AN INTRODUCTION TO
ORDINARY
DIFFERENTIAL EQUATIONS

EARL A. CODDINGTON

PROFESSOR OF MATHEMATICS
UNIVERSITY OF CALIFORNIA, LOS ANGELES

DOVER PUBLICATIONS, INC.
New York

This Dover edition, first published in 1989, is an unabridged, corrected republication of the work first published by Prentice-Hall, Inc., Englewood Cliffs, New Jersey, 1961, in its "Prentice-Hall Mathematics Series."

Library of Congress Cataloging-in-Publication Data

Coddington, Earl A., 1920–
    An introduction to ordinary differential equations / by Earl A. Coddington.—Unabridged, corrected republication.
        p.   cm.
    Reprint. Originally published: Englewood Cliffs, N.J. : Prentice-Hall, 1961. (Prentice-Hall mathematics series)
    Bibliography: p.
    Includes index.
    ISBN-13: 978-0-486-65942-8
    ISBN-10: 0-486-65942-9
    1. Differential equations.   I. Title.
QA372.C59   1989   88-31457
515.3'5—dc19   CIP

Manufactured in the United States by Courier Corporation
65942908
www.doverpublications.com

To Alan, Robert, and Claire

Preface

AS THE TITLE indicates, this book is meant to be a text which can be used for a first course in ordinary differential equations. The student is assumed to have a knowledge of calculus but not what is usually called advanced calculus. During the last four years, I have presented most of this material in a one-semester course.

My aim has been to give an elementary, thorough, systematic introduction to the subject. All significant results are stated as theorems, and careful proofs are given. I have tried to emphasize general properties of equations and their solutions.

The preliminary Chapter 0 contains results from calculus and algebra which are required in the later chapters. Complex numbers and complex-valued functions are introduced here and are used throughout the book. Chapters 1–4 contain material on linear equations. The short Chapter 1 concerns linear equations of the first order. Chapter 2 contains a rather complete discussion of linear equations with constant coefficients, including a uniqueness theorem which is derived from an elementary inequality. In Chapter 3 linear equations with variable coefficients are treated. The early part of this chapter can be covered quite rapidly, since it really amounts to a review of some of the material in Chapter 2. Equations with analytic coefficients, with the Legendre equation as a prime example, are introduced here. Chapter 4 contains a detailed treatment of second order equations with regular singular points; the Bessel equation receives special emphasis. Chapter 5 is concerned with initial value problems for a single—in general nonlinear—equation. Existence and uniqueness of solutions are established. The successive approximation method is used for the existence proof, and general results on uniform convergence (a topic usually taught in advanced calculus) are not used. The required convergence proofs are dealt with by means of explicit inequalities. Both local and non-local existence results are given. In Chapter 6 it is shown how most of the results in Chapter 5 remain valid for systems of equations. Here complex n-dimensional vectors are introduced, and systems are treated as vector equations, with solutions being vector-valued functions.

Several sections are starred. I have usually not devoted any classroom time to these.

Included in the book are many exercises. There are exercises which serve to develop the student's technique in solving equations, and there are many problems which are intended to help sharpen the student's understanding of the mathematical structure of the subject. I have also used the exercises to introduce the student to a variety of topics not treated in the text: for example, stability, equations with periodic coefficients, boundary value problems.

I wish to record here my gratitude to Professor Norman Levinson, from whom I learned so much during the period of our collaboration in writing an earlier advanced work on this subject. Also, I wish to thank Mr. George Biriuk for reading portions of the manuscript, and Professor Richard C. Gilbert, to whom I am particularly indebted for reading all the manuscript and suggesting many interesting exercises. Finally, I would like to express my appreciation to Mr. Richard Hansen, of Prentice-Hall, for his patient understanding.

Earl A. Coddington

Contents

ix

CHAPTER 0

Preliminaries

1. Introduction

In this preliminary chapter we consider briefly some important concepts from calculus and algebra which we shall require for our study of differential equations. Many of these concepts may be familiar to the student, in which case this chapter can serve as a review. First the elementary properties of complex numbers are outlined. This is followed by a discussion of functions which assume complex values, in particular polynomials and power series. Some consequences of the Fundamental Theorem of Algebra are given. The exponential function is defined using power series; it is of central importance for linear differential equations with constant coefficients. The role that determinants play in the solution of systems of linear equations is discussed. Lastly we make a few remarks concerning principles of discovery, and methods of proof, of mathematical results.

2. Complex numbers

It is a fundamental fact about real numbers that the square of any such number is never negative. Thus there is no real x which satisfies the equation

$$x^2 + 1 = 0.$$

We shall use the real numbers to define new numbers which include numbers which satisfy such equations.

A *complex number* z is an ordered pair of real numbers (x, y), and we write

$$z = (x, y).$$

If

$$z_1 = (x_1, y_1), \qquad z_2 = (x_2, y_2),$$

are two such numbers, we define z_1 to be *equal* to z_2, and write $z_1 = z_2$, if $x_1 = x_2$ and $y_1 = y_2$. The *sum* $z_1 + z_2$ is defined to be the complex number given by

$$z_1 + z_2 = (x_1 + x_2, y_1 + y_2).$$

If $z = (x, y)$, the *negative* of z, denoted by $-z$, is defined to be the number

$$-z = (-x, -y).$$

The *zero* complex number, also denoted by 0, is defined by

$$0 = (0, 0).$$

It is clear from these definitions that

     (i) $z_1 + z_2 = z_2 + z_1$

     (ii) $(z_1 + z_2) + z_3 = z_1 + (z_2 + z_3)$

     (iii) $z + 0 = z$

     (iv) $z + (-z) = 0$

for all complex numbers z, z_1, z_2, z_3.

The *difference* $z_1 - z_2$ is defined by

$$z_1 - z_2 = z_1 + (-z_2),$$

and we have

$$z_1 - z_2 = (x_1 - x_2, y_1 - y_2).$$

The *product* $z_1 z_2$ is defined by

$$z_1 z_2 = (x_1 x_2 - y_1 y_2, x_1 y_2 + x_2 y_1).$$

This definition appears curious at first, but we shall soon see a justification for it. It is easy to check that multiplication satisfies

     (v) $z_1 z_2 = z_2 z_1$

     (vi) $(z_1 z_2) z_3 = z_1 (z_2 z_3)$

for all complex numbers z_1, z_2, z_3.

The *unit* complex number, with respect to multiplication, is the number $(1, 0)$ for we see that if $z = (x, y)$ is any complex number

$$z(1, 0) = (x, y)(1, 0) = (x, y) = z.$$

For this reason we denote the number $(1, 0)$ by just 1. Then we have

     (vii) $z1 = z$

for all complex z.

If $z = (x, y) \neq (0, 0)$ there is a unique complex number w such that $zw = 1 (= (1, 0))$. Indeed, if $w = (u, v)$, where u, v are real, the equation $zw = 1$ says that

$$xu - yv = 1$$
$$yu + xv = 0.$$

These equations have the unique solution

$$u = \frac{x}{x^2 + y^2}, \qquad v = \frac{-y}{x^2 + y^2},$$

provided $x^2 + y^2 \neq 0$, which is equivalent to the assumption we made that $z \neq 0$. The number w, such that $zw = 1$, is called the *reciprocal* of z, and we denote it by z^{-1} or $1/z$. Thus

$$z^{-1} = \left(\frac{x}{x^2 + y^2}, \frac{-y}{x^2 + y^2}\right), \qquad \text{if } z \neq 0.$$

Then

$$\text{(viii)} \quad zz^{-1} = 1, \qquad \text{if} \quad z \neq 0.$$

The *quotient* z_1/z_2 is defined when $z_2 \neq 0$ by

$$\frac{z_1}{z_2} = z_1 z_2^{-1}, \qquad \text{if} \quad z_2 \neq 0.$$

The interaction between addition and multiplication is given by the rule

$$\text{(ix)} \quad z_1(z_2 + z_3) = z_1 z_2 + z_1 z_3.$$

The complex numbers of the form $(x, 0)$ are such that the negative and reciprocal of any such number have the same form, for

$$-(x, 0) = (-x, 0),$$
$$(x, 0)^{-1} = (x^{-1}, 0), \qquad \text{if } x \neq 0.$$

Moreover, the sum and product of two such numbers have the same form, since

$$(x_1, 0) + (x_2, 0) = (x_1 + x_2, 0),$$
$$(x_1, 0)(x_2, 0) = (x_1 x_2, 0).$$

The real numbers are in a one-to-one correspondence with the complex numbers of this form, the real number x corresponding to the complex number $z = (x, 0)$. Further, as we have just seen, the numbers corresponding to $-x, x^{-1}, x_1 + x_2, x_1 x_2$ are just $-z, z^{-1}, z_1 + z_2, z_1 z_2$, if $z_1 = (x_1, 0)$, $z_2 = (x_2, 0)$. For this reason it is usual to identify the complex number $(x, 0)$ with the real number x, and we write $x = (x, 0)$. [Notice that this agrees with our earlier identifications $0 = (0, 0)$, $1 = (1, 0)$.] In this sense, the complex numbers contain the real numbers. The properties $(i)-(ix)$, which hold for complex numbers, are also valid for real numbers, and thus we see that we have succeeded in enlarging the set of real numbers without losing any of these algebraic properties. We have gained something also, since there are complex numbers z which satisfy the equation

$$z^2 + 1 = 0.$$

One such number is the *imaginary unit* $i = (0, 1)$, as can be easily checked, and this provides one justification for our definition of multiplication.

If $z = (x, y)$ is a complex number, the real number x is called the *real part* of z, and we write Re $z = x$; whereas y is called the *imaginary part* of z, and we write Im $z = y$. Thus

$$z = (x, y) = x(1, 0) + y(0, 1) = x + iy = \text{Re } z + i(\text{Im } z).$$

Hereafter it will be convenient to denote a complex number (x, y) as $x + iy$.

It is clear that the complex numbers are in a one-to-one correspondence with the points of the (x, y)-plane, the complex number $z = x + iy$ corresponding to the point with coordinates (x, y). Then thought of in this way the x-axis is often called the *real axis*, the y-axis is called the *imaginary axis*, and the plane is called the *complex plane*.

If $z = x + iy$, its mirror image in the real axis is the point $x - iy$. This number is called the *complex conjugate* of z, and is denoted by $\bar{z}$. Thus $\bar{z} = x - iy$ if $z = x + iy$. We see immediately that

$$\bar{\bar{z}} = z, \ \overline{z_1 + z_2} = \bar{z}_1 + \bar{z}_2, \ \overline{z_1 z_2} = \bar{z}_1 \bar{z}_2, \ \overline{z^{-1}} = (\bar{z})^{-1},$$

for any complex numbers z, z_1, z_2.

Introducing polar coordinates (r, θ) in the complex plane via

$$x = r \cos \theta, \quad y = r \sin \theta, \quad (r \geqq 0, 0 \leqq \theta < 2\pi),$$

we see that we may write

$$z = x + iy = r(\cos \theta + i \sin \theta).$$

The *magnitude* of $z = x + iy$, denoted by $|z|$, is defined to be r. Thus

$$|z| = (x^2 + y^2)^{1/2} = (z\bar{z})^{1/2},$$

where the positive square root is understood. Clearly $|\bar{z}| = |z|$. Suppose z is real (that is, Im $z = 0$). Then $z = x + i0$, for some real x, and

$$|z| = (x^2)^{1/2},$$

which is the magnitude of x considered as a real number. In addition the magnitude of a complex number obeys the same rules as the magnitude of a real number, namely:

$$|z| \geqq 0,$$

$$|z| = 0 \quad \text{if and only if} \quad z = 0,$$

$$|-z| = |z|,$$

$$|z_1 + z_2| \leqq |z_1| + |z_2|,$$

$$|z_1 z_2| = |z_1| |z_2|.$$

We show that $|z_1 + z_2| \leqq |z_1| + |z_2|$, for example. First we note that

$$\operatorname{Re} z \leqq |z|$$

for any complex number z. Then

$$\begin{aligned}
|z_1 + z_2|^2 &= (z_1 + z_2)\overline{(z_1 + z_2)} = |z_1|^2 + |z_2|^2 + z_1\bar{z}_2 + \bar{z}_1 z_2 \\
&= |z_1|^2 + |z_2|^2 + 2 \operatorname{Re}(z_1\bar{z}_2) \\
&\leqq |z_1|^2 + |z_2|^2 + 2|z_1\bar{z}_2| \\
&= |z_1|^2 + |z_2|^2 + 2|z_1| |z_2| \\
&= (|z_1| + |z_2|)^2,
\end{aligned}$$

from which it follows that $|z_1 + z_2| \leqq |z_1| + |z_2|$.

From the above rules one can deduce further that

$$||z_1| - |z_2|| \leqq |z_1 + z_2| \leqq |z_1| + |z_2|,$$

$$\left|\frac{z_1}{z_2}\right| = \frac{|z_1|}{|z_2|}.$$

Geometrically we see that $|z_1 - z_2|$ represents the distance between the two points z_1 and z_2 in the complex plane.

EXERCISES

1. Compute the following complex numbers, and express in the form $x + iy$, where x, y are real:

(a) $(2 - i3) + (-1 + i6)$ (b) $(4 + i2) - (6 - i3)$

(c) $(6 - i\sqrt{2})(2 + i4)$ (d) $\dfrac{1 + i}{1 - i}$

(e) $|4 - i5|$ (f) $\operatorname{Re}(4 - i5)$

(g) $\operatorname{Im}(6 + i2)$

2. Express the following complex numbers in the form $r (\cos \theta + i \sin \theta)$ with $r \geqq 0$ and $0 \leqq \theta < 2\pi$:

(a) $1 + i\sqrt{3}$ (b) $(1 + i)^2$

(c) $\dfrac{1 + i}{1 - i}$ (d) $(1 + i)(1 - i)$

3. Indicate graphically the set of all complex numbers z satisfying:

(a) $|z - 2| = 1$ (b) $|z + 2| < 2$
(c) $|\operatorname{Re} z| \leqq 3$ (d) $|\operatorname{Im} z| > 1$
(e) $|z - 1| + |z + 2| = 8$.

4. Prove that:

(a) $z + \bar{z} = 2 \operatorname{Re} z$ (b) $z - \bar{z} = 2i \operatorname{Im} z$
(c) $|\operatorname{Re} z| \leqq |z|$ (d) $|z| \leqq |\operatorname{Re} z| + |\operatorname{Im} z|$

5. If r is a real number, and z complex, show that

$$\operatorname{Re} (rz) = r (\operatorname{Re} z), \qquad \operatorname{Im} (rz) = r (\operatorname{Im} z).$$

6. Prove that

$$|| z_1 | - | z_2 || \leqq | z_1 + z_2 |.$$

(*Hint*: $z_1 = z_1 + z_2 + (-z_2)$, and $z_2 = z_1 + z_2 + (-z_1)$.)

7. Prove that

$$| z_1 + z_2 |^2 + | z_1 - z_2 |^2 = 2 | z_1 |^2 + 2 | z_2 |^2,$$

for all complex z_1, z_2.

8. If $|a| < 1$, what complex z satisfy $\dfrac{|z - a|}{|1 - \bar{a}z|} \leqq 1$?

9. If n is any positive integer, prove that

$$r^n (\cos n\theta + i \sin n\theta) = [r (\cos \theta + i \sin \theta)]^n.$$

(*Hint*: Use induction.)

10. Use the result of Ex. 9 to find
(a) two complex numbers satisfying $z^2 = 2$,
(b) three complex numbers satisfying $z^3 = 1$.

3. Functions

Suppose D is a set whose elements are denoted by $P, Q, \cdots$, which are called the points of the set. Let R be another set. A *function* on D to R is a law f which associates with each point P in D exactly one point in R, which we denote by $f(P)$. The set D is called the *domain* of f. The point $f(P)$ is called the *value* of f at P. We can visualize the concept of a function as

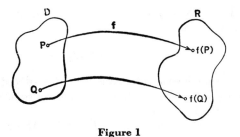

Figure 1

in Fig. 1, where each P in D is connected to a unique $f(P)$ in R by a string according to some rule. This rule, or what amounts to the same thing, the collection of all these strings, is the function f on D to R.

We say that two functions f and g are *equal*, $f = g$, if they have the same domain D, and $f(P) = g(P)$ for all P in D.

The idea of a function is very general, and is a fundamental one in mathematics. We shall consider some examples which are of importance for our study of differential equations.

(a) *Complex-valued functions.* If the set R which contains the values of f is the set of all complex numbers, we say that f is a *complex-valued function*. If f and g are two complex-valued functions with the same domain D, we can define their *sum $f + g$* and *product fg* by

$$(f + g)(P) = f(P) + g(P),$$

$$(fg)(P) = f(P)g(P),$$

for each P in D. Thus $f + g$ and fg are also functions with domain D. If α is any complex number the function which assigns to each P in a domain D the number α is called a *constant function*, and is also denoted by α. Thus if f is any complex-valued function on D we have

$$(\alpha f)(P) = \alpha f(P)$$

for all P in D.

A *real-valued function f* defined on D is one whose values are real numbers. Such a function is a special case of a complex-valued function. Clearly the sum and product of two real-valued functions on D are real-valued functions. Real-valued functions are usually the principal object of study in first courses in calculus.

Every complex-valued function f defined on a domain D gives rise to two real-valued functions $\operatorname{Re} f$, $\operatorname{Im} f$ defined by

$$(\operatorname{Re} f)(P) = \operatorname{Re}[f(P)],$$

$$(\operatorname{Im} f)(P) = \operatorname{Im}[f(P)],$$

for all P in D. Re f and Im f are called the *real* and *imaginary* parts of f respectively and we have

$$f = \operatorname{Re} f + i \operatorname{Im} f.$$

Thus the study of complex-valued functions can be reduced to the study of pairs of real-valued functions. To obtain examples of complex-valued functions we must specify their domains.

(b) *Complex-valued functions with real domains.* Many of the functions we consider in this book have a domain D which is an interval I of the real axis. Recall that an *interval* is a set of real x satisfying one of the nine inequalities

$$a \leqq x \leqq b, \quad a \leqq x < b, \quad a < x \leqq b, \quad a < x < b,$$

$$a \leqq x < \infty, \quad -\infty < x \leqq b, \quad a < x < \infty, \quad -\infty < x < b,$$

$$-\infty < x < \infty,$$

where a, b are distinct real numbers. The calculus of complex-valued functions defined on real intervals is entirely analogous to the calculus of real-valued functions defined on intervals. We sketch the main ideas.

Suppose f is a complex-valued function defined on a real interval I. Then f is said to have the complex number L as a *limit* at x_0 in I, and we write

$$\lim_{x \to x_0} f(x) = L, \quad \text{or} \quad f(x) \to L, \quad (x \to x_0),$$

if

$$|f(x) - L| \to 0, \quad \text{as} \quad 0 < |x - x_0| \to 0.$$

This means that given any $\epsilon > 0$ there is a $\delta > 0$ such that

$$|f(x) - L| < \epsilon, \quad \text{whenever} \quad 0 < |x - x_0| < \delta, \quad x \text{ in } I.$$

Note that here we are using the magnitude of complex numbers. Formally our definition is the same as that for real limits of real-valued functions. Because of this the usual rules for limits, and their proofs, are valid. In particular, if f and g are complex-valued functions defined on I such that for some x_0 in I

$$f(x) \to L, \quad g(x) \to M, \quad (x \to x_0),$$

then

$$(f + g)(x) \to L + M, \quad (fg)(x) \to LM, \quad (x \to x_0).$$

Suppose f has a limit $L = L_1 + iL_2$ at x_0, where L_1, L_2 are real. Then since

$$| (\operatorname{Re} f)(x) - L_1 | = | \operatorname{Re} [f(x) - L] | \leqq | f(x) - L |,$$

and

$$| (\operatorname{Im} f)(x) - L_2 | = | \operatorname{Im}[f(x) - L] | \leqq | f(x) - L |,$$

it follows that

$$(\operatorname{Re} f)(x) \to L_1, \quad (\operatorname{Im} f)(x) \to L_2, \quad (x \to x_0).$$

Conversely, if $\operatorname{Re} f$ and $\operatorname{Im} f$ have limits L_1, L_2 respectively at x_0, then f will have the limit $L = L_1 + iL_2$ at x_0.

We say that a complex-valued function f defined on an interval I is *continuous* at x_0 in I if f has the limit $f(x_0)$ at x_0, that is,

$$| f(x) - f(x_0) | \to 0, \quad \text{as} \quad 0 < | x - x_0 | \to 0.$$

Equivalently, f is continuous at x_0 if both $\operatorname{Re} f$ and $\operatorname{Im} f$ are continuous at x_0. We say f is continuous on I if it is continuous at each point of I. The sum and product of two functions which are continuous at x_0 are continuous there.

The complex-valued function f defined on an interval I is said to be *differentiable* at x_0 in I if the ratio

$$\frac{f(x) - f(x_0)}{x - x_0}, \quad (x \neq x_0),$$

has a limit at x_0. If f is differentiable at x_0 we define its *derivative* at x_0, $f'(x_0)$, to be this limit. Thus, if $f'(x_0)$ exists,

$$\left| \frac{f(x) - f(x_0)}{x - x_0} - f'(x_0) \right| \to 0, \quad \text{as} \quad 0 < | x - x_0 | \to 0.$$

An equivalent definition is: f is differentiable at x_0 if both $\operatorname{Re} f$ and $\operatorname{Im} f$ are differentiable at x_0. The derivative of f at x_0 is given by

$$f'(x_0) = (\operatorname{Re} f)'(x_0) + i(\operatorname{Im} f)'(x_0).$$

Using these definitions one can show that the usual rules for differentiating real-valued functions are valid for complex-valued functions. For example, if f, g are differentiable at x_0 in I, then so are $f + g$ and fg, and

$$(f + g)'(x_0) = f'(x_0) + g'(x_0),$$

$$(fg)'(x_0) = f'(x_0)g(x_0) + f(x_0)g'(x_0).$$

If f is differentiable at every x in an interval I, then f gives rise to a new function f' on I whose value at each x on I is $f'(x)$.

A complex-valued function f with domain the interval $a \leqq x \leqq b$ is said to be *integrable* there if both $\mathrm{Re}\, f$ and $\mathrm{Im}\, f$ are, and in this case we define its integral by

$$\int_a^b f(x)\ dx = \int_a^b (\mathrm{Re}\ f)(x)\ dx + i \int_a^b (\mathrm{Im}\ f)(x)\ dx.$$

Every function f which is continuous on $a \leqq x \leqq b$ is integrable there. This definition implies the usual integration rules. In particular, if f and g are integrable on $a \leqq x \leqq b$, and α, β are two complex numbers,

$$\int_a^b (\alpha f + \beta g)(x)\ dx = \alpha \int_a^b f(x)\ dx + \beta \int_a^b g(x)\ dx.$$

An important inequality connected with the integral of a continuous complex-valued function f defined on $a \leqq x \leqq b$ is

$$\left| \int_a^b f(x)\ dx \right| \leqq \int_a^b |\, f(x)\, |\ dx.^*$$

This inequality is valid if f is real-valued, and the proof for the case when f is complex-valued can be based on this fact. Let

$$F = \int_a^b f(x)\ dx.$$

If $F = 0$ the inequality is obvious. If $F \neq 0$, let

$$F = |F|\, u, \quad u = \cos \theta + i \sin \theta, \qquad (0 \leqq \theta < 2\pi).$$

Then $u\bar{u} = 1$, and we have

$$\left| \int_a^b f(x)\ dx \right| = \bar{u} \int_a^b f(x)\ dx = \mathrm{Re} \left[\bar{u} \int_a^b f(x)\ dx \right]$$

$$= \int_a^b \mathrm{Re}\ [\bar{u} f(x)]\ dx \leqq \int_a^b |\, f(x)\, |dx,$$

*By

$$\int_a^b |f(x)|\, dx$$

is meant the integral of the function $|\, f\, |$ given by $|\, f\, |(x) = |f(x)|$ for $a \leqq x \leqq b$. Thus a more appropriate notation would be

$$\int_a^b |\, f\, |(x)\, dx.$$

We shall use the former notation since it is commonly used, and there will be no chance of confusion.

since

$$\text{Re}\,[\bar{u}f(x)] \leqq |\bar{u}f(x)| = |f(x)|.$$

As particular examples of complex-valued functions let

$$f(x) = x + (1 - i)x^2,$$

$$g(x) = (1 + i)x^2,$$

for all real x. Then

$$(\text{Re}\,f)(x) = x + x^2, \qquad (\text{Im}\,f)(x) = -x^2,$$

$$(f + g)(x) = x + 2x^2,$$

$$(fg)(x) = (1 + i)x^3 + 2x^4,$$

$$f'(x) = 1 + (2 - 2i)x,$$

$$\int_0^1 f(x)\,dx = \int_0^1 x\,dx + (1 - i)\int_0^1 x^2\,dx = \frac{5}{6} - \frac{i}{3}.$$

(c) *Complex-valued functions with complex domains.* We shall need to know a little about complex-valued functions whose domains consist of complex numbers. An example is the function f given by

$$f(z) = z^n,$$

for all complex z, where n is a positive integer.

Let f be a complex-valued function which is defined on some disk

$$D: \quad |z - a| < r$$

with center at the complex number a and radius $r > 0$. Much of the calculus for such functions can be patterned directly after the calculus of complex-valued functions defined on a real interval I. We say that f has the complex number L as a *limit* at z_0 in D if

$$|f(z) - L| \to 0, \quad \text{as} \quad 0 < |z - z_0| \to 0,$$

and we write

$$\lim_{z \to z_0} f(z) = L, \quad \text{or} \quad f(z) \to L, \quad (z \to z_0).$$

If f and g are two complex-valued functions defined on D such that for some z_0 in D

$$f(z) \to L, \quad g(z) \to M, \quad (z \to z_0),$$

then

$$(f + g)(z) \to L + M, \quad (fg)(z) \to LM, \quad (z \to z_0).$$

The proofs are identical to those for functions defined on real intervals.

The function f, defined on the disk D, is said to be *continuous* at z_0 in D if

$$|f(z) - f(z_0)| \to 0, \quad \text{as} \quad 0 < |z - z_0| \to 0.$$

It is said to be continuous on D if it is continuous at each point of D. The sum and product of two functions which are continuous at z_0 are continuous there. Examples of continuous functions on the whole complex plane are

$$f(z) = |z|, \quad g(z) = z^3.$$

Let g be defined on some disk D_1 containing z_0, and let its values be in some disk D_2, where a function f is defined. If g is continuous at z_0, and f is continuous at $g(z_0)$, then "the function of a function" F given by

$$F(z) = f(g(z)), \quad (z \text{ in } D_1), \tag{3.1}$$

is continuous at z_0. The proof follows the same lines as in calculus for real-valued functions defined for real x.

If f is defined on a disk D containing z_0 we say that f is *differentiable* at z_0 if

$$\frac{f(z) - f(z_0)}{z - z_0}, \quad (z \neq z_0),$$

has a limit at z_0. If f is differentiable at z_0 its *derivative* at z_0, $f'(z_0)$, is defined to be this limit. Thus

$$\left| \frac{f(z) - f(z_0)}{z - z_0} - f'(z_0) \right| \to 0, \quad \text{as } 0 < |z - z_0| \to 0.$$

Formally our definition is the same as that for the derivative of a complex-valued function defined on a real interval. For this reason if f and g are functions which have derivatives at z_0 in D then $f + g$, fg have derivatives there, and

$$(f + g)'(z_0) = f'(z_0) + g'(z_0),$$

$$(fg)'(z_0) = f'(z_0)g(z_0) + f(z_0)g'(z_0). \tag{3.2}$$

Also, suppose f and g are two functions as given in (3.1), and that g is differentiable at z_0, whereas f is differentiable at $g(z_0)$. Then F is differentiable at z_0, with

$$F'(z_0) = f'(g(z_0))g'(z_0).$$

It is clear from the definition of a derivative that the function q defined by $q(z) = c$, where c is a complex constant, has a derivative which is zero everywhere, that is, $q'(z) = 0$. Also, if $p_1(z) = z$ for all z, then $p_1'(z) = 1$. Combining these results with the rules (3.2) we obtain the fact that every polynomial has a derivative for all z. A *polynomial* is a function p whose domain is the set of all complex numbers and which has the form

$$p(z) = a_0 z^n + a_1 z^{n-1} + \cdots + a_{n-1} z + a_n,$$

where $a_0, a_1, \cdots, a_n$ are complex constants. The rules (3.2) imply that for such a p

$$p'(z) = a_0 n z^{n-1} + a_1(n - 1)z^{n-2} + \cdots + a_{n-1}.$$

Thus p' is also a polynomial.

It is a rather strong restriction on a function defined on a disk D to demand that it be differentiable at a point z_0 in D. To illustrate this we note that the real-valued function f given by

$$f(x) = |x|,$$

for all real x, is differentiable at all $x \neq 0$. Indeed $f'(x)$ is $+1$ or -1 according as x is positive or negative. However the continuous complex-valued function g given by

$$g(z) = |z|,$$

for all complex z, is not differentiable for any z. Suppose $z_0 = x_0 + y_0 i \neq 0$, for example, and let $z = x + yi$. Then for $z \neq z_0$

$$\frac{|z| - |z_0|}{z - z_0} = \frac{(x^2 + y^2)^{1/2} - (x_0^2 + y_0^2)^{1/2}}{(x - x_0) + i(y - y_0)}$$

$$= \frac{(x^2 + y^2) - (x_0^2 + y_0^2)}{[(x - x_0) + i(y - y_0)][(x^2 + y^2)^{1/2} + (x_0^2 + y_0^2)^{1/2}]}.$$

If we let $|z - z_0| \to 0$ using z of the form $z = x_0 + yi$ (that is $y \to y_0$) we see that

$$\frac{|z| - |z_0|}{z - z_0} \to \frac{y_0}{i(x_0^2 + y_0^2)^{1/2}}, \tag{3.3}$$

whereas if we let $|z - z_0| \to 0$ using z of the form $z = x + y_0 i$ (that is $x \to x_0$) we obtain

$$\frac{|z| - |z_0|}{z - z_0} \to \frac{x_0}{(x_0^2 + y_0^2)^{1/2}}. \tag{3.4}$$

The two limits (3.3) and (3.4) are different. However, in order that g be differentiable at z_0 we must obtain the same limit no matter how $|z - z_0| \to 0$. This shows that g is not differentiable at z_0.

(d) *Other functions.* Other types of functions which are important for our study of differential equations are usually combinations of the types discussed in (b), (c) above. Typical is a complex-valued function f which is defined for real x on some interval $|x - x_0| \leqq a$ (x_0 real, $a > 0$), and for complex z on some disk $|z - z_0| \leqq b$ (z_0 complex, $b > 0$). Thus the domain D of f is given by

$$D: \quad |x - x_0| \leqq a, \quad |z - z_0| \leqq b,$$

and the value of f at (x, z) is denoted by $f(x, z)$. Such a function f is said to be *continuous* at (ξ, η) in D if

$$|f(x, z) - f(\xi, \eta)| \to 0, \quad \text{as} \quad 0 < |x - \xi| + |z - \eta| \to 0.$$

There are two important facts which we shall need in Chap. 5 concerning such continuous functions. The first is that a continuous f on the D given above (*with the equality signs included*) is bounded, that is, there is a positive constant M such that

$$|f(x, z)| \leqq M,$$

for all (x, z) in D. This result is usually proved in advanced calculus courses. The second result relates to "plugging in" a complex-valued function ϕ into f. Suppose ϕ is a complex-valued function defined on

$$|x - x_0| \leqq a,$$

which is continuous there, and has values in $|z - z_0| \leqq b$. Then if f is continuous on D, the function F given by

$$F(x) = f(x, \phi(x)),$$

for all x such that $|x - x_0| \leqq a$, is continuous for such x.

A slightly more complicated type of complex-valued function f is one which is defined for real x and complex $z_1, \cdots, z_n$ on a domain

$$D: \quad |x - x_0| \leqq a, \quad |z_1 - z_{10}| + \cdots + |z_n - z_{n0}| \leqq b.$$

Here x_0 is real, $z_{10}, \cdots, z_{n0}$ are complex, and a, b are positive. The value of f at $x, z_1, \cdots, z_n$ is denoted by $f(x, z_1, \cdots, z_n)$. Continuity of f is defined just as in the case of one z. Thus f is *continuous* at $\xi, \eta_1, \cdots, \eta_n$ in D if

$$\left| f(x, z_1, \cdots, z_n) - f(\xi, \eta_1, \cdots, \eta_n) \right| \to 0,$$

as

$$0 < |x - \xi| + |z_1 - \eta_1| + \cdots + |z_n - \eta_n| \to 0.$$

Such an f is bounded on D, and if $\phi_1, \cdots, \phi_n$ are n continuous complex-valued functions defined on $|x - x_0| \leqq a$, having the property that

$$\left| \phi_1(x) - z_{10} \right| + \cdots + \left| \phi_n(x) - z_{n0} \right| \leqq b$$

for all such x, then the function F given by

$$F(x) = f(x, \phi_1(x), \cdots, \phi_n(x))$$

for $|x - x_0| \leqq a$ is continuous there.

EXERCISES

1. Let $a = 2 + i3$, $b = 1 - i$. If for all real x

$$f(x) = ax + (bx)^2,$$

compute:

   (a) $(\operatorname{Re} f)(x)$ (b) $(\operatorname{Im} f)(x)$

   (c) $f'(x)$ (d) $\displaystyle\int_0^1 f(x)\, dx$

2. If for all real x

$$f(x) = x + ix^2, \qquad g(x) = \frac{x^2}{2},$$

compute:

   (a) The function F given by $F(x) = f(g(x))$ (b) $F'(x)$

3. If a is a real-valued function defined on an interval I, and f is a complex-valued function defined there, show that

$$\operatorname{Re}\,(af) = a(\operatorname{Re} f), \qquad \operatorname{Im}\,(af) = a(\operatorname{Im} f).$$

4. Let $f(z) = z^2$ for all complex z, and let

$$u(x, y) = (\operatorname{Re} f)(x + iy), \quad v(x, y) = (\operatorname{Im} f)(x + iy).$$

   (a) Compute $u(x, y)$ and $v(x, y)$.

   (b) Show that $\dfrac{\partial u}{\partial x} = \dfrac{\partial v}{\partial y}, \qquad \dfrac{\partial u}{\partial y} = -\dfrac{\partial v}{\partial x}.$

(c) Show that

$$\frac{\partial^2 u}{\partial x^2} + \frac{\partial^2 u}{\partial y^2} = 0, \qquad \frac{\partial^2 v}{\partial x^2} + \frac{\partial^2 v}{\partial y^2} = 0.$$

5. Let f be a complex-valued function defined on a disk

$$D: \quad |z| < r \qquad (r > 0),$$

which is differentiable there. Let

$$u(x, y) = (\operatorname{Re} f)(x + iy), \qquad v(x, y) = (\operatorname{Im} f)(x + iy).$$

Show that

$$\frac{\partial u}{\partial x} = \frac{\partial v}{\partial y}, \qquad \frac{\partial u}{\partial y} = -\frac{\partial v}{\partial x}, \qquad (*)$$

for all $z = x + iy$ in D. (*Hint*: If $z_0 = x_0 + iy_0$ is in D, let $0 < |z - z_0| \to 0$, in the definition of $f'(z_0)$, through z of the form $z = x + iy_0$, and then of the form $z = x_0 + iy$, to obtain

$$f'(z_0) = \frac{\partial u}{\partial x}(x_0, y_0) + i \frac{\partial v}{\partial x}(x_0, y_0)$$

$$= \frac{\partial v}{\partial y}(x_0, y_0) - i \frac{\partial u}{\partial y}(x_0, y_0).$$

The equations (*) are called the *Cauchy-Riemann* equations.)

6. Let f be the complex-valued function defined on

$$D: \quad |x| \leqq 1, \qquad |z| \leqq 2,$$

(x real, z complex) by

$$f(x, z) = 3x^2 + xz + z^2,$$

and let ϕ be the function defined on $|x| \leqq 1$ by

$$\phi(x) = x + i.$$

(a) Compute the function F given by

$$F(x) = f(x, \phi(x)), \qquad (|x| \leqq 1).$$

(b) Compute $F'(x)$.

(c) Compute

$$\int_0^1 F(x) \, dx.$$

7. If r is a complex number, and

$$p(z) = (z - r)^n,$$

where n is a positive integer, show that

$$p(r) = p'(r) = \cdots = p^{(n-1)}(r) = 0, \qquad p^{(n)}(r) = n!.$$

4. Polynomials

We have defined a polynomial as a complex-valued function p whose domain is the set of all complex numbers and which has the form

$$p(z) = a_0 z^n + a_1 z^{n-1} + \cdots + a_{n-1} z + a_n,$$

where n is a non-negative integer, and $a_0, a_1 \cdots, a_n$ are complex constants. The highest power of z with non-zero coefficient which appears in the expression defining a polynomial p is called the *degree* of p, and written deg p. A *root* of a polynomial p is a complex number r such that $p(r) = 0$. A root of p is sometimes called a *zero* of p. We shall require, and assume, the following important result.*

Fundamental theorem of algebra. *If* p *is a polynomial such that* deg p ≥ 1, *then* p *has at least one root.*

This is a rather remarkable result, and justifies our introduction of the complex numbers. We have seen that not every polynomial with real coefficients (for example $z^2 + 1$) has a real root, but polynomials of degree greater than zero with complex coefficients always have a complex root. The remarkable fact is that we do not need to invent new numbers, which include the complex numbers, to guarantee a complex root.

We derive some consequences of this fundamental theorem.

Corollary 1. *Let* p *be a polynomial of degree* n ≥ 1, *with leading coefficient* 1 *(the coefficient of* z^n*), and let* r *be a root of* p. *Then*

$$p(z) = (z - r)q(z)$$

where q *is a polynomial of degree* n $- 1$, *with leading coefficient* 1.

Proof. Let $p(z)$ have the form

$$p(z) = z^n + a_1 z^{n-1} + \cdots + a_{n-1} z + a_n,$$

---

* A proof can be found in G. Birkhoff and S. MacLane, *A survey of modern algebra*, New York, rev. ed., 1953, p. 107, and also in K. Knopp, *Theory of functions*, New York, 1945, p. 114.

and let c be any complex number. Then

$$p(z) - p(c) = (z^n - c^n) + a_1(z^{n-1} - c^{n-1}) + \cdots + a_{n-1}(z - c)$$

$$= (z - c)q(z),$$

where q is the polynomial given by

$$q(z) = z^{n-1} + cz^{n-2} + c^2 z^{n-3} + \cdots + c^{n-1}$$

$$+ a_1(z^{n-2} + cz^{n-3} + \cdots + c^{n-2}) + \cdots + a_{n-1}.$$

Clearly deg $q = n - 1$ and q has leading coefficient 1. In particular if $c = r$, a root of p, then we have

$$p(z) = (z - r)q(z),$$

as desired.

If $n - 1 \geqq 1$, the polynomial q has a root, and this root is also a root of p by Corollary 1. Thus applying the Fundamental Theorem of Algebra n times, together with Corollary 1, we obtain

Corollary 2. *If* p *is a polynomial,* deg p $= n \geqq 1$, *with leading coefficient* $a_0 \neq 0$, *then* p *has exactly* n *roots. If* $r_1, r_2, \cdots, r_n$ *are these roots, then*

$$p(z) = a_0(z - r_1)(z - r_2) \cdots (z - r_n). \tag{4.1}$$

Note that $a_0^{-1} p$ is a polynomial which has leading coefficient 1. We remark that the roots need not all be distinct. If r is a root of p, the number of times $z - r$ appears as a factor in (4.1) is called the *multiplicity* of r.

Theorem 1. *If* r *is a root of multiplicity* m *of a polynomial* p, deg p $\geqq 1$, *then*

$$p(r) = p'(r) = \cdots = p^{(m-1)}(r) = 0,$$

and

$$p^{(m)}(r) \neq 0.$$

Proof. Let p have leading coefficient $a_0 \neq 0$, and degree $n \geqq m$. It follows from Corollary 2 that

$$p(z) = a_0(z - r)^m q(z), \tag{4.2}$$

where q is a polynomial of degree $n - m$, and $q(r) \neq 0$. Clearly $p(r) = 0$ by the definition of a root. Also

$$p'(z) = a_0 m(z - r)^{m-1} q(z) + a_0(z - r)^m q'(z),$$

and this implies that, if $m - 1 > 0$, $p'(r) = 0$. If $m = 1$ we have

$$p'(z) = a_0 q(z) + a_0(z - r) q'(z),$$

and thus $p'(r) = a_0 q(r) \neq 0$.

The general argument can be based on (4.2) and the formula

$$(fg)^{(k)} = f^{(k)}g + k f^{(k-1)}g' + \frac{k(k-1)}{2!} f^{(k-2)}g'' + \cdots + fg^{(k)} \quad (4.3)$$

for the k-th derivative of the product fg of two functions having k derivatives. Formula (4.3) can be established by induction. Applying (4.3) to the functions $f(z) = (z - r)^m$, $g(z) = q(z)$ in (4.2), we obtain

$$p^{(k)}(z) = a_0 [m(m - 1) \cdots (m - k + 1)(z - r)^{m-k} q(z)$$

$$+ \text{(terms with higher powers of } (z - r) \text{ as a factor)]}.$$

It is now clear that

$$p(r) = p'(r) = \cdots = p^{(m-1)}(r) = 0,$$

and

$$p^{(m)}(r) = a_0 m! \, q(r) \neq 0,$$

which is the desired result.

EXERCISES

1. Compute the roots, with multiplicities, of the following polynomials:

  (a) $z^2 + z - 6$ (b) $z^2 + z + 1$

  (c) $z^3 - 3z^2 + 4$ (d) $z^3 - (2 + i)z^2 + (1 + i2)z - i$

  (e) $z^4 - 3$

2. If r is such that $r^3 = 1$, and $r \neq 1$, prove that $1 + r + r^2 = 0$.

3. Let p be the polynomial given by

$$p(z) = a_0 z^n + a_1 z^{n-1} + \cdots + a_n,$$

with $a_0, a_1, \cdots, a_n$ all *real*. Show that

$$\overline{p(z)} = p(\bar{z}).$$

As a consequence show that if r is a root of p, then so is $\bar{r}$.

4. Prove that every polynomial of degree 3 with real coefficients has at least one real root.

5. Prove that if p is a polynomial, deg $p \geqq 1$, and r is a complex number such that

$$p(r) = p'(r) = \cdots = p^{(m-1)}(r) = 0, \qquad p^{(m)}(r) \neq 0,$$

then r is a root of p with multiplicity m. This is the converse of Theorem 1.

6. (a) Use the result of Ex. 5 to show that i is a root of the polynomial p given by

$$p(z) = z^5 + (2 - 3i)z^4 + (-1 - 6i)z^3 + (-6 - 5i)z^2 + (-6 + 2i)z + 2i,$$

and compute the multiplicity of i.

(b) Find the other roots of the polynomial p in (a).

7. Prove the formula (4.3). This can be written in the form

$$(fg)^{(k)} = f^{(k)}g + \binom{k}{1}f^{(k-1)}g' + \binom{k}{2}f^{(k-2)}g''$$

$$+ \cdots + \binom{k}{l}f^{(k-l)}g^{(l)} + \cdots + fg^{(k)},$$

where

$$\binom{k}{l} = \frac{k!}{l!(k-l)!}$$

is a binomial coefficient. *Hint:* Use induction, and show that

$$\binom{k+1}{l} = \binom{k}{l-1} + \binom{k}{l}.$$

5. Complex series and the exponential function

If x is a real number, and e is the base for the natural logarithms, the number e^x exists, and

$$e^x = \sum_{k=0}^{\infty} \frac{x^k}{k!}, \qquad (0! = 1),$$

where the series converges for all real x. Indeed, this series may be taken as the definition of e^x. We shall need to know what e^z is for *complex z*. One way is to *define* e^z by

$$e^z = \sum_{k=0}^{\infty} \frac{z^k}{k!}. \tag{5.1}$$

Now we have to prove that this series converges for all complex z, and in fact there is the problem of defining what we mean by a convergent series with complex terms. The method is the same as that used to define convergent series with real terms.

A series

$$\sum_{k=0}^{\infty} c_k, \tag{5.2}$$

where all c_k are complex numbers, is said to be *convergent* if the sequence of partial sums

$$s_n = \sum_{k=0}^{n} c_k, \qquad (n = 0, 1, 2, \cdots),$$

tends to a limit s, as $n \to \infty$. That is, s is a complex number such that

$$|s_n - s| \to 0, \qquad (n \to \infty),$$

where the magnitude is the magnitude for complex numbers. If the series (5.2) is convergent, and $s_n \to s$, we call s the *sum* of the series, and write

$$s = \sum_{k=0}^{\infty} c_k.$$

If the series is not convergent we say that it is *divergent*.

The series (5.2) with complex terms c_k gives rise to two series with real terms, namely

$$\sum_{k=0}^{\infty} \operatorname{Re} c_k, \qquad \sum_{k=0}^{\infty} \operatorname{Im} c_k, \tag{5.3}$$

and it is not difficult to see that the series (5.2) is convergent with sum $s = \operatorname{Re} s + i \operatorname{Im} s$ if, and only if, the two real series in (5.3) are convergent with sums $\operatorname{Re} s$ and $\operatorname{Im} s$ respectively. In principle, therefore, the study of series with complex terms is the study of pairs of real series.

The series (5.2) is said to be *absolutely convergent* if the series

$$\sum_{k=0}^{\infty} |c_k| \tag{5.4}$$

is convergent. It can be shown that every absolutely convergent series is convergent. Since the series (5.4) has terms which are real and non-negative, any condition which implies the convergence of such series can be applied to guarantee the convergence of the series (5.2). One of the most important tests for convergence is the *ratio test*. One version of this is the following.

Ratio test. *Consider the series*

where the c_k are complex. If $|c_k| > 0$ for all k beyond a certain positive integer, and

$$\frac{|c_{k+1}|}{|c_k|} \to L, \qquad (k \to \infty), \tag{5.5}$$

then the series is convergent if $L < 1$, and divergent for $L > 1$.

Thus the series (5.2) is convergent if (5.5) is valid for an $L < 1$.

An immediate application of this result is to the series

$$\sum_{k=0}^{\infty} \frac{z^k}{k!}.$$

Here $c_k = z^k/k!$ and

$$\frac{|c_{k+1}|}{|c_k|} = \left| \frac{z^{k+1}}{(k+1)!} \cdot \frac{k!}{z^k} \right| = \frac{|z|}{k+1} \to 0, \qquad (k \to \infty).$$

Thus this series converges for every z such that $|z| < \infty$, that is, for all complex z. Hence our definition (5.1) of e^z as the sum of this series makes sense. The function which associates with each z the complex number e^z is called the *exponential function*.

The series defining e^z is an example of a *power series*

$$\sum_{k=0}^{\infty} a_k (z - z_0)^k \tag{5.6}$$

about some point z_0, the a_k being complex. Many of the properties of a power series of the type

$$\sum_{k=0}^{\infty} a_k (x - x_0)^k,$$

where the a_k, x, x_0 are real, remain true for series of the form (5.6), and the proofs are identical. In particular, if a series (5.6) is convergent on a disk $D : |z - z_0| < r$ $(r > 0)$, then the function f defined by

$$f(z) = \sum_{k=0}^{\infty} a_k (z - z_0)^k, \qquad (z \text{ in } D),$$

has all derivatives in D, and these may be computed by differentiating term by term. Thus

$$f'(z) = \sum_{k=0}^{\infty} k a_k (z - z_0)^{k-1} = \sum_{k=1}^{\infty} k a_k (z - z_0)^{k-1},$$

where the last series converges in D. Applying this result to (5.1) we find that

$$(e^z)' = \sum_{k=1}^{\infty} k \frac{z^{k-1}}{k!} = \sum_{k=1}^{\infty} \frac{z^{k-1}}{(k-1)!} = \sum_{k=0}^{\infty} \frac{z^k}{k!} = e^z.$$

Another important property of the exponential function is that

$$e^{z_1+z_2} = e^{z_1}e^{z_2} \tag{5.7}$$

for every complex z_1, z_2. This can be proved by justifying the following steps

$$e^{z_1}e^{z_2} = \left(\sum_{k=0}^{\infty} \frac{z_1^k}{k!}\right)\left(\sum_{k=0}^{\infty} \frac{z_2^k}{k!}\right) = \sum_{k=0}^{\infty} c_k.$$

Here

$$c_k = \sum_{n=0}^{k} \frac{z_1^{k-n}}{(k-n)!} \frac{z_2^n}{n!}$$

$$= \frac{1}{k!} \sum_{n=0}^{k} \frac{k!}{(k-n)!\, n!} z_1^{k-n} z_2^n$$

$$= \frac{1}{k!} (z_1 + z_2)^k.$$

Thus formally we have the product of the series defining e^{z_1} and e^{z_2} is the series defining $e^{z_1+z_2}$, and these steps can be justified to give a proof of the equality (5.7). A consequence of (5.7) is that

$$(e^z)^n = e^{nz}$$

for every integer n. In particular $1/e^z = e^{-z}$.

Another property of the exponential function is that for all real θ,

$$e^{i\theta} = \cos\theta + i\sin\theta, \tag{5.8}$$

and the proof results from adding the series involved. Indeed, $i^2 = -1$, $i^3 = -i$, $i^4 = 1$, etc., and thus

$$\cos\theta = 1 - \frac{\theta^2}{2!} + \frac{\theta^4}{4!} - \cdots$$

$$= 1 + \frac{(i\theta)^2}{2!} + \frac{(i\theta)^4}{4!} + \cdots,$$

$$\sin\theta = \theta - \frac{\theta^3}{3!} + \frac{\theta^5}{5!} - \cdots,$$

$$i\sin\theta = i\theta + \frac{(i\theta)^3}{3!} + \frac{(i\theta)^5}{5!} + \cdots.$$

Hence

$$\cos \theta + i \sin \theta = 1 + i\theta + \frac{(i\theta)^2}{2!} + \frac{(i\theta)^3}{3!} + \cdots$$

$$= \sum_{k=0}^{\infty} \frac{(i\theta)^k}{k!} = e^{i\theta}.$$

A consequence of (5.8) is that

$$e^{-i\theta} = \cos \theta - i \sin \theta, \tag{5.9}$$

since $\cos(-\theta) = \cos \theta$, and $\sin(-\theta) = -\sin \theta$. Using (5.8) and (5.9) we can solve for $\cos \theta$ and $\sin \theta$, obtaining

$$\cos \theta = \frac{e^{i\theta} + e^{-i\theta}}{2},$$

$$\sin \theta = \frac{e^{i\theta} - e^{-i\theta}}{2i}.$$

If z is a complex number with polar coordinates (r, θ), then

$$z = r(\cos \theta + i \sin \theta), \qquad (r \geq 0, 0 \leq \theta < 2\pi),$$

and we have, using (5.8),

$$z = re^{i\theta}. \tag{5.10}$$

Note that $|z| = r$, $|e^{i\theta}| = 1$ for every real θ. The relation (5.10) can be employed to find the roots of polynomials p of the form

$$p(z) = z^n - c, \tag{5.11}$$

where c is a complex constant. Suppose $c = |c|e^{i\alpha}$, where α is real, $0 \leq \alpha < 2\pi$, and $re^{i\theta}$ is a root. Then

$$r^n e^{in\theta} = |c|e^{i\alpha},$$

and taking magnitudes of both sides we see that

$$r^n = |c|, \quad \text{or} \quad r = |c|^{1/n},$$

where the positive n-th root is understood. Further

$$e^{in\theta} = e^{i\alpha}, \quad \text{or} \quad e^{i(n\theta - \alpha)} = 1.$$

There are exactly n distinct values of θ satisfying this relation and $0 \leqq \theta < 2\pi$, namely, those for which

$$n\theta - \alpha = 2\pi k,$$

or

$$\theta = \frac{\alpha + 2\pi k}{n}, \qquad (k = 0, 1, \cdots, n - 1).$$

Thus the roots $z_1, \cdots, z_n$ of the polynomial p in (5.11) are given by

$$z_{k+1} = |c|^{1/n}e^{i(\alpha+2\pi k)/n}$$

$$= |c|^{1/n}\left[\cos\left(\frac{\alpha + 2\pi k}{n}\right) + i\sin\left(\frac{\alpha + 2\pi k}{n}\right)\right], \quad (k = 0, 1, \cdots, n - 1).$$

Geometrically we can describe the roots of p as follows. All roots lie on a circle about the origin with radius $|c|^{1/n}$. One root has an angle α/n with the real axis, if c has angle α with the real axis. The remainder of the roots are located by cutting the circle into n even parts, with the first cut being at the root at angle α/n.

As a particular example let us find the three cube roots of $4i$. Thus we want the roots of $z^3 - 4i$. Here $c = 4i$, and hence the cube roots will all have a magnitude of $|4i|^{1/3} = 4^{1/3}$. If we write $c = |c|e^{i\alpha}$, we see that $\alpha = \pi/2$ in this case. Thus the three cube roots of $4i$ are given by

$$z_1 = 4^{1/3}e^{i\pi/6}, \qquad z_2 = 4^{1/3}e^{i5\pi/6}, \qquad z_3 = 4^{1/3}e^{i9\pi/6},$$

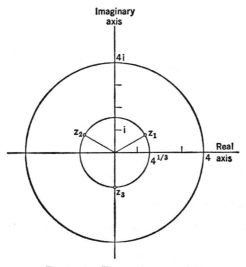

Figure 2. Three cube roots of $4i$

or since $\pi/6$ represents $30°$,

$$z_1 = 4^{1/3}\left(\frac{\sqrt{3}}{2} + \frac{i}{2}\right), \qquad z_2 = 4^{1/3}\left(-\frac{\sqrt{3}}{2} + \frac{i}{2}\right),$$

$$z_3 = -4^{1/3}i.$$

These roots are sketched in Fig. 2.

EXERCISES

1. Find the three cube roots of 1.

2. Find the two square roots of i.

3. Find all roots of the polynomials:
(a) $z^3 + 24$ (b) $z^4 + i64$
(c) $z^4 + 4z^2 + 4$ (d) $z^{100} - 1$

4. If $z = x + iy$, where x, y are real, show that $|e^z| = e^x$. As a consequence show that there is no complex z such that $e^z = 0$.

5. If a, b, x are real show that:
(a) $\mathrm{Re}\,[e^{(a+ib)x}] = e^{ax} \cos bx$ (b) $\mathrm{Im}\,[e^{(a+ib)x}] = e^{ax} \sin bx$

6. (a) If $r = a + ib \neq 0$, where a, b are real, show that $(e^{rx})' = re^{rx}$.
(b) Using (a) compute:

   (i) $\displaystyle\int_0^1 e^{rx}\, dx$

   (ii) $\displaystyle\int_0^1 e^{ax} \cos bx\, dx$

   (iii) $\displaystyle\int_0^1 e^{ax} \sin bx\, dx$

7. (a) If $\phi(x) = e^{rx}$, where r is a complex constant, and x is real, show that $\phi'(x) - r\phi(x) = 0$.
(b) If $\phi(x) = e^{iax}$, where a is a real constant, show that:
   (i) $\phi'(x) - ia\phi(x) = 0$
   (ii) $\phi''(x) + a^2\phi(x) = 0$

8. For what values of the constant r will the function ϕ given by $\phi(x) = e^{rx}$ satisfy

$$\phi''(x) + 3\phi'(x) - 2\phi(x) = 0$$

for all real x?

9. Let $a_k = k! + (i/k!)$. For what real x are the following series convergent?

(a) $\displaystyle\sum_{k=0}^{\infty} (\text{Re } a_k)x^k$

(b) $\displaystyle\sum_{k=0}^{\infty} (\text{Im } a_k)x^k$

(c) $\displaystyle\sum_{k=0}^{\infty} a_k x^k$

10. Consider the series

$$\sum_{k=0}^{\infty} z^k, \qquad (*)$$

where z is complex.

(a) Show that the partial sum

$$s_n(z) = \sum_{k=0}^{n} z^k = \frac{1 - z^{n+1}}{1 - z},$$

if $z \neq 1$.

(b) Show that the series (*) converges absolutely for $|z| < 1$.

(c) Compute the sum $s(z)$ of the series (*) for $|z| < 1$.

6. Determinants

We shall need to know the connection between determinants and the solution of systems of linear equations. Suppose we have such a system of n equations

$$a_{11}z_1 + a_{12}z_2 + \cdots + a_{1n}z_n = c_1$$

$$a_{21}z_1 + a_{22}z_2 + \cdots + a_{2n}z_n = c_2 \qquad (6.1)$$

$$\vdots$$

$$a_{n1}z_1 + a_{n2}z_2 + \cdots + a_{nn}z_n = c_n,$$

where the a_{ij} and c_i are given complex constants. The problem is to find complex numbers $z_1, \cdots, z_n$ satisfying these equations. Such a set of n numbers is called a *solution* of (6.1). We say that two solutions $z_1, \cdots, z_n$ and $z_1', \cdots, z_n'$ of (6.1) are *equal* if $z_1 = z_1', \cdots, z_n = z_n'$. If $c_1 = c_2 = \cdots = c_n = 0$ we say that the system is a *homogeneous system* of n linear equations, otherwise we say (6.1) is a *non-homogeneous* system. The determinant Δ of the coefficients in (6.1) is denoted by

$$\Delta = \begin{vmatrix} a_{11} & a_{12} & \cdots & a_{1n} \\ a_{21} & a_{22} & \cdots & a_{2n} \\ & & \bullet & \\ & & \bullet & \\ & & \bullet & \\ a_{n1} & a_{n2} & \cdots & a_{nn} \end{vmatrix},$$

and is shorthand for the number Δ given by

$$\Delta = \sum (\pm) a_{1 i_1} a_{2 i_2} \cdots a_{n i_n},$$

where the sum is over all indices $i_1, \cdots, i_n$ such that $i_1, \cdots, i_n$ is a permutation of $1, \cdots, n$ and each term occurs with a $+$ or $-$ sign according as $i_1, \cdots, i_n$ is an even or odd permutation of $1, \cdots, n$. Thus

$$\begin{vmatrix} a_{11} & a_{12} \\ a_{21} & a_{22} \end{vmatrix} = a_{11}a_{22} - a_{12}a_{21},$$

and

$$\begin{vmatrix} a_{11} & a_{12} & a_{13} \\ a_{21} & a_{22} & a_{23} \\ a_{31} & a_{32} & a_{33} \end{vmatrix} = a_{11}a_{22}a_{33} - a_{11}a_{23}a_{32} + a_{12}a_{23}a_{31} \\ - a_{12}a_{21}a_{33} + a_{13}a_{21}a_{32} - a_{13}a_{22}a_{31}.$$

The principal results we require concerning determinants are contained in the following theorems. They are usually proved in elementary texts on linear algebra.

Theorem 2. *If the determinant Δ of the coefficients in* (6.1) *is not zero there is a unique solution of the system for* $z_1, \cdots, z_n$. *It is given by*

$$z_k = \frac{\Delta_k}{\Delta}, \qquad (k = 1, \cdots, n),$$

where Δ_k is the determinant obtained from Δ by replacing its kth column $a_{1k}, \cdots, a_{nk}$ *by* $c_1, \cdots, c_n$.

Proof for the case $n = 2$. In this case suppose z_1, z_2 satisfy

$$\begin{aligned} a_{11}z_1 + a_{12}z_2 &= c_1 \\ a_{21}z_1 + a_{22}z_2 &= c_2. \end{aligned} \tag{6.2}$$

Multiply the first equation by a_{22}, the second equation by $-a_{12}$, and add. There results

$$z_1\Delta = a_{22}c_1 - a_{12}c_2 = \begin{vmatrix} c_1 & a_{12} \\ c_2 & a_{22} \end{vmatrix} = \Delta_1.$$

Multiply the first equation by $-a_{21}$, and the second by a_{11}, and add, obtaining

$$z_2\Delta = -a_{21}c_1 + a_{11}c_2 = \begin{vmatrix} a_{11} & c_1 \\ a_{21} & c_2 \end{vmatrix} = \Delta_2.$$

Thus if $\Delta \neq 0$, z_k must be Δ_k/Δ $(k = 1, 2)$, and it is readily verified that these values satisfy (6.2).

We note that for a homogeneous system $(c_1 = c_2 = \cdots = c_n = 0$ in (6.1)) there is always the solution

$$z_1 = z_2 = \cdots = z_n = 0.$$

This solution is called the *trivial solution*.

Theorem 3. *If $c_1 = c_2 = \cdots = c_n = 0$ in* (6.1), *and the determinant of the coefficients $\Delta = 0$, there is a solution of* (6.1) *such that not all the z_k are* 0.

Proof for the case $n = 2$. We are dealing with the case

$$a_{11}z_1 + a_{12}z_2 = 0$$

$$a_{21}z_1 + a_{22}z_2 = 0,$$

where

$$a_{11}a_{22} - a_{21}a_{12} = 0.$$

If $a_{11} \neq 0$,

$$z_1 = \frac{-a_{12}}{a_{11}}, \qquad z_2 = 1,$$

is a solution. If $a_{11} = 0$, and $a_{21} \neq 0$,

$$z_1 = \frac{-a_{22}}{a_{21}}, \qquad z_2 = 1,$$

is a solution. If $a_{11} = 0$, and $a_{21} = 0$,

$$z_1 = 1, \qquad z_2 = 0,$$

is a solution.

Combining Theorem 3 with Theorem 2 we obtain

Theorem 4. *The system of equations* (6.1) *has a unique solution if, and only if, the determinant Δ of the coefficients is not zero.*

Proof. If $\Delta \neq 0$ Theorem 2 says that there is a unique solution. Conversely, suppose there is a unique solution $z_1, \cdots, z_n$ of (6.1). If $\Delta = 0$, by

Theorem 3 there is a solution $\zeta_1, \cdots, \zeta_n$ of the corresponding homogeneous system, which is not the trivial solution. Then it is easy to check that $z_1 + \zeta_1, \cdots, z_n + \zeta_n$ is a solution of (6.1) distinct from $z_1, \cdots, z_n$, and forces us to conclude that $\Delta \neq 0$.

EXERCISES

1. Consider the system of equations

$$iz_1 + z_2 = 1 + i$$
$$2z_1 + (2 - i)z_2 = 1.$$

   (a) Compute the determinant of the coefficients.
   (b) Solve the system for z_1 and z_2.

2. Solve the following system for z_1, z_2 and z_3:

$$3z_1 + z_2 - z_3 = 0$$
$$2z_1 \qquad - z_3 = 1$$
$$z_2 + 2z_3 = 2$$

3. Does the following system of equations have any solution other than $z_1 = z_2 = z_3 = 0$? If so find one.

$$4z_1 + 2z_2 + 2z_3 = 0$$
$$3z_1 + 7z_2 + 2z_3 = 0$$
$$2z_1 + z_2 + z_3 = 0$$

4. Consider the homogeneous system corresponding to (6.1) (the case $c_1 = c_2 = \cdots = c_n = 0$). Show that if the determinant of the coefficients $\Delta = 0$, there are an infinite number of solutions. (*Hint:* If $z_1, \cdots, z_n$ is a non-trivial solution, show that $\alpha z_1, \cdots, \alpha z_n$ is also a solution for any complex number α.)

5. Prove that if the determinant Δ of the coefficients in (6.1) is zero then either there is no solution of (6.1), or there are an infinite number of solutions. (*Hint:* Use Ex. 4.)

7. Remarks on methods of discovery and proof

Often a student studying mathematics has difficulty in understanding why or how a particular result, or method of proof, was ever conceived in the first place. Sometimes ideas seem to appear from nowhere. Now it is true that mathematical geniuses do invent radically new results, and methods for proving old results, which often appear quite strange. The most that ordinary people can do is to accept these brilliant ideas for what they are, try to understand their consequences, and build on them to obtain further information. However, there are a few general principles which, if followed, can lead to a better understanding of mathematical discovery and proof. Concerning discovery, we mention two principles:

(a) *use simple examples as a basis for conjecturing general results,*
(b) *argue in reverse.*

Both of these principles are illustrated in the proof we gave of Theorem 2 for the case $n = 2$. We were faced with trying to find out whether the system (6.1) of n linear equations has a solution or not, and what condition, or conditions, would guarantee a unique solution. We looked at the simplest example, which occurs for $n = 2$ (using (a)). Then we *assumed* that we had a solution (principle (b)), and found out what must be true for a solution, namely, that

$$z_1\Delta = \Delta_1, \qquad z_2\Delta = \Delta_2.$$

We immediately saw that if $\Delta \neq 0$, then

$$z_1 = \frac{\Delta_1}{\Delta}, \qquad z_2 = \frac{\Delta_2}{\Delta}. \tag{7.1}$$

Note that at this point we have not yet shown that there *is* a solution. All we have shown is that *if* z_1, z_2 is a solution, and $\Delta \neq 0$, it must be given by (7.1). We can now guess that if $\Delta \neq 0$, then z_1, z_2 given by (7.1) is a solution. This can be readily verified by substituting (7.1) into the given equations. An alternate procedure is to check that the steps leading to (7.1) can be reversed, if $\Delta \neq 0$. Once we have discovered the right condition for the case $n = 2$, it is natural to conjecture that a similar condition will work for a general n.

Three important methods of proving mathematical results are:

(i) *a constructive method,*
(ii) *method of contradiction,*
(iii) *method of induction.*

A typical example of a constructive method appears in the proof of Theorem 3 for the case $n = 2$. We wanted to show that nontrivial solutions of the two homogeneous equations exist if $\Delta = 0$. To do this we constructed solutions explicitly. An example of the method of contradiction appears in the proof of Theorem 4. We supposed that the system (6.1) had a unique solution. We assumed that $\Delta = 0$, and, using logical arguments, we arrived at the fact that (6.1) does not have a unique solution. This is a contradiction, and the only thing that can be wrong is our assumption that $\Delta = 0$.

The only other alternative is that $\Delta \neq 0$, which is the conclusion we desired.

The method of induction is concerned with proving an infinite number of statements $s_1, s_2, \cdots$, one for each positive integer n. If s_1 is true, and if for any positive integer k the statement s_k implies the statement s_{k+1}, then all the statements $s_1, s_2, \cdots$, are true. An example of a result which can be proved using induction is the formula

$$(fg)^{(k)} = \sum_{l=0}^{k} \binom{k}{l} f^{(k-l)}g^{(l)}, \qquad \binom{k}{l} = \frac{k!}{l!(k-l)!},$$

for the k-th derivative of the product of two complex-valued functions f, g which have k derivatives; see (4.3). The proof is the same as the induction used to prove the binomial formula

$$(a + b)^k = \sum_{l=0}^{k} \binom{k}{l} a^{k-l}b^l, \qquad (k = 1, 2, \cdots),$$

for the powers of the sum of two complex numbers a, b. The method of induction is equivalent to a property of the positive integers, and consequently we assume that this method is a valid method of proof.

The principles of discovery (a), (b), and the methods of proof (i), (ii), (iii), will be used many times throughout this book. It will be instructive for the student to identify which principles and methods are being used in any particular situation.

CHAPTER 1

Introduction—Linear Equations of the First Order

1. Introduction

In Sec. 2 we discuss what is meant by an ordinary differential equation and its solutions. Various problems which arise in connection with differential equations are considered in Sec. 3, notably initial value problems, boundary value problems, and the qualitative behavior of solutions. In a succession of easy steps we solve the linear equation of the first order in Secs. 4–7.

2. Differential equations

Suppose f is a complex-valued function defined for all real x in an interval I, and for complex y in some set S. The value of f at (x, y) is denoted by $f(x, y)$. An important problem associated with f is to find a (complex-valued) function ϕ on I, which is differentiable there, such that for all x on I,

$$\text{(i)} \quad \phi(x) \text{ is in } S,$$

$$\text{(ii)} \quad \phi'(x) = f(x, \phi(x)).$$

This problem is called an *ordinary differential equation of the first order*, and is denoted by

$$y' = f(x, y). \tag{2.1}$$

The *ordinary* refers to the fact that only ordinary derivatives enter into the problem, and not partial derivatives. If such a function ϕ exists on I satisfying (i) and (ii) there, then ϕ is called a *solution* of (2.1) on I.

As an example consider the case when f is independent of y, that is, we have the equation

$$y' = f(x), \tag{2.2}$$

where f is defined on some interval I. The problem is to find a function ϕ on I such that ϕ' exists there, and $\phi'(x) = f(x)$. This is one of the most important problems considered in the study of calculus. Indeed, if f is *continuous* on I, we know that the indefinite integral function ϕ_0 defined by

$$\phi_0(x) = \int_{x_0}^{x} f(t) \, dt,$$

where x_0 is some fixed point in I, is a solution of (2.2). Moreover, if ϕ is any solution of (2.2), then there is a constant c such that

$$\phi(x) = \phi_0(x) + c$$

for all x in I; and every constant c gives rise to a solution in this way. Thus all solutions of (2.2) are known in case f is continuous on I, and the study of (2.2) reduces to the study of integration.

For a second example, suppose that $\phi(x)$ denotes the amount of a certain substance at time x, and we know that the substance increases at a rate proportional to the amount present at any time x. Then we must have

$$\phi'(x) = k\phi(x),$$

where k is some constant. Thus ϕ is a solution of the differential equation

$$y' = ky. \tag{2.3}$$

Conventional examples of processes described by this equation are population growth $(k > 0)$ and radioactive decay $(k < 0)$. A solution of (2.3) is given by

$$\phi(x) = e^{kx},$$

which exists for all real x.

The problem posed by the equation $y' = f(x, y)$ has a simple geometrical interpretation in case f is real-valued, and y is defined on a set S of real numbers. Then for each x in I and y in S we are given a number $f(x, y)$, which may be thought of as the slope of a straight line through the point (x, y). A solution of $y' = f(x, y)$ on I is a function ϕ whose graph (the set of points $(x, \phi(x))$, x in I) is a curve whose tangent at $(x, \phi(x))$ has the slope $\phi'(x)$, which is the same as the given slope $f(x, \phi(x))$ at this point. Thus, geometrically we are given a set of directions, and the differential equation is the problem of finding curves having these directions as tangents. The set of directions $\{f(x, y)\}$ is called a *direction field*. Fig. 3 shows such a field for $f(x, y) = -xy$, and the curve sketched is the solution $\phi(x) = 2e^{-(x^2/2)}$ of the equation $y' = -xy$.

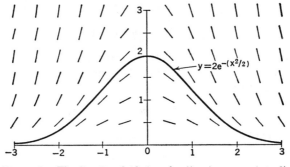

Figure 3. The direction field given by $f(x, y) = -xy$ $(y > 0)$

Sometimes a differential equation occurs in a slightly more general form, where the derivative y' is not by itself on one side of the equation. Thus it might be necessary to consider an equation of the form

$$F(x, y, y') = 0. \tag{2.4}$$

Here F is some function defined for real x in an interval I, and complex y_1, y_2 in sets S_1, S_2 respectively. Then (2.4) is the problem of finding a (complex-valued) function ϕ on I, which is differentiable there, such that for all x on I,

  (i) $\phi(x)$ is in S_1, $\phi'(x)$ is in S_2,

    (so that $F(x, \phi(x), \phi'(x))$ is defined),

  (ii) $F(x, \phi(x), \phi'(x)) = 0$.

This problem is also called an *ordinary differential equation of the first order*. The equation (2.1) is the special case when

$$F(x, y, y') = y' - f(x, y).$$

Usually we shall consider equations in the form (2.1), since it can be shown that (2.4) can be reduced to the form (2.1) under rather general conditions on F.

  More general differential equations involve higher order derivatives. Let F now be a function defined for real x in an interval I, and for complex $y_1, y_2, \cdots, y_{n+1}$ in sets $S_1, S_2, \cdots, S_{n+1}$ respectively. The problem of finding a function ϕ on I, having n derivatives there, and such that for all x in I,

  (i) $\phi^{(k-1)}(x)$ is in S_k $(k = 1, 2, \cdots, n + 1)$,

    $(\phi^{(0)}(x) = \phi(x))$,

  (ii) $F(x, \phi(x), \cdots, \phi^{(n)}(x)) = 0$,

is called an *ordinary differential equation of the nth order*, and is denoted by

$$F(x, y, y', \cdots, y^{(n)}) = 0. \tag{2.5}$$

A function ϕ on I, with n derivatives, satisfying (i) and (ii) is called a *solution* of (2.5) on I. Again it will be the usual situation to consider those equations of the form

$$y^{(n)} = f(x, y, y', \cdots, y^{(n-1)}).$$

An example of a second order equation is

$$y'' + y = 0, \tag{2.6}$$

which arises naturally in the study of electrical and mechanical oscillations. Two solutions ϕ_1, ϕ_2 which exist for all real x are given by

$$\phi_1(x) = \cos x, \qquad \phi_2(x) = \sin x.$$

3. Problems associated with differential equations

When presented with a differential equation our first impulse might be to try to find *all* solutions of it. Ideally we would like to write down these solutions in terms of well-known functions. This can be done for a large number of very important equations. For example, we indicated in Sec. 2 that all solutions of

$$y' = f(x), \qquad (f \text{ continuous}), \tag{3.1}$$

are given by

$$\phi(x) = \int_{x_0}^{x} f(t) \, dt + c, \tag{3.2}$$

where x_0 is some point in the interval where f is defined, and c is any constant. All solutions of

$$y' = ky \tag{3.3}$$

are of the form

$$\phi(x) = ce^{kx}, \tag{3.4}$$

and c can be any constant. We shall prove this elementary fact in Sec. 5. Also every solution of

$$y'' + y = 0 \tag{3.5}$$

has the form

$$\phi(x) = c_1 \cos x + c_2 \sin x, \tag{3.6}$$

where c_1, c_2 may be arbitrary constants. The proof of this will occur in Chap. 2.

Frequently we are not interested in *all* solutions of an equation, but only those satisfying certain other conditions. These conditions may take many forms, but two of the most important types are *initial conditions* and *boundary conditions*. An initial condition is a condition on the solution at one point. For example, the solution of (3.3) having the property that $\phi(0) = 2$ (the initial condition) is readily seen from (3.4) to be given by

$$\phi(x) = 2e^{kx}.$$

Such an *initial value problem* would be denoted by

$$y' = ky, \qquad y(0) = 2.$$

Similarly, the solution ϕ of (3.5) satisfying

$$\phi(0) = 1, \qquad \phi'(0) = 2,$$

is given by

$$\phi(x) = \cos x + 2 \sin x.$$

This problem would be denoted by

$$y'' + y = 0, \qquad y(0) = 1, \qquad y'(0) = 2.$$

A boundary condition is a condition on the solution at two or more points. For example, the solution ϕ of (3.5) satisfying

$$\phi(0) = 1, \qquad \phi'(2\pi) = -1,$$

is given by

$$\phi(x) = \cos x - \sin x.$$

There are many equations for which it is not obvious that solutions exist at all; and if they do, it might not be possible to write down "nice" formulas for them. For example, consider the equation

$$y'' + y' + \sin y = 0, \tag{3.7}$$

which is encountered in the study of the motion of a pendulum. It can be shown that (3.7) has solutions, satisfying any given real initial condition, which exist for all real x, although we can not express them in terms of functions we meet in calculus. How do we *solve* equations such as (3.7), that is, find the solutions? One method is to develop mathematical procedures which would allow us to compute the value of a solution at any given x to any desired degree of accuracy. This method should be sufficiently general to cover a large number of equations. Such a procedure is developed in Chap. 5, where a general method for computing solutions to initial value problems is given.

Even though it is impossible to express solutions of some equations in nice formulas, it is often the case that we can say a good deal about the *properties* of the solutions. In many situations it is just some property of the solutions which we wish to investigate. For example, without solving (3.7) we can show that any solution ϕ for which

$$-\pi < \phi(0) < \pi, \qquad \phi'(0) = 0,$$

will tend to zero as $x \to \infty$. This corresponds to the fact that the oscillations of the pendulum are damped, and eventually the pendulum will stay arbitrarily close to its equilibrium position $y = 0$.

EXERCISES

1. Find all solutions of the following equations on $-\infty < x < \infty$:
   (a) $y' = e^{3x} + \sin x$ (b) $y'' = 2 + x$
   (c) $y^{(k)} = 0$, (k a positive integer)   (d) $y''' = x^2$

2. Verify that the following are solutions of the differential equations given:
   (a) $\phi(x) = e^{-\sin x}$, for $y' + (\cos x)y = 0$
   (b) $\phi(x) = \sin x - 1$, for $y' + (\cos x)y = \sin x \cos x$
   (c) $\phi(x) = 1$, for $y'' - y' = 0$
   (d) $\phi(x) = e^x$, for $y'' - y' = 0$
   (e) $\phi(x) = c_1 + c_2 e^x$, for $y'' - y' = 0$, (c_1, c_2 any constants)
   (f) $\phi(x) = \sin 2x$, for $y'' + 4y = 0$
   (g) $\phi(x) = e^{2ix}$, for $y'' + 4y = 0$
   (h) $\phi(x) = c_1 \cos kx + c_2 \sin kx$, for $y'' + k^2 y = 0$, (k a positive constant, and c_1, c_2 any constants)

3. Consider the equation $y' + 5y = 2$.
   (a) Show that the function ϕ given by

$$\phi(x) = \tfrac{2}{5} + ce^{-5x}$$

   is a solution, where c is any constant.
   (b) Assuming every solution has this form, find that solution satisfying $\phi(1) = 2$.
   (c) Find that solution satisfying $\phi(1) = 3\phi(0)$.

4. Consider the equation $y'' = 3x + 1$.
   (a) Find all solutions on the interval $0 \leq x \leq 1$.
   (b) Find that solution ϕ which satisfies $\phi(0) = 1, \phi'(0) = 2$.
   (c) Find that solution ϕ which satisfies $\phi(0) = 0, \phi(1) = 3$.

5. Consider the equation $y' = ky$ on $-\infty < x < \infty$, where k is some constant.
   (a) Show that if ϕ is any solution, and $\psi(x) = \phi(x)e^{-kx}$, then $\psi(x) = c$, where c is a constant. (*Hint*: Show that $\psi'(x) = 0$ for all x.)
   (b) Prove that if Re $k < 0$ then every solution tends to zero as $x \to \infty$.
   (c) Prove that if Re $k > 0$ then the magnitude of every non-trivial (not identically zero) solution tends to ∞ as $x \to \infty$.
   (d) What can you say about the magnitudes of the solutions if Re $k = 0$?

4. Linear equations of the first order

We initiate our study of differential equations by considering the simple case of a *linear equation of the first order*. This is an equation of the form

$$y' + a(x)y = b(x), \tag{4.1}$$

where a, b are certain functions defined on an interval I. Writing this in the form $y' = f(x, y)$ we see that

$$f(x, y) = -a(x)y + b(x). \tag{4.2}$$

If $b(x) = 0$ for all x in I, the corresponding equation

$$y' + a(x)y = 0$$

is called a *homogeneous equation*, whereas if b is not identically zero on I, (4.1) is called a *non-homogeneous equation*.

We note that if $b(x) = 0$ for all x in I, then the f of (4.2) is linear in y, that is,

$$f(x, y_1 + y_2) = f(x, y_1) + f(x, y_2),$$

and homogeneous in y, that is,

$$f(x, cy) = cf(x, y),$$

where c is any constant.

We first solve the simple case of (4.1) when a is a constant, and then treat the more general case.

5. The equation $y' + ay = 0$

If a is a constant and ϕ is a solution of

$$y' + ay = 0, \tag{5.1}$$

then $\phi' + a\phi = 0$, and this implies that

$$e^{ax}(\phi' + a\phi) = 0,$$

or

$$(e^{ax}\phi)' = 0.$$

Therefore there is a constant c such that $e^{ax}\phi(x) = c$, or

$$\phi(x) = ce^{-ax}. \tag{5.2}$$

We have shown that any solution ϕ of (5.1) must have the form (5.2), where c is some constant. Conversely, if c is any constant, the function ϕ defined by (5.2) is a solution of (5.1), for

$$\phi'(x) + a\phi(x) = -ace^{-ax} + ace^{-ax} = 0.$$

We have proved a small theorem.

Theorem 1. *Consider the equation*

$$y' + ay = 0,$$

where a *is a complex constant. If* c *is any complex number, the function* ϕ *defined by*

$$\phi(x) = ce^{-ax}$$

is a solution of this equation, and moreover every solution has this form.

Notice that all solutions exist for all real x, that is, for $-\infty < x < \infty$. Also note that the constant c is the value of ϕ at 0, that is, $c = \phi(0)$.

6. The equation $y' + ay = b(x)$

Let a be a constant and let b be a continuous function on some interval I. We consider the equation

$$y' + ay = b(x), \tag{6.1}$$

and try to solve it using the same method as in Sec. 5. If ϕ is a solution of (6.1), then

$$e^{ax}(\phi' + a\phi) = e^{ax}b,$$

or

$$(e^{ax}\phi)' = e^{ax}b.$$

Let B be a function such that $B'(x) = e^{ax}b(x)$, for example,

$$B(x) = \int_{x_0}^{x} e^{at}b(t)\ dt,$$

where x_0 is some fixed point in I. Since $e^{ax}\phi$ is another function whose derivative is $e^{ax}b$, it follows that

$$e^{ax}\phi(x) = B(x) + c$$

for some constant c. Therefore

$$\phi(x) = e^{-ax}B(x) + ce^{-ax}. \tag{6.2}$$

It is easy to see that our steps can be retraced to prove that if ϕ is defined by (6.2), where c is *any* constant, then ϕ is a solution of (6.1).

We observe from (6.2) that the function ψ defined by $\psi(x) = e^{-ax}B(x)$ is a particular solution of (6.1), since it is the solution corresponding to $c = 0$. We summarize our result.

Theorem 2. *Consider the equation*

$$y' + ay = b(x),$$

where a *is a constant, and* b *is a continuous function on an interval* I. *If* x_0 *is a point in* I *and* c *is any constant, the function* ϕ *defined by*

$$\phi(x) = e^{-ax} \int_{x_0}^{x} e^{at}b(t)\ dt + ce^{-ax}$$

is a solution of this equation. Every solution has this form.

EXERCISES

1. Find all solutions of the following equations:
   (a) $y' - 2y = 1$ (b) $y' + y = e^x$
   (c) $y' - 2y = x^2 + x$ (d) $3y' + y = 2e^{-x}$
   (e) $y' + 3y = e^{ix}$

2. Let ϕ be the solution of $y' + iy = x$ such that $\phi(0) = 2$. Find $\phi(\pi)$.

3. Consider the equation

$$Ly' + Ry = E,$$

where L, R, E are positive constants.
   (a) Solve this equation.
   (b) Find the solution ϕ satisfying $\phi(0) = I_0$, where I_0 is a given positive constant.
   (c) Sketch a graph of the solution given in (b) for the case $I_0 > E/R$.
   (d) Show that every solution tends to E/R as $x \to \infty$.

4. Consider the equation

$$Ly' + Ry = E \sin \omega x,$$

where L, R, E, ω are positive constants.
   (a) Compute the solution ϕ satisfying $\phi(0) = 0$.
   (b) Show that this solution may be written in the form

$$\phi(x) = \frac{E\omega L}{R^2 + \omega^2 L^2} e^{-(R/L)x} + \frac{E}{\sqrt{R^2 + \omega^2 L^2}} \sin(\omega x - \alpha),$$

   where α is the angle satisfying

$$\cos \alpha = \frac{R}{\sqrt{R^2 + \omega^2 L^2}}, \quad \sin \alpha = \frac{\omega L}{\sqrt{R^2 + \omega^2 L^2}}.$$

   (c) Sketch the graph of the solution given in (b).

5. Consider the equation

$$Ly' + Ry = Ee^{i\omega x},$$

where L, R, E, ω are positive constants.

(a) Compute the solution ϕ which satisfies $\phi(0) = 0$.

(b) Using the differential equation show that $\phi_1 = \text{Re } \phi$ satisfies

$$Ly' + Ry = E \cos \omega x.$$

Compute ϕ_1.

(c) Using the differential equation show that $\phi_2 = \text{Im } \phi$ satisfies

$$Ly' + Ry = E \sin \omega x.$$

Compute ϕ_2.

6. Let ϕ satisfy the equation

$$y' + ay = b_1(x),$$

and let ψ satisfy the equation

$$y' + ay = b_2(x),$$

where b_1, b_2 are defined on the same interval I, and a is a constant.

(a) Show that $\chi = \phi + \psi$ satisfies

$$y' + ay = b_1(x) + b_2(x)$$

on I.

(b) Apply the result of (a) to find the solution of

$$y' + y = \sin x + 3 \cos 2x$$

whose graph passes through the origin.

7. Consider the equation

$$y' + ay = b(x),$$

where a is a constant, and b is a continuous function on $0 \leqq x < \infty$, satisfying there $|b(x)| \leqq k$, where k is some positive number.

(a) Find the solution ϕ satisfying $\phi(0) = 0$.

(b) If $\text{Re } a \neq 0$, show that this solution satisfies

$$|\phi(x)| \leqq \frac{k}{\text{Re } a} [1 - e^{-(\text{Re } a)x}].$$

(c) Show that the right side of the inequality in (b) is the solution of

$$y' + (\text{Re } a)y = k, \qquad (\text{Re } a \neq 0),$$

whose graph passes through the origin.

8. Let a be a constant, and let b_1, b_2 be two continuous functions on $0 \leqq x < \infty$ such that

$$|b_1(x) - b_2(x)| \leqq k, \qquad (0 \leqq x < \infty), \tag{*}$$

for some constant $k > 0$. Let ϕ be a solution of $y' + ay = b_1(x)$, and ψ a solution of $y' + ay = b_2(x)$. Assume that $\phi(0) = \psi(0)$. Show that

$$| \phi(x) - \psi(x) | \leq \frac{k}{\text{Re } a} [1 - e^{-(\text{Re } a)x}] \qquad (**)$$

for $0 \leq x < \infty$.

(*Note*: If b_2 approximates b_1 with an error at most k, in the sense of (*), then (**) gives an estimate for the difference between the solutions. If k is small ψ will be close to ϕ.)

9. Consider the equation $y' + ay = b(x)$, where a is a constant such that Re $a > 0$, and b is a continuous function on $0 \leq x < \infty$ which tends to the constant β as $x \to \infty$. Prove that every solution of this equation tends to β/a as $x \to \infty$.

7. The general linear equation of the first order

We now consider the equation

$$y' + a(x)y = b(x), \qquad (7.1)$$

where a and b are continuous functions on some interval I. If we are given an equation

$$\alpha(x)y' + \beta(x)y = \gamma(x)$$

and $\alpha(x) \neq 0$ on I, we may divide by $\alpha(x)$ to obtain an equation of the form (7.1). The points where $\alpha(x) = 0$, called *singular points*, are frequently troublesome. We postpone a discussion of these difficulties until later; see Chap. 4.

We try to solve (7.1) in the same way we solved the case when a was constant. Suppose ϕ is a solution of (7.1). We try to find a function u such that

$$u(\phi' + a\phi) = (u\phi)'.$$

If A is a function whose derivative is a, for example

$$A(x) = \int_{x_0}^{x} a(t) \, dt,$$

where x_0 is a fixed point in I, then such a function u is given by $u = e^A$ since

$$(e^A\phi)' = e^A\phi' + ae^A\phi = e^A(\phi' + a\phi).$$

Therefore $\phi' + a\phi = b$ if and only if

$$(e^A\phi)' = e^A b,$$

and this is valid if and only if

$$e^A\phi = B + c, \tag{7.2}$$

where c is a constant, and B is a function whose derivative is $e^A b$. For example we can choose B to be given by

$$B(x) = \int_{x_0}^{x} e^{A(t)}b(t)\, dt.$$

Now (7.2) holds if and only if

$$\phi(x) = e^{-A(x)}B(x) + ce^{-A(x)}. \tag{7.3}$$

We have thus shown that every solution of (7.1) has the form (7.3), and conversely, if c is any constant, the function ϕ defined by (7.3) is a solution of (7.1).

We remark that the function $\psi = e^{-A}B$ is a particular solution of (7.1) (the case $c = 0$), and that $\phi_1 = e^{-A}$ is a solution of the homogeneous equation $y' + a(x)y = 0$.

Theorem 3. *Suppose* a *and* b *are continuous functions on an interval* I. *Let* A *be a function such that* $A' = a$. *Then the function* ψ *given by*

$$\psi(x) = e^{-A(x)} \int_{x_0}^{x} e^{A(t)}b(t)\, dt,$$

where x_0 *is in* I, *is a solution of the equation*

$$y' + a(x)y = b(x) \tag{7.1}$$

on I. *The function* ϕ_1 *given by*

$$\phi_1(x) = e^{-A(x)}$$

is a solution of the homogeneous equation

$$y' + a(x)y = 0.$$

If c *is any constant,* $\phi = \psi + c\phi_1$ *is a solution of* (7.1), *and every solution of* (7.1) *has this form.*

In solving a particular linear equation a person with a good memory could remember (7.3), but it is probably easier to remember that multiplication of $\phi' + a\phi = b$ by e^A yields $(e^A\phi)' = e^A b$, which can be immediately integrated to give (7.3). As an example consider the equation

$$y' + (\cos x)y = \sin x \cos x.$$

Here $a(x) = \cos x$, $b(x) = \sin x \cos x$, and a choice for A is $A(x) = \sin x$. Thus, if ϕ is any solution,

$$(e^{\sin x}\phi)' = e^{\sin x} \sin x \cos x,$$

and an integration gives

$$e^{\sin x}\phi(x) = (\sin x - 1)e^{\sin x} + c,$$

or

$$\phi(x) = (\sin x - 1) + ce^{-\sin x},$$

where c is an arbitrary constant.

EXERCISES

1. Find all solutions of the following equations:

(a) $y' + 2xy = x$

(b) $xy' + y = 3x^3 - 1$ (for $x > 0$)

(c) $y' + e^x y = 3e^x$

(d) $y' - (\tan x)y = e^{\sin x}$ (for $0 < x < \pi/2$)

(e) $y' + 2xy = xe^{-x^2}$

2. Consider the equation $y' + (\cos x)y = e^{-\sin x}$.

(a) Find the solution ϕ which satisfies $\phi(\pi) = \pi$.

(b) Show that any solution ϕ has the property that

$$\phi(\pi k) - \phi(0) = \pi k,$$

where k is any integer.

3. Consider the equation $x^2 y' + 2xy = 1$ on $0 < x < \infty$.

(a) Show that every solution tends to zero as $x \to \infty$.

(b) Find that solution ϕ which satisfies $\phi(2) = 2\phi(1)$.

4. Consider the homogeneous equation

$$y' + a(x)y = 0, \qquad\qquad (*)$$

where a is continuous on an interval I.

(a) Show that the function ϕ given by $\phi(x) = 0$ for all x in I (the identically zero function) satisfies this equation. This solution is called the *trivial* solution.

(b) If ϕ is any solution of (*), and $\phi(x_0) = 0$ for some x_0 in I, show that ϕ is the trivial solution.

(c) If ϕ, ψ are two solutions of (*) satisfying $\phi(x_0) = \psi(x_0)$ for some x_0 in I, show that $\phi(x) = \psi(x)$ for *all* x in I.

(d) If ϕ is not the trivial solution, and ψ is any other solution, show that there is a constant c such that $\psi = c\phi$, that is, $\psi(x) = c\phi(x)$ for all x in I.

5. The equation

$$y' + \alpha(x)y = \beta(x)y^k, \qquad (k \text{ constant}),$$

is called *Bernoulli's equation.*

(a) Show that the formal substitution $z = y^{1-k}$ transforms this into the linear equation

$$z' + (1 - k)\alpha(x)z = (1 - k)\beta(x).$$

(b) Find all solutions of $y' - 2xy = xy^2$.

6. Consider the homogeneous equation $y' + a(x)y = 0$, where a is a continuous function on $-\infty < x < \infty$ which is periodic with period $\xi > 0$, that is, $a(x + \xi) = a(x)$ for all x.

(a) Let ϕ be a non-trivial solution, and let $\psi(x) = \phi(x + \xi)$. Show that ψ is a solution.

(b) Show that there is a constant c such that $\phi(x + \xi) = c\,\phi(x)$ for all x. (*Hint*: Ex. 4 (d)). Show that

$$c = \exp\left(-\int_0^\xi a(t)\,dt\right).$$

(*Note*: exp u is an alternate notation for e^u.)

(c) What condition must a satisfy in order that there exist a non-trivial solution of period ξ; of period 2ξ? If a is real-valued, what is the condition?

(d) If a is a constant, what must this constant be in order that a non-trivial solution of period 2ξ exist?

7. Consider the non-homogeneous equation $y' + a(x)y = b(x)$, where a, b are continuous real-valued functions on $-\infty < x < \infty$ which are of period $\xi > 0$, and b is not identically zero.

(a) Show that a solution ϕ is periodic of period ξ if, and only if, $\phi(0) = \phi(\xi)$.

(b) Show that there exists a unique solution of period ξ if there is no non-trivial solution of the homogeneous equation of period ξ.

(c) Suppose there is a non-trivial periodic solution of the homogeneous equation of period ξ. Show that there are periodic solutions of period ξ of the non-homogeneous equation if, and only if,

$$\int_0^\xi e^{A(t)}b(t)\,dt = 0,$$

where $A(t) = \int_0^t a(s)\,ds$.

(d) Find solutions of period 2π for the equations:

(i) $y' + 3y = \cos x$

(ii) $y' + (\cos x)y = \sin 2x$

8. Find all solutions of the equation

$$y' + 2y = b(x), \qquad (-\infty < x < \infty),$$

where $b(x) = 1 - |x|$ for $|x| \le 1$, and $b(x) = 0$ for $|x| > 1$.

9. The formula

$$\psi(x) = e^{-A(x)} \int_{x_0}^{x} e^{A(t)} b(t) \, dt$$

for a solution ψ of the equation

$$y' + a(x)y = b(x)$$

makes sense for some functions b which are not continuous. It is sometimes convenient to consider such b, and this ψ is called a solution even in this case. Of course ψ satisfies the differential equation at the continuity points of b. Find a solution of the equation

$$y' + ay = b(x), \qquad (a \text{ constant}),$$

where $b(x) = 1$ for $0 \leq x \leq \xi$, and $b(x) = 0$ for $x > \xi$. Here ξ is some positive constant.

10. Suppose ϕ is a function with a continuous derivative on $0 \leq x \leq 1$ satisfying there $\phi'(x) - 2\phi(x) \leq 1$, and $\phi(0) = 1$. Show that

$$\phi(x) \leq \tfrac{3}{2} e^{2x} - \tfrac{1}{2}.$$

11. Let ϕ, ψ be solutions of $y' + a(x)y = b(x)$ on an interval I containing x_0. Show that for x in I,

$$\psi(x) - \phi(x) = [\psi(x_0) - \phi(x_0)] \exp\left[-\int_{x_0}^{x} a(t) \, dt \right],$$

and consequently that

$$|\psi(x) - \phi(x)| = |\psi(x_0) - \phi(x_0)| \exp\left[-\int_{x_0}^{x} \operatorname{Re} a(t) \, dt \right]$$

12. Consider the boundary value problem

$$iy' = ly, \qquad y(1) = e^{i\alpha} y(0),$$

where α is a fixed real number, and l is a complex number.

(a) Show that this problem has a non-trivial solution if, and only if,

$$l = \lambda_k = 2\pi k - \alpha,$$

where $k = 0, \pm 1, \pm 2, \cdots$.

(b) Compute a solution ϕ_k of the problem for $l = \lambda_k$ which satisfies

$$\int_{0}^{1} |\phi_k(x)|^2 \, dx = 1.$$

(c) If ϕ_j, ϕ_k are the solutions determined in (b) for $l = \lambda_j$, $l = \lambda_k$ respectively, show that

$$\int_{0}^{1} \phi_j(x) \overline{\phi_k(x)} \, dx = 0$$

if $\lambda_j \neq \lambda_k$.

(d) If f is a function having the form

$$f = A_1\phi_1 + \cdots + A_n\phi_n,$$

where the ϕ_k are as in (b), and the A_k are constants, show that

$$A_k = \int_0^1 f(x)\overline{\phi_k(x)}\ dx.$$

(*Hint*: Use (b) and (c).)

13. Let f be any continuous function on $0 \leq x \leq 1$, and consider the problem

$$iy' - ly = f(x), \qquad y(1) = e^{i\alpha}y(0),$$

where α is real, and l is a complex number not equal to any of the λ_k in Ex. 12, (a). Find a solution ψ of this problem, and show that it can be expressed in the form

$$\psi(x) = \int_0^1 g(x, y)f(y)\ dy,$$

where g has a discontinuity at $y = x$.

14. (a) Find the solution ϕ of the linear equation

$$y' = 1 + y$$

satisfying $\phi(0) = 0$. Observe that this solution exists for all real x.
(b) Find the real-valued solution ψ of the nonlinear equation

$$y' = 1 + y^2$$

satisfying $\psi(0) = 0$. Observe that this solution exists only for $-(\pi/2) < x < (\pi/2)$. (*Hint*: For any t for which such a ψ exists we must have

$$\frac{\psi'(t)}{1 + [\psi(t)]^2} = (\tan^{-1}\psi)'(t) = 1.$$

Integrating from 0 to x we obtain $\tan^{-1}\psi(x) = x$, or $\psi(x) = \tan x$. Check that this ψ is the solution desired.)

(*Note*: This illustrates one of the differences between linear and nonlinear equations. General techniques for solving equations such as in (b) will be considered in Chap. 5.)

CHAPTER 2

Linear Equations with Constant Coefficients

1. Introduction

A linear differential equation of order n with constant coefficients is an equation of the form

$$a_0 y^{(n)} + a_1 y^{(n-1)} + a_2 y^{(n-2)} + \cdots + a_n y = b(x),$$

where $a_0 \neq 0$, $a_1, \cdots, a_n$ are complex constants, and b is some complex-valued function on an interval I. By dividing by a_0 we can arrive at an equation of the same form with a_0 replaced by 1. Therefore we can always assume $a_0 = 1$, and our equation becomes

$$y^{(n)} + a_1 y^{(n-1)} + a_2 y^{(n-2)} + \cdots + a_n y = b(x). \tag{1.1}$$

It will be convenient to denote the differential expression on the left of the equality (1.1) by $L(y)$. Thus

$$L(y) = y^{(n)} + a_1 y^{(n-1)} + a_2 y^{(n-2)} + \cdots + a_n y,$$

and the equation (1.1) becomes simply $L(y) = b(x)$. If $b(x) = 0$ for all x in I the corresponding equation $L(y) = 0$ is called a *homogeneous equation*, whereas if $b(x) \neq 0$ for some x in I, $L(y) = b(x)$ is called a *non-homogeneous equation*.

We give a meaning to L itself as a *differential operator* which operates on functions which have n derivatives on I, and transforms such a function ϕ into a function $L(\phi)$ whose value at x is given by

$$L(\phi)(x) = \phi^{(n)}(x) + a_1 \phi^{(n-1)}(x) + \cdots + a_n \phi(x).$$

Thus

$$L(\phi) = \phi^{(n)} + a_1 \phi^{(n-1)} + \cdots + a_n \phi.$$

49

A solution of $L(y) = b(x)$ is therefore a function ϕ having n derivatives on I such that $L(\phi) = b$.

From a theoretical standpoint, if b is continuous on I, it is possible to find *all* solutions of $L(y) = b(x)$. We have done this for the case $n = 1$ in Chap. 1, Sec. 6. In this chapter we first consider the simple case of the second order equation $(n = 2)$. All solutions of the homogeneous equation can be found by a simple device which reduces the problem to the algebraic one of locating roots of a polynomial. The solutions of the non-homogeneous equation can be generated by using the solutions of the corresponding homogeneous equation, together with an integration involving the function b. Secondly we show how the methods which work for the second order case can be extended to solve the n-th order equation. Finally we indicate a method of solving the non-homogeneous equation that works for a large class of b, and which is often quicker than the general method to apply.

2. The second order homogeneous equation

Here we are concerned with the equation

$$L(y) = y'' + a_1 y' + a_2 y = 0, \tag{2.1}$$

where a_1 and a_2 are constants. We recall that the first order equation with constant coefficients $y' + ay = 0$ has a solution e^{-ax}. The constant $-a$ in this solution is the solution of the equation $r + a = 0$. Since differentiating an exponential e^{rx} any number of times, where r is a constant, always yields a constant times e^{rx}, it is reasonable to expect that for some appropriate constant r, e^{rx} will be a solution of the equation (2.1). We have seen that this works for equations of the first order. Let us try it for (2.1). We find

$$L(e^{rx}) = (r^2 + a_1 r + a_2)e^{rx},$$

and e^{rx} will be a solution of $L(y) = 0$, i.e. $L(e^{rx}) = 0$, if r satisfies $r^2 + a_1 r + a_2 = 0$. We let

$$p(r) = r^2 + a_1 r + a_2,$$

and call p the *characteristic polynomial* of L, or of the equation (2.1). Note that $p(r)$ can be obtained from $L(y)$ by replacing $y^{(k)}$ everywhere by r^k, where we use the conventions that the zero-th derivative of y, $y^{(0)}$, is y itself, and that $r^0 = 1$. From the Fundamental Theorem of Algebra we know that the polynomial p always has two complex roots r_1, r_2 (which may be real). If $r_1 \neq r_2$, we see that $e^{r_1 x}$ and $e^{r_2 x}$ are two distinct solutions of $L(y) = 0$.

It is possible to find two distinct solutions in the case $r_1 = r_2$ also. We have

$$L(e^{rx}) = p(r)e^{rx} \tag{2.2}$$

for all r and x. We recall that if r_1 is a repeated root of p, then not only $p(r_1) = 0$, but $p'(r_1) = 0$. This suggests differentiating the equation (2.2) with respect to r. In doing this we observe that, since L involves only differentiation with respect to x,

$$\frac{\partial}{\partial r}L(e^{rx}) = L\left(\frac{\partial}{\partial r}e^{rx}\right) = L(xe^{rx}),$$

and therefore

$$L(xe^{rx}) = [p'(r) + xp(r)]e^{rx}.$$

Now setting $r = r_1$ in this equation we see that $L(xe^{r_1x}) = 0$, thus showing that xe^{r_1x} is another solution in case $r_1 = r_2$. We formulate our results so far as a theorem.

Theorem 1. *Let* a_1, a_2 *be constants, and consider the equation*

$$L(y) = y'' + a_1y' + a_2y = 0.$$

If r_1, r_2 *are distinct roots of the characteristic polynomial* p, *where*

$$p(r) = r^2 + a_1r + a_2,$$

then the functions ϕ_1, ϕ_2 *defined by*

$$\phi_1(x) = e^{r_1x}, \qquad \phi_2(x) = e^{r_2x}, \tag{2.3}$$

are solutions of $L(y) = 0$. *If* r_1 *is a repeated root of* p, *then the functions* ϕ_1, ϕ_2 *defined by*

$$\phi_1(x) = e^{r_1x}, \qquad \phi_2(x) = xe^{r_1x} \tag{2.4}$$

are solutions of $L(y) = 0$.

We now turn to the problem of finding *all* solutions of $L(y) = 0$. It is a remarkable fact that every solution of this equation is a linear combination, with constant coefficients, of the two functions ϕ_1, ϕ_2 given by (2.3) in case $r_1 \neq r_2$, and by (2.4) in case $r_1 = r_2$. This will be shown in Sec. 3 (Theorem 5).

First we verify the interesting fact that if ϕ_1, ϕ_2 are any two solutions of $L(y) = 0$, and c_1, c_2 are any two constants, then the function $\phi = c_1\phi_1 + c_2\phi_2$ is also a solution of $L(y) = 0$. Indeed

$$L(\phi) = (c_1\phi_1 + c_2\phi_2)'' + a_1(c_1\phi_1 + c_2\phi_2)' + a_2(c_1\phi_1 + c_2\phi_2)$$
$$= c_1\phi_1'' + c_2\phi_2'' + c_1a_1\phi_1' + c_2a_1\phi_2' + c_1a_2\phi_1 + c_2a_2\phi_2$$
$$= c_1L(\phi_1) + c_2L(\phi_2) = 0.$$

The function ϕ which is zero for all x is also a solution, the *trivial solution* of $L(y) = 0$.

The result of Theorem 1, together with Theorem 5, allows us to solve all homogeneous linear equations of the second order with constant coefficients. We have only to compute the two solutions ϕ_1, ϕ_2 and then form all linear combinations $c_1\phi_1 + c_2\phi_2$ of these two, where c_1, c_2 are any constants.

As an example consider the equation

$$y'' + y' - 2y = 0. \tag{2.5}$$

The characteristic polynomial is

$$p(r) = r^2 + r - 2,$$

whose roots are -2 and 1. Therefore every solution ϕ has the form

$$\phi(x) = c_1 e^{-2x} + c_2 e^x, \tag{2.6}$$

where c_1, c_2 are constants. Moreover, if c_1, c_2 are *any* two constants the ϕ given by (2.6) is a solution.

As a second example consider

$$y'' + \omega^2 y = 0, \tag{2.7}$$

where ω is a positive constant. The characteristic polynomial is

$$p(r) = r^2 + \omega^2,$$

whose roots are $i\omega$ and $-i\omega$. Consequently all solutions of (2.7) are of the form

$$c_1 e^{i\omega x} + c_2 e^{-i\omega x},$$

where c_1, c_2 may be any two constants. Taking $c_1 = \frac{1}{2}$, $c_2 = \frac{1}{2}$, we see that $\cos \omega x$ is a solution; and letting $c_1 = 1/2i, c_2 = -1/2i$, we find that $\sin \omega x$ is a solution. The equation (2.7) is important in the study of oscillatory behavior in many physical situations, and is called the *harmonic oscillator equation*.

EXERCISES

1. Find all solutions of the following equations:
   (a) $y'' - 4y = 0$ (b) $3y'' + 2y' = 0$
   (c) $y'' + 16y = 0$ (d) $y'' = 0$
   (e) $y'' + 2iy' + y = 0$ (f) $y'' - 4y' + 5y = 0$
   (g) $y'' + (3i - 1)y' - 3iy = 0$

2. Consider the equation $y'' + y' - 6y = 0$.
   (a) Compute the solution ϕ satisfying $\phi(0) = 1, \phi'(0) = 0$.
   (b) Compute the solution ψ satisfying $\psi(0) = 0, \psi'(0) = 1$.
   (c) Compute $\phi(1)$ and $\psi(1)$.

3. Find all solutions ϕ of $y'' + y = 0$ satisfying:
(a) $\phi(0) = 1, \phi(\pi/2) = 2$ (b) $\phi(0) = 0, \phi(\pi) = 0$
(c) $\phi(0) = 0, \phi'(\pi/2) = 0$ (d) $\phi(0) = 0, \phi(\pi/2) = 0$

4. Consider the equation

$$y'' + a_1 y' + a_2 y = 0,$$

where the constants a_1, a_2 are real. Suppose $\alpha + i\beta$ is a complex root of the characteristic polynomial, where α, β are real, $\beta \neq 0$.
(a) Show that $\alpha - i\beta$ is also a root.
(b) Show that any solution ϕ may be written in the form

$$\phi(x) = e^{\alpha x}(d_1 \cos \beta x + d_2 \sin \beta x),$$

where d_1, d_2 are constants.
(c) Show that $\alpha = -a_1/2$, $\beta^2 = a_2 - (a_1^2/4)$.
(d) Show that every solution tends to zero as $x \to +\infty$ if $a_1 > 0$.
(e) Show that the magnitude of every non-trivial solution assumes arbitrarily large values as $x \to +\infty$ if $a_1 < 0$.

5. Consider the equation

$$Ly'' + Ry' + \frac{1}{C}y = 0,$$

where L, R, and C are positive constants. (*Note:* L is not a differential operator here.)
(a) Compute all solutions for the three cases:

(i) $\dfrac{R^2}{L^2} - \dfrac{4}{LC} > 0$

(ii) $\dfrac{R^2}{L^2} - \dfrac{4}{LC} = 0$

(iii) $\dfrac{R^2}{L^2} - \dfrac{4}{LC} < 0$

(b) Show that all solutions tend to zero as $x \to \infty$ for each of the cases (i), (ii), (iii) of (a).
(c) Sketch the solution ϕ satisfying $\phi(0) = 1, \phi'(0) = 0$ in the case (iii).
(d) Show that any solution ϕ in case (iii) may be written in the form

$$\phi(x) = Ae^{\alpha x} \cos (\beta x - \omega),$$

where A, α, β, ω are constants. Determine α, β.

6. Show that every solution of the constant coefficient equation

$$y'' + a_1 y' + a_2 y = 0$$

tends to zero as $x \to \infty$ if, and only if, the real parts of the roots of the characteristic polynomial are negative. (*Note:* In this case the solutions are often called *transients*.)

7. Show that every solution of the constant coefficient equation

$$y'' + a_1y' + a_2y = 0$$

is bounded on $0 \leq x < \infty$ if, and only if, the real parts of the roots of the characteristic polynomial are non-positive and the roots with zero real part have multiplicity one.

8. Consider the equation $y'' + k^2y = 0$, where k is a non-negative constant.
(a) For what values of k will there exist non-trivial solutions ϕ satisfying
  (i) $\phi(0) = 0, \phi(\pi) = 0$,
  (ii) $\phi'(0) = 0, \phi'(\pi) = 0$,
  (iii) $\phi(0) = \phi(\pi), \phi'(0) = \phi'(\pi)$,
  (iv) $\phi(0) = -\phi(\pi), \phi'(0) = -\phi'(\pi)$?
(b) Find the non-trivial solutions for each of the cases (i)–(iv) in (a).

9. Let ϕ be a solution of the equation

$$y'' + a_1y' + a_2y = 0,$$

where a_1, a_2 are constants. If

$$\psi(x) = e^{(a_1/2)x}\phi(x),$$

show that ψ satisfies an equation $y'' + ky = 0$, where k is some constant. Compute k.

3. Initial value problems for second order equations

The demonstration that every solution of the equation

$$L(y) = y'' + a_1y' + a_2y = 0$$

is a linear combination of the solutions (2.3) or (2.4) will depend on showing that the initial value problems for this equation have unique solutions. An *initial value problem* for $L(y) = 0$ is a problem of finding a solution ϕ satisfying

$$\phi(x_0) = \alpha, \qquad \phi'(x_0) = \beta, \tag{3.1}$$

where x_0 is some real number, and α, β are two given constants. Thus we specify ϕ and its first derivative at some initial point x_0. This problem is denoted by

$$L(y) = 0, \qquad y(x_0) = \alpha, \qquad y'(x_0) = \beta. \tag{3.2}$$

Theorem 2. (*Existence Theorem*) *For any real* x_0, *and constants* α, β, *there exists a solution* ϕ *of the initial value problem* (3.2) *on* $-\infty < x < \infty$.

Proof. We show that there are unique constants c_1, c_2 such that $\phi = c_1\phi_1 + c_2\phi_2$ satisfies (3.1), where ϕ_1, ϕ_2 are the solutions given by (2.3), or (2.4). In order to satisfy the relations (3.1) we must have

$$c_1\phi_1(x_0) + c_2\phi_2(x_0) = \alpha$$
$$c_1\phi_1'(x_0) + c_2\phi_2'(x_0) = \beta \tag{3.3}$$

and these equations will have a unique solution c_1, c_2 if the determinant

$$\Delta = \begin{vmatrix} \phi_1(x_0) & \phi_2(x_0) \\ \phi_1'(x_0) & \phi_2'(x_0) \end{vmatrix} = \phi_1(x_0)\phi_2'(x_0) - \phi_1'(x_0)\phi_2(x_0) \neq 0.$$

In case $r_1 \neq r_2$,

$$\phi_1(x) = e^{r_1 x}, \qquad \phi_2(x) = e^{r_2 x},$$

and

$$\Delta = r_2 e^{r_1 x_0} e^{r_2 x_0} - r_1 e^{r_1 x_0} e^{r_2 x_0} = (r_2 - r_1) e^{(r_1 + r_2)x_0},$$

which is not zero, since $e^{(r_1 + r_2)x_0} \neq 0$. If $r_1 = r_2$,

$$\phi_1(x) = e^{r_1 x}, \qquad \phi_2(x) = x e^{r_1 x},$$

and

$$\Delta = e^{r_1 x_0}(e^{r_1 x_0} + x_0 r_1 e^{r_1 x_0}) - r_1 x_0 e^{r_1 x_0} e^{r_1 x_0} = e^{2r_1 x_0} \neq 0.$$

Therefore the determinant condition is satisfied in either case. Thus, if c_1, c_2 are the unique constants satisfying (3.3), the function

$$\phi = c_1\phi_1 + c_2\phi_2$$

will be the desired solution satisfying (3.1).

We have shown that there is a unique linear combination of ϕ_1 and ϕ_2 which is a solution of (3.2). Although it is not quite obvious, it turns out that this solution is the only one. Before proving this we give an estimate for the rate of growth of any solution ϕ of $L(y) = 0$, and its first derivative ϕ', in terms of the coefficients 1, a_1, a_2 appearing in $L(y)$. As a measure of the "size" of ϕ and ϕ' we take*

$$\| \phi(x) \| = [|\phi(x)|^2 + |\phi'(x)|^2]^{1/2},$$

where the positive square root is understood. The "size" of L will be measured by

$$k = 1 + |a_1| + |a_2|.$$

---

* Note that $\| \phi(x) \|$ is just the magnitude, or length, of the vector with components $\phi(x)$, $\phi'(x)$.

In the course of the proof we shall require the elementary fact that if b and c are any two constants, then

$$2|b||c| \leqq |b|^2 + |c|^2. \tag{3.4}$$

This inequality results by noticing that

$$0 \leqq (|b| - |c|)^2 = |b|^2 + |c|^2 - 2|b||c|.$$

Theorem 3. *Let ϕ be any solution of*

$$L(y) = y'' + a_1 y' + a_2 y = 0$$

on an interval I *containing a point* x_0. *Then for all* x *in* I

$$||\phi(x_0)||\, e^{-k|x-x_0|} \leqq ||\phi(x)|| \leqq ||\phi(x_0)||\, e^{k|x-x_0|} \tag{3.5}$$

where

$$||\phi(x)|| = [|\phi(x)|^2 + |\phi'(x)|^2]^{1/2}, \qquad k = 1 + |a_1| + |a_2|.$$

Remark. Geometrically the inequality (3.5) says that $||\phi(x)||$ always remains between the two curves

$$y = ||\phi(x_0)||\, e^{k(x-x_0)} \quad \text{and} \quad y = ||\phi(x_0)||\, e^{-k(x-x_0)};$$

the shaded area in Fig. 4.

Proof of Theorem 3. We let $u(x) = ||\phi(x)||^2$. Thus

$$u = \phi\bar{\phi} + \phi'\bar{\phi'},$$

where $\bar{\phi}(x) = \overline{\phi(x)}$, $\bar{\phi'}(x) = \overline{\phi'(x)}$. Then

$$u' = \phi'\bar{\phi} + \phi\bar{\phi'} + \phi''\bar{\phi'} + \phi'\bar{\phi''}$$

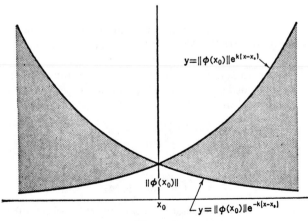

Figure 4

and therefore*

$$|u'(x)| \leq 2|\phi(x)||\phi'(x)| + 2|\phi'(x)||\phi''(x)|. \tag{3.6}$$

Since ϕ satisfies $L(\phi) = 0$ we have

$$\phi'' = -a_1\phi' - a_2\phi,$$

and hence

$$|\phi''(x)| \leq |a_1||\phi'(x)| + |a_2||\phi(x)|. \tag{3.7}$$

Using (3.7) in (3.6) we obtain

$$|u'(x)| \leq 2(1+|a_2|)|\phi(x)||\phi'(x)| + 2|a_1||\phi'(x)|^2.$$

Now applying (3.4) to $b = \phi(x)$, $c = \phi'(x)$, this gives

$$|u'(x)| \leq (1+|a_2|)|\phi(x)|^2 + (1+2|a_1|+|a_2|)|\phi'(x)|^2$$
$$\leq 2(1+|a_1|+|a_2|)[|\phi(x)|^2 + |\phi'(x)|^2],$$

or

$$|u'(x)| \leq 2ku(x).$$

This is equivalent to

$$-2ku(x) \leq u'(x) \leq 2ku(x) \tag{3.8}$$

and these inequalities lead directly to (3.5). Indeed, consider the right inequality which can be written as

$$u' - 2ku \leq 0.$$

If this were an equality, it would be a linear differential equation for u of the first order. We "integrate" this inequality using the same procedure we used in Chap. 1. It is equivalent to

$$e^{-2kx}(u' - 2ku) = (e^{-2kx}u)' \leq 0.$$

If $x > x_0$ we integrate from x_0 to x obtaining

$$e^{-2kx}u(x) - e^{-2kx_0}u(x_0) \leq 0,$$

or

$$u(x) \leq u(x_0)e^{2k(x-x_0)},$$

yielding†

$$\|\phi(x)\| \leq \|\phi(x_0)\| e^{k(x-x_0)}, \qquad (x > x_0).$$

The left inequality in (3.8) similarly implies

$$\|\phi(x_0)\| e^{-k(x-x_0)} \leq \|\phi(x)\|, \qquad (x > x_0),$$

and therefore

$$\|\phi(x_0)\| e^{-k(x-x_0)} \leq \|\phi(x)\| \leq \|\phi(x_0)\| e^{k(x-x_0)}, \qquad (x > x_0),$$

---

* From the definition of a derivative it follows that $\overline{\phi'} = \overline{\phi}'$. Also $|\phi(x)| = |\overline{\phi(x)}|$.
† If $0 \leq b \leq c$ then $0 \leq b^{\frac{1}{2}} \leq c^{\frac{1}{2}}$, where the positive square root is understood.

which is just (3.5) for $x > x_0$. A consideration of (3.8) for $x < x_0$, together with an integration from x to x_0, yields

$$\| \phi(x_0) \| e^{k(x-x_0)} \leq \| \phi(x) \| \leq \| \phi(x_0) \| e^{-k(x-x_0)}, \qquad (x < x_0),$$

which is (3.5) for $x < x_0$.

Theorem 4. (*Uniqueness Theorem*) *Let α, β be any two constants, and let x_0 be any real number. On any interval I containing x_0 there exists at most one solution ϕ of the initial value problem*

$$L(y) = 0, \qquad y(x_0) = \alpha, \qquad y'(x_0) = \beta.$$

Proof. Suppose ϕ, ψ were two solutions. Let $\chi = \phi - \psi$. Then

$$L(\chi) = L(\phi) - L(\psi) = 0, \quad \text{and} \quad \chi(x_0) = 0, \quad \chi'(x_0) = 0.$$

Thus $\| \chi(x_0) \| = 0$, and applying the inequalities (3.5) to χ we see that

$$\| \chi(x) \| = 0 \quad \text{for all } x \text{ in } I.$$

This implies $\chi(x) = 0$ for all x in I, or $\phi = \psi$, proving our result.

Theorems 2 and 4 now imply the result we promised in Sec. 2.

Theorem 5. *Let ϕ_1, ϕ_2 be the two solutions of $L(y) = 0$ given by (2.3) in case $r_1 \neq r_2$, and by (2.4) in case $r_1 = r_2$. If c_1, c_2 are any two constants the function $\phi = c_1\phi_1 + c_2\phi_2$ is a solution of*

$$L(y) = 0 \text{ on } -\infty < x < \infty.$$

Conversely, if ϕ is any solution of

$$L(y) = 0 \text{ on } -\infty < x < \infty,$$

there are unique constants c_1, c_2 such that

$$\phi = c_1\phi_1 + c_2\phi_2.$$

Proof. The first part of the theorem follows, as we have seen, from the fact that

$$L(\phi) = c_1 L(\phi_1) + c_2 L(\phi_2).$$

If ϕ is a solution and x_0 is real, let $\phi(x_0) = \alpha$, $\phi'(x_0) = \beta$. In the proof of Theorem 2 we showed that there is a solution ψ of $L(y) = 0$, satisfying $\psi(x_0) = \alpha$, $\psi'(x_0) = \beta$, of the form

$$\psi = c_1\phi_1 + c_2\phi_2,$$

where c_1, c_2 are uniquely determined by α, β. By uniqueness (Theorem 4) $\phi = \psi$.

EXERCISES

1. Find the solutions of the following initial value problems:
(a) $y'' - 2y' - 3y = 0$, $y(0) = 0$, $y'(0) = 1$
(b) $y'' + (4i + 1)y' + y = 0$, $y(0) = 0$, $y'(0) = 0$
(c) $y'' + (3i - 1)y' - 3iy = 0$, $y(0) = 2$, $y'(0) = 0$
(d) $y'' + 10y = 0$, $y(0) = \pi$, $y'(0) = \pi^2$

2. Suppose ϕ is a function having a continuous derivative on $0 \leqq x < \infty$, such that $\phi'(x) + 2\phi(x) \leqq 1$ for all such x, and $\phi(0) = 0$. Show that $\phi(x) < \frac{1}{2}$ for $x \geqq 0$.

3. Find a function ϕ which has a continuous derivative on $0 \leqq x \leqq 2$ which satisfies

$$\phi(0) = 0, \qquad \phi'(0) = 1,$$

and

$$y'' - y = 0, \qquad \text{for } 0 \leqq x \leqq 1,$$

and

$$y'' - 9y = 0, \qquad \text{for } 1 \leqq x \leqq 2.$$

4. Suppose ϕ, ψ are two solutions of the constant coefficient equation

$$L(y) = y'' + a_1 y' + a_2 y = 0$$

on a finite interval I including a point x_0. Let

$$\phi(x_0) = \alpha_1, \qquad \phi'(x_0) = \beta_1,$$

$$\psi(x_0) = \alpha_2, \qquad \psi'(x_0) = \beta_2,$$

and suppose

$$(\alpha_1 - \alpha_2)^2 + (\beta_1 - \beta_2)^2 = \epsilon^2.$$

(a) If $\chi = \phi - \psi$ show that χ satisfies $L(y) = 0$, and

$$\chi(x_0) = \alpha_1 - \alpha_2, \qquad \chi'(x_0) = \beta_1 - \beta_2.$$

(b) Show that

$$|\phi(x) - \psi(x)| \leqq \epsilon e^{k|I|},$$

for all x in I, where $k = 1 + |a_1| + |a_2|$, and $|I|$ is the length of I. (*Note:* This result implies that if $\alpha_2 \to \alpha_1$, $\beta_2 \to \beta_1$, then $\epsilon \to 0$, and hence $\psi(x) \to \phi(x)$ on I.)

5. Consider the constant coefficient equation

$$L(y) = y'' + a_1 y' + a_2 y = 0.$$

Let ϕ_1 be the solution satisfying

$$\phi_1(x_0) = 1, \qquad \phi_1'(x_0) = 0,$$

and let ϕ_2 be the solution satisfying

$$\phi_2(x_0) = 0, \qquad \phi_2'(x_0) = 1.$$

If ϕ is a solution satisfying

$$\phi(x_0) = \alpha, \qquad \phi'(x_0) = \beta,$$

show that

$$\phi(x) = \alpha\phi_1(x) + \beta\phi_2(x)$$

for all x. (*Note*: This shows that every solution ϕ is a linear combination of ϕ_1 and ϕ_2, and that ϕ is a linear function of its initial conditions α, β.)

6. Let I be the interval $0 < x < 1$. Find a function ϕ which has a continuous derivative on $-\infty < x < \infty$, which satisfies

$$y'' = 0 \quad \text{in } I,$$

$$y'' + k^2y = 0 \quad \text{outside } I, \qquad (k > 0),$$

and which has the form

$$\phi(x) = e^{ikx} + Ae^{-ikx}, \qquad (x \leq 0),$$

and

$$\phi(x) = Be^{ikx}, \qquad (x \geq 1).$$

Determine ϕ by computing the constants A and B, and its values in I.

4. Linear dependence and independence

Two functions ϕ_1, ϕ_2 defined on an interval I are said to be *linearly dependent* on I if there exist two constants c_1, c_2, not both zero, such that

$$c_1\phi_1(x) + c_2\phi_2(x) = 0$$

for all x in I. The functions ϕ_1, ϕ_2 are said to be *linearly independent* on I if they are not linearly dependent there. Thus ϕ_1, ϕ_2 are linearly independent on I if the only constants c_1, c_2 such that $c_1\phi_1(x) + c_2\phi_2(x) = 0$ for all x in I are the constants $c_1 = 0$, $c_2 = 0$.

The functions defined by (2.3) are linearly independent on any interval I. For suppose

$$c_1e^{r_1x} + c_2e^{r_2x} = 0 \tag{4.1}$$

for all x in I. Then, multiplying by e^{-r_1x}, we obtain

$$c_1 + c_2e^{(r_2-r_1)x} = 0,$$

and differentiating there results

$$c_2(r_2 - r_1)e^{(r_2-r_1)x} = 0.$$

Since $r_1 \neq r_2$, and $e^{(r_2-r_1)x}$ is never zero, this implies $c_2 = 0$. But if $c_2 = 0$, the relation (4.1) gives $c_1e^{r_1x} = 0$, or $c_1 = 0$ also.

Similarly the functions ϕ_1, ϕ_2 defined by (2.4) are linearly independent on any interval I. The proof is the same. If

$$c_1 e^{r_1 x} + c_2 x e^{r_1 x} = 0$$

on I, by multiplying by $e^{-r_1 x}$ we get

$$c_1 + c_2 x = 0,$$

and differentiating we obtain $c_2 = 0$, and this implies $c_1 = 0$.

There is a simple test which enables us to tell whether two solutions ϕ_1, ϕ_2 of $L(y) = 0$ are linearly independent or not. It involves the determinant

$$W(\phi_1, \phi_2) = \begin{vmatrix} \phi_1 & \phi_2 \\ \phi_1' & \phi_2' \end{vmatrix} = \phi_1 \phi_2' - \phi_1' \phi_2,$$

which is called the *Wronskian* of ϕ_1, ϕ_2. It is a function, and its value at x is denoted by $W(\phi_1, \phi_2)(x)$.

Theorem 6. *Two solutions ϕ_1, ϕ_2 of* $L(y) = 0$ *are linearly independent on an interval* I *if, and only if,*

$$W(\phi_1, \phi_2)(x) \neq 0$$

for all x *in* I.

Proof. First suppose $W(\phi_1, \phi_2)(x) \neq 0$ for all x in I, and let c_1, c_2 be constants such that

$$c_1 \phi_1(x) + c_2 \phi_2(x) = 0 \qquad (4.2)$$

for all x in I. Then also

$$c_1 \phi_1'(x) + c_2 \phi_2'(x) = 0 \qquad (4.3)$$

for all x in I. For a fixed x the equations (4.2), (4.3) are linear homogeneous equations satisfied by c_1, c_2. The determinant of the coefficients is just $W(\phi_1, \phi_2)(x)$ which is not zero. Therefore (Theorem 2, Chap. 0) $c_1 = c_2 = 0$ is the only solution of (4.2), (4.3). This proves that ϕ_1, ϕ_2 are linearly independent on I.

Conversely, assume ϕ_1, ϕ_2 are linearly independent on I. Suppose that there is an x_0 in I such that $W(\phi_1, \phi_2)(x_0) = 0$. This implies that the system of two equations

$$c_1 \phi_1(x_0) + c_2 \phi_2(x_0) = 0,$$
$$c_1 \phi_1'(x_0) + c_2 \phi_2'(x_0) = 0, \qquad (4.4)$$

has a solution c_1, c_2, where at least one of these numbers is not zero (Theorem 3, Chap. 0). Let c_1, c_2 be such a solution and consider the function $\psi = c_1 \phi_1 + c_2 \phi_2$. Now $L(\psi) = 0$, and from (4.4) we see that

$$\psi(x_0) = 0, \qquad \psi'(x_0) = 0.$$

From the uniqueness theorem (Theorem 4) we infer that $\psi(x) = 0$ for all x in I, and thus

$$c_1\phi_1(x) + c_2\phi_2(x) = 0$$

for all x in I. But this contradicts the fact that ϕ_1, ϕ_2 are linearly independent on I. Thus the supposition that there was a point x_0 in I such that

$$W(\phi_1, \phi_2)(x_0) = 0$$

must be false. We have consequently proved that

$$W(\phi_1, \phi_2)(x) \neq 0$$

for all x in I.

It is easy to see that we need compute $W(\phi_1, \phi_2)$ at only one convenient point to test the linear independence of the solutions ϕ_1, ϕ_2.

Theorem 7. *Let ϕ_1, ϕ_2 be two solutions of* $L(y) = 0$ *on an interval* I, *and let* x_0 *be any point in* I. *Then ϕ_1, ϕ_2 are linearly independent on* I *if and only if*

$$W(\phi_1, \phi_2)(x_0) \neq 0.$$

Proof. If ϕ_1, ϕ_2 are linearly independent on I then

$$W(\phi_1, \phi_2)(x) \neq 0$$

for all x in I, by Theorem 6. In particular

$$W(\phi_1, \phi_2)(x_0) \neq 0.$$

Conversely, suppose $W(\phi_1, \phi_2)(x_0) \neq 0$, and suppose c_1, c_2 are constants such that

$$c_1\phi_1(x) + c_2\phi_2(x) = 0$$

for all x in I. Then we see that

$$c_1\phi_1(x_0) + c_2\phi_2(x_0) = 0,$$
$$c_1\phi_1'(x_0) + c_2\phi_2'(x_0) = 0,$$

and since the determinant of the coefficients is $W(\phi_1, \phi_2)(x_0) \neq 0$, we obtain $c_1 = c_2 = 0$. Thus ϕ_1, ϕ_2 are linearly independent on I.

Using the concept of linear independence we can show any two linearly independent solutions of $L(y) = 0$ determine all solutions, in the sense of the following theorem.

Theorem 8. *Let ϕ_1, ϕ_2 be any two linearly independent solutions of* $L(y) = 0$ *on an interval* I. *Every solution ϕ of* $L(y) = 0$ *can be written uniquely as*

$$\phi = c_1\phi_1 + c_2\phi_2,$$

where c_1, c_2 are constants.

Proof. Let x_0 be a point in I. Since ϕ_1, ϕ_2 are linearly independent on I we know that $W(\phi_1, \phi_2)(x_0) \neq 0$. Let $\phi(x_0) = \alpha$, $\phi'(x_0) = \beta$, and consider the two equations

$$c_1\phi_1(x_0) + c_2\phi_2(x_0) = \alpha$$
$$c_1\phi_1'(x_0) + c_2\phi_2'(x_0) = \beta$$

for the constants c_1, c_2. Since the determinant of the coefficients of c_1, c_2 is just

$$W(\phi_1, \phi_2)(x_0) \neq 0,$$

there is a unique pair of constants c_1, c_2 satisfying these equations. Choose c_1, c_2 to be these constants. Then the function $\psi = c_1\phi_1 + c_2\phi_2$ is such that

$$\psi(x_0) = \phi(x_0), \quad \psi'(x_0) = \phi'(x_0), \quad \text{and} \quad L(\psi) = 0.$$

From the uniqueness theorem (Theorem 4) it follows that $\psi = \phi$ on I, that is,

$$\phi = c_1\phi_1 + c_2\phi_2.$$

The importance of Theorem 8 is that we need only to find *any* two linearly independent solutions of $L(y) = 0$ (not necessarily the ones we found in Sec. 2) in order to obtain all solutions of $L(y) = 0$. For example, the equation

$$y'' + y = 0$$

has the two solutions e^{ix}, e^{-ix}, which are linearly independent, but it also has the two linearly independent solutions $\cos x$, $\sin x$. Sometimes it is more convenient to express a solution in terms of the latter set of functions, especially when we want to observe the oscillatory character of a real-valued solution.

EXERCISES

1. The functions ϕ_1, ϕ_2 defined below exist for $-\infty < x < \infty$. Determine whether they are linearly dependent or independent there.
    (a) $\phi_1(x) = x$, $\phi_2(x) = e^{rx}$, r is a complex constant
    (b) $\phi_1(x) = \cos x$, $\phi_2(x) = \sin x$
    (c) $\phi_1(x) = x^2$, $\phi_2(x) = 5x^2$
    (d) $\phi_1(x) = \sin x$, $\phi_2(x) = e^{ix}$
    (e) $\phi_1(x) = \cos x$, $\phi_2(x) = 3(e^{ix} + e^{-ix})$
    (f) $\phi_1(x) = x$, $\phi_2(x) = |x|$

2. Are the following statements true or false? If the statement is true, prove it; if it is false, give a counterexample showing it is false.
    (a) "If ϕ_1, ϕ_2 are linearly independent functions on an interval I, they are linearly independent on any interval J contained inside I."

(b) "If ϕ_1, ϕ_2 are linearly dependent on an interval I, they are linearly dependent on any interval J contained inside I."

(c) "If ϕ_1, ϕ_2 are linearly independent solutions of $L(y) = 0$ on an interval I, they are linearly independent on any interval J contained inside I."

(d) "If ϕ_1, ϕ_2 are linearly dependent solutions of $L(y) = 0$ on an interval I, they are linearly dependent on any interval J contained inside I."

3. (a) Show that the functions ϕ_1, ϕ_2 defined by

$$\phi_1(x) = x^2, \qquad \phi_2(x) = x \mid x \mid,$$

are linearly independent for $-\infty < x < \infty$.

(b) Compute the Wronskian of these functions.

(c) Do the results of parts (a) and (b) contradict Theorem 6? Explain your answer.

4. Consider the equation

$$y'' + a_1 y' + a_2 y = 0,$$

where a_1, a_2 are real constants such that $4a_2 - a_1^2 > 0$. Let

$$\alpha + i\beta, \qquad \alpha - i\beta \qquad (\alpha, \beta \text{ real})$$

be the roots of the characteristic polynomial.

(a) Show that ϕ_1, ϕ_2 defined by

$$\phi_1(x) = e^{\alpha x} \cos \beta x, \qquad \phi_2(x) = e^{\alpha x} \sin \beta x$$

are solutions of the equation.

(b) Compute $W(\phi_1, \phi_2)$, and show that ϕ_1, ϕ_2 are linearly independent on any interval I. (*Hint*: See Ex. 4, Sec. 2.)

5. (a) Let ϕ_n be any function satisfying the boundary value problem

$$y'' + n^2 y = 0, \qquad y(0) = y(2\pi), \qquad y'(0) = y'(2\pi), \qquad (*)$$

where $n = 0, 1, 2 \cdots$. Show that

$$\int_0^{2\pi} \phi_n(x)\phi_m(x) \, dx = 0$$

if $n \neq m$. (*Hint*: $-\phi_n'' = n^2\phi_n$, and $-\phi_m'' = m^2\phi_m$. Thus

$$(n^2 - m^2)\phi_n\phi_m = \phi_n\phi_m'' - \phi_m\phi_n'' = [\phi_n\phi_m' - \phi_m\phi_n']'.$$

Integrate this equality from 0 to 2π, and use the boundary conditions satisfied by ϕ_n and ϕ_m.)

(b) Show that $\cos nx$ and $\sin nx$ are functions satisfying the boundary value problem (*). The result of (a) then implies that

$$\int_0^{2\pi} \cos nx \cos mx \, dx = 0, \qquad \int_0^{2\pi} \cos nx \sin mx \, dx = 0,$$

$$\int_0^{2\pi} \sin nx \sin mx \, dx = 0, \qquad (n \neq m).$$

6. (a) Show that $\phi_n(x) = \sin nx$ satisfies the boundary value problem

$$y'' + n^2 y = 0, \qquad y(0) = 0, \qquad y(\pi) = 0,$$

where $n = 1, 2, \cdots$.

(b) Using (a) show that

$$\int_0^\pi \sin nx \sin mx \, dx = 0$$

if $n \neq m$. (*Hint*: See Ex. 5 (a).)

(c) Prove that for any positive integer n, $\phi_1, \cdots, \phi_n$ are linearly independent on $0 \leq x \leq \pi$. (*Hint*: Suppose $a_1 \phi_1 + \cdots + a_n \phi_n = 0$. Multiply both sides of this equality by ϕ_k (k fixed between 1 and n) and integrate from 0 to π. Use (b).)

7. Determine all complex numbers l for which the problem

$$-y'' = ly, \qquad y(0) = 0, \qquad y(1) = 0,$$

has a non-trivial solution, and compute such a solution for each of these l.

8. Suppose ϕ_1, ϕ_2 are linearly independent solutions of the constant coefficient equation

$$y'' + a_1 y' + a_2 y = 0,$$

and let $W(\phi_1, \phi_2)$ be abbreviated to W. Show that W is a constant if and only if $a_1 = 0$. (*Hint*: Compute W'.)

9. Let ϕ_1, ϕ_2 be two differentiable functions on an interval I, which are not necessarily solutions of an equation $L(y) = 0$. Prove the following:

(a) If ϕ_1, ϕ_2 are linearly dependent on I, then $W(\phi_1, \phi_2)(x) = 0$ for all x in I.

(b) If $W(\phi_1, \phi_2)(x_0) \neq 0$ for some x_0 in I, then ϕ_1, ϕ_2 are linearly independent on I.

(c) $W(\phi_1, \phi_2)(x) = 0$ for all x in I does not imply that ϕ_1, ϕ_2 are linearly dependent on I. (*Hint*: Ex. 3.)

(d) $W(\phi_1, \phi_2)(x) = 0$ for all x in I, and $\phi_2(x) \neq 0$ on I, imply that ϕ_1, ϕ_2 are linearly dependent on I. (*Hint*: Compute $(\phi_1/\phi_2)'$.)

5. A formula for the Wronskian

There is a convenient formula for the Wronskian of two solutions of $L(y) = 0$, which results from the fact that $W(\phi_1, \phi_2)$ satisfies a first order linear equation.

Theorem 9. *If ϕ_1, ϕ_2 are two solutions of* $\mathrm{L(y)} = 0$ *on an interval* I *containing a point* x_0, *then*

$$W(\phi_1, \phi_2)(x) = e^{-a_1(x-x_0)} W(\phi_1, \phi_2)(x_0). \tag{5.1}$$

Proof. We have

$$\phi_1'' + a_1\phi_1' + a_2\phi_1 = 0,$$

$$\phi_2'' + a_1\phi_2' + a_2\phi_2 = 0,$$

and multiplying the first equation by $-\phi_2$, the second by ϕ_1, and adding we obtain

$$(\phi_1\phi_2'' - \phi_1''\phi_2) + a_1(\phi_1\phi_2' - \phi_1'\phi_2) = 0.$$

We notice that if $W = W(\phi_1, \phi_2)$,

$$W = \phi_1\phi_2' - \phi_1'\phi_2, \quad \text{and} \quad W' = \phi_1\phi_2'' - \phi_1''\phi_2.$$

Thus W satisfies the first order equation

$$W' + a_1W = 0.$$

Hence

$$W(x) = ce^{-a_1x},$$

where c is some constant. Setting $x = x_0$ we see that

$$W(x_0) = ce^{-a_1x_0},$$

or

$$c = e^{a_1x_0}W(x_0),$$

and thus

$$W(x) = e^{-a_1(x-x_0)}W(x_0),$$

which was to be proved.

6. The non-homogeneous equation of order two

We turn now to the problem of finding all solutions of the equation

$$L(y) = y'' + a_1y' + a_2y = b(x),$$

where b is some continuous function on an interval I. Suppose we know that ψ_p is a particular solution of this equation, and that ψ is any other solution. Then

$$L(\psi - \psi_p) = L(\psi) - L(\psi_p) = b - b = 0$$

on I. This shows that $\psi - \psi_p$ is a solution of the homogeneous equation $L(y) = 0$. Therefore if ϕ_1, ϕ_2 are linearly independent solutions of $L(y) = 0$, there are unique constants c_1, c_2 such that

$$\psi - \psi_p = c_1\phi_1 + c_2\phi_2.$$

In other words every solution ψ of $L(y) = b(x)$ can be written in the form

$$\psi = \psi_p + c_1\phi_1 + c_2\phi_2,$$

and we see that the problem of finding all solutions of $L(y) = b(x)$ reduces to finding a particular one ψ_p, and two linearly independent solutions ϕ_1, ϕ_2 of $L(y) = 0$. Note that if

$$L(\psi_p) = b, \quad \text{and} \quad L(\phi_1) = L(\phi_2) = 0,$$

and c_1, c_2 are any constants, then

$$\psi = \psi_p + c_1\phi_1 + c_2\phi_2$$

satisfies $L(\psi) = b$.

To find a particular solution of $L(y) = b(x)$ we reason in the following way. Every solution of $L(y) = 0$ is of the form $c_1\phi_1 + c_2\phi_2$ where c_1, c_2 are constants, and ϕ_1, ϕ_2 are linearly independent solutions. Such a function $c_1\phi_1 + c_2\phi_2$ can not be a solution of $L(y) = b(x)$ unless $b(x) = 0$ on I. However, suppose we allow c_1, c_2 to become *functions* u_1, u_2 (not necessarily constants) on I, and then ask whether there is a solution of $L(y) = b(x)$ of the form $u_1\phi_1 + u_2\phi_2$ on I. This procedure is known as the *variation of constants*. The remarkable thing is that it works. We argue in reverse. Suppose we have a solution of $L(y) = b(x)$ of the form $u_1\phi_1 + u_2\phi_2$, where u_1, u_2 are functions. Then

$$(u_1\phi_1 + u_2\phi_2)'' + a_1(u_1\phi_1 + u_2\phi_2)' + a_2(u_1\phi_1 + u_2\phi_2)$$

$$= u_1 L(\phi_1) + u_2 L(\phi_2) + (\phi_1 u_1'' + \phi_2 u_2'') + 2(\phi_1' u_1' + \phi_2' u_2')$$

$$+ a_1(\phi_1 u_1' + \phi_2 u_2')$$

$$= (\phi_1 u_1'' + \phi_2 u_2'') + 2(\phi_1' u_1' + \phi_2' u_2') + a_1(\phi_1 u_1' + \phi_2 u_2') = b,$$

and we notice that if

$$\phi_1 u_1' + \phi_2 u_2' = 0 \tag{6.1}$$

then

$$0 = (\phi_1 u_1' + \phi_2 u_2')' = (\phi_1' u_1' + \phi_2' u_2') + (\phi_1 u_2'' + \phi_2 u_2''),$$

and we must have

$$\phi_1' u_1' + \phi_2' u_2' = b. \tag{6.2}$$

Looking at this reasoning in reverse we see that if we can find two functions u_1, u_2 satisfying (6.1), (6.2), then indeed $u_1\phi_1 + u_2\phi_2$ will satisfy $L(y) = b(x)$.

The equations (6.1), (6.2) are two linear equations for u_1', u_2', with a determinant which is just the Wronskian $W(\phi_1, \phi_2)$. Since we assumed ϕ_1, ϕ_2 to be linearly independent this determinant is never zero on I, and there exist unique solutions u_1', u_2'. Indeed, a little calculation shows that

$$u_1' = \frac{-\phi_2 b}{W(\phi_1, \phi_2)}, \qquad u_2' = \frac{\phi_1 b}{W(\phi_1, \phi_2)}.$$

In order to obtain u_1, u_2 all we have to do is integrate. For example, if x_0 is in I we may take for u_1, u_2

$$u_1(x) = -\int_{x_0}^{x} \frac{\phi_2(t)b(t)}{W(\phi_1, \phi_2)(t)} \, dt, \qquad u_2(x) = \int_{x_0}^{x} \frac{\phi_1(t)b(t)}{W(\phi_1, \phi_2)(t)} \, dt.$$

The solution $\psi_p = u_1\phi_1 + u_2\phi_2$ then takes the form

$$\psi_p(x) = \int_{x_0}^{x} \frac{[\phi_1(t)\phi_2(x) - \phi_1(x)\phi_2(t)]b(t)}{W(\phi_1, \phi_2)(t)} \, dt. \tag{6.3}$$

We summarize our results.

Theorem 10. *Let* b *be continuous on an interval* I. *Every solution* ψ *of* $L(y) = b(x)$ *on* I *can be written as*

$$\psi = \psi_p + c_1\phi_1 + c_2\phi_2,$$

where ψ_p *is a particular solution,* ϕ_1, ϕ_2 *are two linearly independent solutions of* $L(y) = 0$, *and* c_1, c_2 *are constants. A particular solution* ψ_p *is given by* (6.3). *Conversely every such* ψ *is a solution of* $L(y) = b(x)$.

As an example let us solve $L(y) = b(x)$ in the case $p(r) = r^2 + a_1 r + a_2$ has two distinct roots r_1, r_2. We may take

$$\phi_1(x) = e^{r_1 x}, \qquad \phi_2(x) = e^{r_2 x},$$

and then

$$W(\phi_1, \phi_2)(x) = (r_2 - r_1)e^{(r_1+r_2)x}.$$

Also

$$\phi_1(t)\phi_2(x) - \phi_1(x)\phi_2(t) = e^{r_1 t}e^{r_2 x} - e^{r_1 x}e^{r_2 t}.$$

Thus every solution ψ of $L(y) = b(x)$ in this case has the form

$$\psi(x) = c_1 e^{r_1 x} + c_2 e^{r_2 x} + \frac{1}{r_1 - r_2} \int_{x_0}^{x} [e^{r_1(x-t)} - e^{r_2(x-t)}]b(t) \, dt,$$

where x_0 is a real number, and c_1, c_2 are constants.

For a more concrete illustration of this method of solving a non-homogeneous equation consider the equation

$$y'' - y' - 2y = e^{-x}.$$

The characteristic polynomial is

$$r^2 - r - 2 = (r + 1)(r - 2),$$

and therefore two linearly independent solutions ϕ_1, ϕ_2 of the homogeneous equation are

$$\phi_1(x) = e^{-x}, \qquad \phi_2(x) = e^{2x}.$$

A particular solution ψ_p of the non-homogeneous equation is of the form

$$\psi_p(x) = u_1(x)e^{-x} + u_2(x)e^{2x},$$

where u_1', u_2' satisfy the equations (6.1), (6.2), that is,

$$u_1'(x)e^{-x} + u_2'(x)e^{2x} = 0,$$

$$-u_1'(x)e^{-x} + 2u_2'(x)e^{2x} = e^{-x}.$$

Solving these for u_1', u_2' we find that

$$u_1'(x) = -\tfrac{1}{3}, \qquad u_2'(x) = \tfrac{1}{3}e^{-3x},$$

and for u_1, u_2 we can take

$$u_1(x) = -\frac{x}{3}, \qquad u_2(x) = -\tfrac{1}{9}e^{-3x}.$$

Thus ψ_p is given by

$$\psi_p(x) = -\frac{x}{3}e^{-x} - \tfrac{1}{9}e^{-x}.$$

We note that $-(e^{-x}/9)$ is a solution of the homogeneous equation, so that we may take $-(xe^{-x}/3)$ as a simpler particular solution of the non-homogeneous equation. The most general solution ψ of the non-homogeneous equation then has the form

$$\psi(x) = -\frac{x}{3}e^{-x} + c_1e^{-x} + c_2e^{2x},$$

where c_1, c_2 are any two constants.

EXERCISES

1. Find all solutions of the following equations:
  (a) $y'' + 4y = \cos x$
  (b) $y'' + 9y = \sin 3x$
  (c) $y'' + y = \tan x$, $(-\pi/2 < x < \pi/2)$
  (d) $y'' + 2iy' + y = x$
  (e) $y'' - 4y' + 5y = 3e^{-x} + 2x^2$
  (f) $y'' - 7y' + 6y = \sin x$
  (g) $y'' + y = 2\sin x \sin 2x$
  (h) $y'' + y = \sec x$, $(-\pi/2 < x < \pi/2)$
  (i) $4y'' - y = e^x$
  (j) $6y'' + 5y' - 6y = x$

2. Let $L(y) = y'' + a_1y' + a_2y$, where a_1, a_2 are constants, and let p be the characteristic polynomial $p(r) = r^2 + a_1r + a_2$.

(a) If A, α are constants, and $p(\alpha) \neq 0$, show that there is a solution ϕ of $L(y) = Ae^{\alpha x}$ of the form $\phi(x) = Be^{\alpha x}$, where B is a constant. (*Hint*: Compute $L(Be^{\alpha x})$.)

(b) Compute a particular solution of $L(y) = Ae^{\alpha x}$ in case $p(\alpha) = 0$. (*Hint*: If B, r are constants compute $L(Bxe^{rx})$, and then let $r = \alpha$.)

(c) If ϕ, ψ are solutions of

$$L(y) = b_1(x), \qquad L(y) = b_2(x),$$

respectively, on some interval I, show that $\chi = \phi + \psi$ is a solution of

$$L(y) = b_1(x) + b_2(x) \quad \text{on} \quad I.$$

(d) Suppose A_1, A_2, α_1, α_2 are constants, and $p(\alpha_1) \neq 0$, $p(\alpha_2) \neq 0$. Find a solution of

$$L(y) = A_1e^{\alpha_1 x} + A_2e^{\alpha_2 x}.$$

3. Consider

$$L(y) = y'' + a_1y' + a_2y,$$

where a_1, a_2 are real constants. Let A, ω be real constants such that $p(i\omega) \neq 0$, where p is the characteristic polynomial.

(a) Show that the equation $L(y) = Ae^{i\omega x}$ has a solution ϕ given by

$$\phi(x) = \frac{A}{|p(i\omega)|}\, e^{i(\omega x - \alpha)},$$

where $p(i\omega) = |p(i\omega)|\, e^{i\alpha}$. (*Hint*: Ex. 2 (a).)

(b) If ϕ is any solution of $L(y) = Ae^{i\omega x}$, show that $\phi_1 = \operatorname{Re} \phi$, $\phi_2 = \operatorname{Im} \phi$, are solutions of

$$L(y) = A \cos \omega x, \qquad L(y) = A \sin \omega x,$$

respectively.

(c) Using (a), (b) show that there is a particular solution ϕ of

$$Ly'' + Ry' + \frac{1}{C}y = E \cos \omega x,$$

where L, R, C, E, ω are positive constants, which has the form $\phi(x) = B \cos (\omega x - \alpha)$. (*Note*: L is a constant here, and not a differential operator.)

(d) Suppose that $R^2C < 2L$ in (c). For what value of ω is B a maximum? (*Note*: This ω is often referred to as the *resonance* ω.)

4. Consider the equation

$$y'' + \omega^2 y = A \cos \omega x,$$

where A, ω are positive constants.

(a) Find all solutions on $0 \leq x < \infty$.

(b) Show that every solution ϕ is such that $|\phi(x)|$ assumes arbitrarily large values as $x \to \infty$.

(c) Sketch the graph of that solution ϕ satisfying $\phi(0) = 0$, $\phi'(0) = 1$.

5. Consider the equation

$$L(y) = y'' + a_1 y' + a_2 y = b(x),$$

where a_1, a_2 are constants, and b is a continuous function on $0 \leq x < \infty$. Suppose that the roots r_1, r_2 of the characteristic polynomial

$$p(r) = r^2 + a_1 r + a_2$$

are distinct, and Re $r_1 < 0$, Re $r_2 < 0$.

(a) Suppose b is bounded on $0 \leq x < \infty$, that is, there is a constant $k > 0$ such that

$$|b(x)| \leq k, \qquad (0 \leq x < \infty).$$

Show that every solution of $L(y) = b(x)$ is bounded on $0 \leq x < \infty$. (*Hint:* Use the formula for a solution ψ which was developed just after Theorem 10.)

(b) If $b(x) \to 0$, as $x \to \infty$, show that every solution of $L(y) = b(x)$ tends to zero as $x \to \infty$.

7. The homogeneous equation of order n

Everything we have done for the second order equation can be carried over to the case of the equation of order n. Now let $L(y)$ be given by

$$L(y) = y^{(n)} + a_1 y^{(n-1)} + a_2 y^{(n-2)} + \cdots + a_n y,$$

where a_1, a_2, $\cdots$, a_n are constants. We try to solve $L(y) = 0$ as before by trying an exponential e^{rx}. We see that

$$L(e^{rx}) = p(r) e^{rx}, \tag{7.1}$$

where

$$p(r) = r^n + a_1 r^{n-1} + a_2 r^{n-2} + \cdots + a_n.$$

We call p the *characteristic polynomial* of L. If r_1 is a root of p, then clearly $L(e^{r_1 x}) = 0$, and we have a solution $e^{r_1 x}$. If r_1 is a root of multiplicity m_1 of p, then

$$p(r_1) = 0, \qquad p'(r_1) = 0, \cdots, p^{(m_1-1)}(r_1) = 0.$$

If we differentiate the equation (7.1) k times with respect to r, we obtain*

$$\frac{\partial^k}{\partial r^k} L(e^{rx}) = L\left(\frac{\partial^k}{\partial r^k} e^{rx}\right) = L(x^k e^{rx})$$

$$= \left[p^{(k)}(r) + k p^{(k-1)}(r) x + \frac{k(k-1)}{2!} p^{(k-2)}(r) x^2 + \cdots + p(r) x^k \right] e^{rx}.$$

* If f, g are two functions having k derivatives, then

$$(fg)^{(k)} = f^{(k)} g + k f^{(k-1)} g' + \frac{k(k-1)}{2!} f^{(k-2)} g'' + \cdots + f g^{(k)}.$$

Thus for $k = 0, 1, \cdots, m_1 - 1$, we see that $x^k e^{r_1 x}$ is a solution of $L(y) = 0$. Repeating this process for each root of p we arrive at the following result.

Theorem 11. *Let* $r_1, \cdots, r_s$ *be the distinct roots of the characteristic polynomial* p, *and suppose* r_i *has multiplicity* m_i *(thus* $m_1 + m_2 + \cdots + m_s = n$*). Then the* n *functions*

$$e^{r_1 x}, \ xe^{r_1 x}, \ \cdots, \ x^{m_1-1}e^{r_1 x};$$

$$e^{r_2 x}, \ xe^{r_2 x}, \ \cdots, \ x^{m_2-1}e^{r_2 x}; \quad \cdots;$$

$$e^{r_s x}, \ xe^{r_s x}, \ \cdots, \ x^{m_s-1}e^{r_s x}$$

are solutions of $L(y) = 0$.

The n functions $\phi_1, \cdots, \phi_n$ on an interval I are said to be *linearly dependent* on I if there are constants $c_1, \cdots, c_n$ not all zero, such that

$$c_1\phi_1(x) + \cdots + c_n\phi_n(x) = 0$$

for all x in I. The functions $\phi_1, \cdots, \phi_n$ are said to be *linearly independent* on I if they are not linearly dependent on I.

Theorem 12. *The* n *solutions of* $L(y) = 0$ *given in Theorem* 11 *are linearly independent on any interval* I.

Proof. Suppose we have n constants

$$c_{ij} \qquad (i = 1, \cdots, s; j = 0, \cdots, m_i - 1)$$

such that

$$\sum_{i=1}^{s} \sum_{j=0}^{m_i-1} c_{ij}x^j e^{r_i x} = 0 \tag{7.2}$$

on I. Summing over j for fixed i, we let

$$P_i(x) = \sum_{j=0}^{m_i-1} c_{ij}x^j$$

be the polynomial coefficient of $e^{r_i x}$ in (7.2). Thus we have

$$P_1(x)e^{r_1 x} + P_2(x)e^{r_2 x} + \cdots + P_s(x)e^{r_s x} = 0 \tag{7.3}$$

on I. Assume that not all the constants c_{ij} are 0. Then there will be at least one of the polynomials P_i which is not identically zero on I. By relabeling the roots r_i if necessary we can assume that P_s is not identically zero on I. Now (7.3) implies that

$$P_1(x) + P_2(x)e^{(r_2-r_1)x} + \cdots + P_s(x)e^{(r_s-r_1)x} = 0 \tag{7.4}$$

on I. Upon differentiating (7.4) sufficiently many times (at most m_1 times) we can reduce $P_1(x)$ to 0. In this process the degrees of the polynomials multiplying $e^{(r_i - r_1)x}$ remain unchanged, as well as the non-identically vanishing character of any of these polynomials. We obtain an expression of the form

$$Q_2(x)e^{(r_2-r_1)x} + \cdots + Q_s(x)e^{(r_s-r_1)x} = 0,$$

or

$$Q_2(x)e^{r_2x} + \cdots + Q_s(x)e^{r_sx} = 0$$

on I, where the Q_i are polynomials, deg Q_i = deg P_i, and Q_s does not vanish identically. Continuing this process we finally arrive at a situation where

$$R_s(x)e^{r_sx} = 0 \qquad (7.5)$$

on I, and R_s is a polynomial, deg R_s = deg P_s, which does not vanish identically on I. But (7.5) implies that $R_s(x) = 0$ for all x on I. This contradiction forces us to abandon the supposition that P_s is not identically zero. Thus $P_s(x) = 0$ for all x in I, and we have shown that all the constants $c_{ij} = 0$, proving that the n solutions given in Theorem 11 are linearly independent on any interval I.

If $\phi_1, \cdots, \phi_m$ are any m solutions of $L(y) = 0$ on an interval I, and $c_1, \cdots, c_m$ are any m constants, then

$$\phi = c_1\phi_1 + \cdots + c_m\phi_m$$

is also a solution since

$$L(\phi) = c_1 L(\phi_1) + \cdots + c_m L(\phi_m) = 0.$$

As in the case $n = 2$ every solution of $L(y) = 0$ is a linear combination of n linearly independent solutions. The proof of this fact depends on the uniqueness of solutions to initial value problems, which we shall establish in Sec. 8, Theorem 17.

As an example consider the equation

$$y''' - 3y' + 2y = 0.$$

The characteristic polynomial is

$$p(r) = r^3 - 3r + 2,$$

and its roots are 1, 1, -2. Thus three linearly independent solutions are given by

$$e^x, \qquad xe^x, \qquad e^{-2x},$$

and any solution ϕ has the form

$$\phi(x) = (c_1 + c_2x)e^x + c_3e^{-2x},$$

where c_1, c_2, c_3 are any constants.

EXERCISES

1. Are the following sets of functions defined on $-\infty < x < \infty$ linearly independent or dependent there? Why?

   (a) $\phi_1(x) = 1$, $\phi_2(x) = x$, $\phi_3(x) = x^3$

   (b) $\phi_1(x) = e^{ix}$, $\phi_2(x) = \sin x$, $\phi_3(x) = 2 \cos x$

   (c) $\phi_1(x) = x$, $\phi_2(x) = e^{2x}$, $\phi_3(x) = |x|$

2. Prove that if p_1, p_2, p_3, p_4 are polynomials of degree two, they are linearly dependent on $-\infty < x < \infty$.

3. Are the following statements true or false? If the statement is true, prove it; otherwise give a counterexample.

   (a) "If ϕ_1, $\cdots$, ϕ_n are linearly independent functions on an interval I, then any subset of them forms a linearly independent set of functions on I."

   (b) "If ϕ_1, $\cdots$, ϕ_n are linearly dependent functions on an interval I, then any subset of them forms a linearly dependent set of functions on I."

4. Find all solutions of the following equations:

   (a) $y''' - 8y = 0$ (b) $y^{(4)} + 16y = 0$

   (c) $y''' - 5y'' + 6y' = 0$ (d) $y''' - iy'' + 4y' - 4iy = 0$

   (e) $y^{(100)} + 100y = 0$ (f) $y^{(4)} + 5y'' + 4y = 0$

   (g) $y^{(4)} - 16y = 0$ (h) $y''' - 3y' - 2y = 0$

   (i) $y''' - 3iy'' - 3y' + iy = 0$

5. (a) Compute the Wronskian of four linearly independent solutions of the equation $y^{(4)} + 16y = 0$.

   (b) Compute that solution ϕ of this equation which satisfies

$$\phi(0) = 1, \quad \phi'(0) = 0, \quad \phi''(0) = 0, \quad \phi'''(0) = 0.$$

6. Find four linearly independent solutions of the equation

$$y^{(4)} + \lambda y = 0,$$

in case:

   (a) $\lambda = 0$ (b) $\lambda > 0$ (c) $\lambda < 0$

7. Suppose the constants a_1, $\cdots$, a_n in

$$L(y) = y^{(n)} + a_1 y^{(n-1)} + \cdots + a_n y$$

are all real.

   (a) Show that if ϕ is a solution of $L(y) = 0$ then so are

$$\phi_1 = \operatorname{Re} \phi \quad \text{and} \quad \phi_2 = \operatorname{Im} \phi.$$

   (b) If $\alpha + i\beta$ (α, β real) is a root of the characteristic polynomial of L show that the functions

$$e^{\alpha x} \cos \beta x, \qquad e^{\alpha x} \sin \beta x$$

are solutions of $L(y) = 0$.

8. Suppose all roots of the characteristic polynomial for the equation

$$y^{(n)} + a_1 y^{(n-1)} + \cdots + a_n y = 0$$

have negative real parts. Show that every solution tends to zero as $x \to +\infty$.

9. Suppose all roots of the characteristic polynomial for the equation

$$y^{(n)} + a_1 y^{(n-1)} + \cdots + a_n y = 0$$

have non-positive real parts, and those roots with zero real parts have multiplicity one. Show that all solutions are bounded on $0 \leq x < \infty$. (*Note*: A solution ϕ is bounded on $0 \leq x < \infty$ if there is a constant $k > 0$ such that

$$|\phi(x)| \leq k \quad \text{for} \quad 0 \leq x < \infty.)$$

8. Initial value problems for n-th order equations

An *initial value problem* for $L(y) = 0$ is a problem of finding a solution ϕ which has prescribed values for it, and its first $n - 1$ derivatives, at some point x_0 (the initial point). If $\alpha_1, \cdots, \alpha_n$ are given constants, and x_0 is some real number, the problem of finding a solution ϕ of $L(y) = 0$ satisfying

$$\phi(x_0) = \alpha_1, \quad \phi'(x_0) = \alpha_2, \quad \cdots, \quad \phi^{(n-1)}(x_0) = \alpha_n,$$

is denoted by

$$L(y) = 0, \quad y(x_0) = \alpha_1, \quad y'(x_0) = \alpha_2, \quad \cdots, \quad y^{(n-1)}(x_0) = \alpha_n.$$

There is only one solution to such an initial value problem, and the demonstration of this will depend on an estimate for the rate of growth of a solution ϕ of $L(y) = 0$, together with its derivatives $\phi', \cdots, \phi^{(n-1)}$. We define $\|\phi(x)\|$ by

$$\|\phi(x)\| = [|\phi(x)|^2 + \cdots + |\phi^{(n-1)}(x)|^2]^{1/2},$$

the positive square root being understood, and give the analogue of Theorem 3.

Theorem 13. *Let ϕ be any solution of*

$$L(y) = y^{(n)} + a_1 y^{(n-1)} + \cdots + a_n y = 0$$

on an interval I *containing a point* x_0. *Then for all* x *in* I

$$\|\phi(x_0)\| e^{-k|x-x_0|} \leq \|\phi(x)\| \leq \|\phi(x_0)\| e^{k|x-x_0|}, \tag{8.1}$$

where

$$k = 1 + |a_1| + \cdots + |a_n|.$$

Proof. Letting $u(x) = \| \phi(x) \|^2$ we have

$$u = \phi\bar{\phi} + \phi'\bar{\phi'} + \cdots + \phi^{(n-1)}\overline{\phi^{(n-1)}}.$$

Hence

$$u' = \phi'\bar{\phi} + \phi''\bar{\phi'} + \cdots + \phi^{(n)}\overline{\phi^{(n-1)}}$$
$$+ \phi\bar{\phi'} + \phi'\bar{\phi''} + \cdots + \phi^{(n-1)}\overline{\phi^{(n)}},$$

and therefore

$$| u'(x) | \leq 2 | \phi(x) || \phi'(x) | + 2 | \phi'(x) || \phi''(x) | + \cdots$$
$$+ 2 | \phi^{(n-1)}(x) || \phi^{(n)}(x) |. \quad (8.2)$$

Since ϕ satisfies $L(\phi) = 0$ we have

$$\phi^{(n)} = - [a_1\phi^{(n-1)} + \cdots + a_n\phi],$$

and

$$| \phi^{(n)}(x) | \leq | a_1 || \phi^{(n-1)}(\dot{x}) | + \cdots + | a_n || \phi(x) |. \quad (8.3)$$

Using (8.3) in (8.2) there results

$$| u'(x) | \leq 2 | \phi(x) || \phi'(x) | + 2 | \phi'(x) || \phi''(x) | + \cdots$$
$$+ 2 | \phi^{(n-2)}(x) || \phi^{(n-1)}(x) | + 2 | a_1 || \phi^{(n-1)}(x) |^2$$
$$+ 2 | a_2 | | \phi^{(n-2)}(x) || \phi^{(n-1)}(x) | + \cdots + 2 | a_n || \phi(x) || \phi^{(n-1)}(x) |.$$

We now apply the elementary inequality

$$2 | b || c | \leq | b |^2 + | c |^2$$

to obtain

$$| u'(x) | \leq (1 + | a_n |) | \phi(x) |^2 + (2 + | a_{n-1} |) | \phi'(x) |^2$$
$$+ \cdots + (2 + | a_2 |) | \phi^{(n-2)}(x) |^2$$
$$+ (1 + 2 | a_1 | + | a_2 | + \cdots + | a_n |) | \phi^{(n-1)}(x) |^2.$$

Therefore

$$| u'(x) | \leq 2ku(x),$$

and the remainder of the proof is the same as the steps following (3.8) in the proof of Theorem 3.

Theorem 14. (*Uniqueness Theorem*) *Let* $\alpha_1, \cdots, \alpha_n$ *be any* n *constants, and let* x_0 *be any real number. On any interval* I *containing* x_0 *there exists at most one solution* ϕ *of* $L(y) = 0$ *satisfying*

$$\phi(x_0) = \alpha_1, \quad \phi'(x_0) = \alpha_2, \quad \cdots, \quad \phi^{(n-1)}(x_0) = \alpha_n.$$

Proof. The proof is the same as that of Theorem 4. Suppose ϕ, ψ were two solutions of $L(y) = 0$ on I satisfying the above conditions at x_0. Then $\chi = \phi - \psi$ satisfies $L(\chi) = 0$ and

$$\chi(x_0) = \chi'(x_0) = \cdots = \chi^{(n-1)}(x_0) = 0.$$

Thus $\| \chi(x_0) \| = 0$, and applying (8.1) to χ we obtain $\| \chi(x) \| = 0$ for all x in I. This implies $\chi(x) = 0$ for all x in I, or $\phi = \psi$.

The *Wronskian* $W(\phi_1, \cdots, \phi_n)$ of n functions $\phi_1, \cdots, \phi_n$ having $n - 1$ derivatives on an interval I is defined to be the determinant function

$$W(\phi_1, \cdots, \phi_n) = \begin{vmatrix} \phi_1 & \cdots & \phi_n \\ \phi_1' & \cdots & \phi_n' \\ \cdot & & \cdot \\ \cdot & & \cdot \\ \cdot & & \cdot \\ \phi_1^{(n-1)} & \cdots & \phi_n^{(n-1)} \end{vmatrix},$$

its value at any x in I being $W(\phi_1, \cdots, \phi_n)(x)$.

Theorem 15. *If $\phi_1, \cdots, \phi_n$ are n solutions of* $L(y) = 0$ *on an interval* I, *they are linearly independent there if, and only if,* $W(\phi_1, \cdots, \phi_n)(x) \neq 0$ *for all* x *in* I.

The proof is entirely similar to the proof of Theorem 6 (the case $n = 2$), and so will be omitted. The result and the proof do not depend on the fact that L has *constant* coefficients. See the proof for a more general case in Chap. 3, Theorem 6.

Theorem 16. (*Existence Theorem*) *Let $\alpha_1, \cdots, \alpha_n$ be any n constants, and let x_0 be any real number. There exists a solution ϕ of* $L(y) = 0$ *on* $-\infty < x < \infty$ *satisfying*

$$\phi(x_0) = \alpha_1, \quad \phi'(x_0) = \alpha_2, \quad \cdots, \quad \phi^{(n-1)}(x_0) = \alpha_n. \tag{8.4}$$

Proof. Let $\phi_1, \cdots, \phi_n$ be any set of n linearly independent solutions of $L(y) = 0$ on $-\infty < x < \infty$, for example these could be the solutions obtained in Theorem 11. It will be shown that there exist unique constants $c_1, \cdots, c_n$ such that

$$\phi = c_1\phi_1 + \cdots + c_n\phi_n$$

is a solution of $L(y) = 0$ satisfying (8.4). Such constants would have to satisfy

$$
\begin{aligned}
c_1\phi_1(x_0) + \cdots + c_n\phi_n(x_0) &= \alpha_1 \\
c_1\phi_1'(x_0) + \cdots + c_n\phi_n'(x_0) &= \alpha_2 \\
&\vdots \\
c_1\phi_1^{(n-1)}(x_0) + \cdots + c_n\phi_n^{(n-1)}(x_0) &= \alpha_n,
\end{aligned}
\tag{8.5}
$$

which is a system of n linear equations for $c_1, \cdots, c_n$. The determinant of the coefficients is just $W(\phi_1, \cdots, \phi_n)(x_0)$ which is not zero, by Theorem 15.

Therefore there is a unique set of constants $c_1, \cdots, c_n$ satisfying (8.5). For this choice of $c_1, \cdots, c_n$ the function

$$\phi = c_1\phi_1 + \cdots + c_n\phi_n$$

will be the desired solution.

The results of Theorems 14 and 16 allow us to describe all solutions of the homogeneous n-th order equation with constant coefficients $L(y) = 0$.

Theorem 17. *Let* $\phi_1, \cdots, \phi_n$ *be* n *linearly independent solutions of* $L(y) = 0$ *on an interval* I. *If* $c_1, \cdots, c_n$ *are any constants*

$$\phi = c_1\phi_1 + \cdots + c_n\phi_n \qquad (8.6)$$

is a solution, and every solution may be represented in this form.

Proof. We have already seen that

$$L(\phi) = c_1 L(\phi_1) + \cdots + c_n L(\phi_n) = 0.$$

Now, let ϕ be any solution of $L(y) = 0$, and let x_0 be in I. Suppose

$$\phi(x_0) = \alpha_1, \quad \phi'(x_0) = \alpha_2, \quad \cdots, \quad \phi^{(n-1)}(x_0) = \alpha_n.$$

In the proof of Theorem 16 we showed that there exist unique constants $c_1, \cdots, c_n$ such that $\psi = c_1\phi_1 + \cdots + c_n\phi_n$ is a solution of $L(y) = 0$ on I satisfying

$$\psi(x_0) = \alpha_1, \quad \psi'(x_0) = \alpha_2, \quad \cdots, \quad \psi^{(n-1)}(x_0) = \alpha_n.$$

The uniqueness theorem (Theorem 14) implies that $\phi = \psi$, proving that ϕ may be represented as in (8.6).

A simple formula exists for the Wronskian, as in the case $n = 2$.

Theorem 18. *Let* $\phi_1, \cdots, \phi_n$ *be* n *solutions of* $L(y) = 0$ *on an interval* I *containing a point* x_0. *Then*

$$W(\phi_1, \cdots, \phi_n)(x) = e^{-a_1(x-x_0)} W(\phi_1, \cdots, \phi_n)(x_0). \qquad (8.7)$$

This result is a corollary of a more general result concerning the Wronskian of n solutions of a linear homogeneous equation with variable coefficients; see Theorem 8, Chap. 3. We therefore omit the proof here.

Corollary to Theorem 18. *Let* $\phi_1, \cdots, \phi_n$ *be* n *solutions of* $L(y) = 0$ *on an interval* I *containing* x_0. *Then they are linearly independent on* I *if and only if* $W(\phi_1, \cdots, \phi_n)(x_0) \neq 0$.

Proof. The proof is an immediate consequence of Theorem 15 and the formula (8.7).

As a simple illustration of the use of (8.7) consider a homogeneous equation of order 3 which has a root r_1 with multiplicity 3. Its characteristic polynomial is

$$p(r) = (r - r_1)^3 = r^3 - 3r_1r^2 + 3r_1^2r - r_1^3.$$

Hence

$$L(y) = y''' - 3r_1y'' + 3r_1^2y' - r_1^3y,$$

and we have $a_1 = -3r_1$. We take

$$\phi_1(x) = e^{r_1x}, \qquad \phi_2(x) = xe^{r_1x}, \qquad \phi_3(x) = x^2e^{r_1x},$$

and then obtain

$$W(\phi_1, \phi_2, \phi_3)(x) = \begin{vmatrix} e^{r_1x} & xe^{r_1x} & x^2e^{r_1x} \\ r_1e^{r_1x} & (1 + r_1x)e^{r_1x} & (2x + r_1x^2)e^{r_1x} \\ r_1^2e^{r_1x} & (2r_1 + r_1^2x)e^{r_1x} & (2 + 4r_1x + r_1^2x^2)e^{r_1x} \end{vmatrix}$$

This becomes a little involved to evaluate directly, but using (8.7) with $x_0 = 0$ we obtain

$$W(\phi_1, \phi_2, \phi_3)(0) = \begin{vmatrix} 1 & 0 & 0 \\ r_1 & 1 & 0 \\ r_1^2 & 2r_1 & 2 \end{vmatrix} = 2,$$

and hence

$$W(\phi_1, \phi_2, \phi_3)(x) = 2e^{3r_1x}.$$

EXERCISES

1. Consider the equation

$$y''' - 4y' = 0.$$

(a) Compute three linearly independent solutions.

(b) Compute the Wronskian of the solutions found in (a).

(c) Find that solution ϕ satisfying

$$\phi(0) = 0, \qquad \phi'(0) = 1, \qquad \phi''(0) = 0.$$

2. Consider the equation

$$y^{(5)} - y^{(4)} - y' + y = 0.$$

(a) Compute five linearly independent solutions.

(b) Compute the Wronskian of the solutions found in (a), using Theorem 18.

(c) Find that solution ϕ satisfying

$$\phi(0) = 1, \qquad \phi'(0) = \phi''(0) = \phi'''(0) = \phi^{(4)}(0) = 0.$$

3. Suppose ϕ is a solution of

$$y^{(n)} + a_1 y^{(n-1)} + \cdots + a_n y = 0,$$

and $\psi(x) = \phi(x) \exp(a_1 x/n)$. Show that ψ satisfies a linear homogeneous equation with constant coefficients

$$y^{(n)} + b_1 y^{(n-1)} + \cdots + b_n y = 0,$$

with $b_1 = 0$. (*Note*: The Wronskian of any n linearly independent solutions of the latter equation is a constant; see (8.7).)

4. Consider the constant coefficient equation

$$y^{(n)} + a_1 y^{(n-1)} + \cdots + a_n y = 0,$$

and suppose $\phi_1, \cdots, \phi_n$ are solutions satisfying for some real x_0

$$\phi_i^{(j-1)}(x_0) = \delta_{ij}, \qquad (i, j = 1, \cdots, n),$$

where $\delta_{ij} = 1$ if $i = j$, and $\delta_{ij} = 0$ if $i \neq j$.
(a) Show that $\phi_1, \cdots, \phi_n$ are linearly independent.
(b) If ϕ is a solution satisfying

$$\phi^{(j-1)}(x_0) = \alpha_j, \qquad (j = 1, \cdots, n),$$

show that

$$\phi = \alpha_1 \phi_1 + \alpha_2 \phi_2 + \cdots + \alpha_n \phi_n.$$

(*Note*: This shows that ϕ is a *linear* function of its initial conditions $\alpha_1, \cdots, \alpha_n$.)

9. Equations with real constants

Suppose that the constants $a_1, \cdots, a_n$ in

$$L(y) = y^{(n)} + a_1 y^{(n-1)} + \cdots + a_n y$$

are all *real* numbers. The characteristic polynomial

$$p(r) = r^n + a_1 r^{n-1} + \cdots + a_n$$

then has all real coefficients. This implies that

$$\overline{p(r)} = p(\bar{r}) \tag{9.1}$$

for all r, since

$$\begin{aligned}
\overline{p(r)} &= \overline{r^n + a_1 r^{n-1} + \cdots + a_n} \\
&= \overline{r^n} + \overline{a_1 r^{n-1}} + \cdots + \overline{a_n} \\
&= \bar{r}^n + \bar{a}_1 \bar{r}^{n-1} + \cdots + \overline{a_n} \\
&= \bar{r}^n + a_1 \bar{r}^{n-1} + \cdots + a_n \\
&= p(\bar{r}).
\end{aligned}$$

From (9.1) it follows that if r_1 is a root of p, then so is $\bar{r}_1$. Thus the roots of p whose imaginary parts do not vanish occur in conjugate pairs. A slight extension of this argument shows that if r_1 is a root of multiplicity m_1 then $\bar{r}_1$ is a root with the same multiplicity m_1. If there are s distinct roots of p, let us enumerate them as follows:

$$r_1, \bar{r}_1, r_2, \bar{r}_2, \cdots, r_j, \bar{r}_j, r_{2j+1}, \cdots, r_s,$$

where

$$r_k = \sigma_k + i\tau_k, \quad (k = 1, \cdots, j; \sigma_k, \tau_k \text{ real}; \tau_k \neq 0),$$

and $r_{2j+1}, \cdots, r_s$ are real. Suppose that r_k has multiplicity m_k. Then we have

$$2(m_1 + \cdots + m_j) + m_{2j+1} + \cdots + m_s = n.$$

Corresponding to these roots we have the n linearly independent solutions

$$e^{r_1x}, xe^{r_1x}, \cdots, x^{m_1-1}e^{r_1x}; e^{\bar{r}_1x}, xe^{\bar{r}_1x}, \cdots, x^{m_1-1}e^{\bar{r}_1x}; \cdots; e^{r_sx}, xe^{r_sx}, \cdots, x^{m_s-1}e^{r_sx}$$

$$\text{(9.2)}$$

of $L(y) = 0$. Every solution is a linear combination, with constant coefficients, of these. We now note that if $1 \leq k \leq j, 0 \leq h \leq m_k - 1$,

$$x^h e^{r_k x} = x^h e^{(\sigma_k + i\tau_k)x} = x^h e^{\sigma_k x}(\cos \tau_k x + i \sin \tau_k x),$$
$$x^h e^{\bar{r}_k x} = x^h e^{(\sigma_k - i\tau_k)x} = x^h e^{\sigma_k x}(\cos \tau_k x - i \sin \tau_k x).$$

$$\text{(9.3)}$$

Thus every solution is a linear combination, with constant coefficients, of the n functions

$$e^{\sigma_1 x} \cos \tau_1 x, xe^{\sigma_1 x} \cos \tau_1 x, \cdots, x^{m_1-1}e^{\sigma_1 x} \cos \tau_1 x;$$

$$e^{\sigma_1 x} \sin \tau_1 x, xe^{\sigma_1 x} \sin \tau_1 x, \cdots, x^{m_1-1}e^{\sigma_1 x} \sin \tau_1 x;$$

$$\cdot$$
$$\cdot \qquad \text{(9.4)}$$
$$\cdot$$

$$e^{r_s x}, xe^{r_s x}, \cdots, x^{m_s-1}e^{r_s x}.$$

Each of the functions in (9.4) is a solution of $L(y) = 0$ since, from (9.3),

$$x^h e^{\sigma_k x} \cos \tau_k x = \frac{1}{2}x^h(e^{r_k x} + e^{\bar{r}_k x}),$$

$$\text{(9.5)}$$

$$x^h e^{\sigma_k x} \sin \tau_k x = \frac{1}{2i}x^h(e^{r_k x} - e^{\bar{r}_k x}).$$

The solutions in (9.4) are all *real-valued*, and they are linearly independent. For suppose we have a linear combination of these functions equal to zero. Let us denote the terms in this sum which involve

$$x^h e^{\sigma_k x} \cos \tau_k x, \qquad x^h e^{\sigma_k x} \sin \tau_k x$$

by

$$c x^h e^{\sigma_k x} \cos \tau_k x + d x^h e^{\sigma_k x} \sin \tau_k x,$$

where c and d are constants. Using (9.5) we find that we have a linear combination of the functions (9.2) equal to zero, and the terms involving $x^h e^{r_k x}$, $x^h e^{\bar{r}_k x}$ will be

$$\frac{(c - id)}{2} x^h e^{r_k x} + \frac{(c + id)}{2} x^h e^{\bar{r}_k x}.$$

Since the functions (9.2) are linearly independent we must have all the coefficients in this sum equal to zero. In particular

$$c + id = 0, \qquad c - id = 0,$$

from which it follows that $c = 0$, $d = 0$. Thus the solutions (9.4) are linearly independent.

If ϕ is any *real-valued solution* of $L(y) = 0$, then ϕ is a linear combination of the real solutions (9.4) with *real* coefficients. Indeed, if we denote the solutions in (9.4) by $\phi_1, \cdots, \phi_n$, we have

$$\phi = c_1 \phi_1 + \cdots + c_n \phi_n,$$

for some constants $c_1, \cdots, c_n$. Since $\phi, \phi_1, \cdots, \phi_n$ are all real-valued, we have

$$0 = \operatorname{Im} \phi = (\operatorname{Im} c_1) \phi_1 + \cdots + (\operatorname{Im} c_n) \phi_n,$$

and since $\phi_1, \cdots, \phi_n$ are linearly independent we must have

$$\operatorname{Im} c_1 = \operatorname{Im} c_2 = \cdots = \operatorname{Im} c_n = 0.$$

This shows that $c_1, \cdots, c_n$ are all real numbers.

We remark that if ϕ is a solution of $L(y) = 0$ which is such that

$$\phi(x_0) = \alpha_1, \quad \phi'(x_0) = \alpha_2, \quad \cdots, \quad \phi^{(n-1)}(x_0) = \alpha_n, \qquad (9.6)$$

where $\alpha_1, \cdots, \alpha_n$ are *real* constants, then ϕ is real-valued. One way to see this is to note that since

$$L(\bar{\phi}) = \overline{L(\phi)} = 0,$$

$\bar{\phi}$ is also a solution, and hence so is

$$\psi = (1/2i)(\phi - \bar{\phi}) = \operatorname{Im} \phi.$$

But, from (9.6) we see that

$$\psi(x_0) = 0, \quad \psi'(x_0) = 0, \quad \cdots, \quad \psi^{(n-1)}(x_0) = 0.$$

The uniqueness theorem implies that $\psi(x) = 0$ for all x, or Im $\phi = 0$, showing that ϕ is real-valued.

We summarize in the following theorem.

Theorem 19. *Suppose the constants* $a_1, \cdots, a_n$ *in the equation*

$$L(y) = y^{(n)} + a_1 y^{(n-1)} + \cdots + a_n y = 0$$

are all real. There exists a set of n *linearly independent real-valued solutions* (9.4), *and every real-valued solution is a linear combination of these with real coefficients. If a solution satisfies real initial conditions, it is real-valued.*

The importance of Theorem 19 is that in many practical problems differential equations are encountered with real coefficients, and the real solutions are the ones sought. For example, the equation

$$y^{(4)} + y = 0 \tag{9.7}$$

arises in the study of the deflection of beams. The characteristic polynomial is given by

$$p(r) = r^4 + 1,$$

and its roots are

$$\frac{1}{\sqrt{2}}(1 + i), \quad \frac{1}{\sqrt{2}}(1 - i), \quad \frac{1}{\sqrt{2}}(-1 + i), \quad \frac{1}{\sqrt{2}}(-1 - i).$$

Thus every real solution ϕ of (9.7) has the form

$$\phi(x) = e^{x/\sqrt{2}}[c_1 \cos (x/\sqrt{2}) + c_2 \sin (x/\sqrt{2})]$$

$$+ e^{-x/\sqrt{2}}[c_3 \cos (x/\sqrt{2}) + c_4 \sin (x/\sqrt{2})],$$

where $c_1, \cdots, c_4$ are real constants.

EXERCISES

1. Find all real-valued solutions of the following equations:
(a) $y'' + y = 0$ (b) $y'' - y = 0$
(c) $y^{(4)} - y = 0$ (d) $y^{(5)} + 2y = 0$
(e) $y^{(4)} - 5y'' + 4y = 0$

2. Find the solution ϕ of the initial-value problem

$$y''' + y = 0, \quad y(0) = 0, \quad y'(0) = 1, \quad y''(0) = 0.$$

3. Determine all real-valued solutions of the equations:
(a) $y''' - iy'' + y' - iy = 0$ (b) $y'' - 2iy' - y = 0$

4. Show that if there exists a non-trivial solution of the problem

$$y^{(n)} + a_1 y^{(n-1)} + \cdots + a_n y = 0, \qquad (a_1, \cdots, a_n \text{ real}),$$

$$y^{(k-1)}(0) = y^{(k-1)}(1), \qquad (k = 1, \cdots, n),$$

then there exists a non-trivial real-valued solution.

5. Consider the equation

$$y^{(4)} - k^4 y = 0,$$

where k is a real constant.

(a) Show that $\cos kx$, $\sin kx$, $\cosh kx$, $\sinh kx$ are solutions if $k \neq 0$. (*Note:* $\cosh u = (e^u + e^{-u})/2$, $\sinh u = (e^u - e^{-u})/2$.)

(b) Show that there are non-trivial solutions ϕ satisfying

$$\phi(0) = 0, \quad \phi'(0) = 0, \quad \phi(1) = 0, \quad \phi'(1) = 0,$$

if and only if $\cos k \cosh k = 1$ and $k \neq 0$.

(c) Compute all non-trivial solutions satisfying the conditions in (b).

(d) For what values of k will there exist non-trivial solutions satisfying

$$\phi^{(j)}(0) = \phi^{(j)}(1), \qquad (j = 0, 1, 2, 3)?$$

(e) Compute all non-trivial solutions satisfying the conditions in (d).

6. Suppose the characteristic polynomial p of

$$L(y) = y^{(n)} + a_1 y^{(n-1)} + \cdots + a_n y = 0$$

has a real root r with multiplicity m, and $-r$ is also a root of multiplicity m. Show that

$$\cosh rx, \quad x \cosh rx, \quad \cdots, \quad x^{m-1} \cosh rx,$$

$$\sinh rx, \quad x \sinh rx, \quad \cdots, \quad x^{m-1} \sinh rx$$

are $2m$ linearly independent solutions which can be used to replace

$$e^{rx}, \quad xe^{rx}, \quad \cdots, \quad x^{m-1}e^{rx}$$

$$e^{-rx}, \quad xe^{-rx}, \quad \cdots, \quad x^{m-1}e^{-rx}$$

in a set of n linearly independent solutions of $L(y) = 0$.

10. The non-homogeneous equation of order n

Let b be a continuous function on an interval I, and consider the equation

$$L(y) = y^{(n)} + a_1 y^{(n-1)} + a_2 y^{(n-2)} + \cdots + a_n y = b(x),$$

where $a_1, a_2, \cdots, a_n$ are constants. If ψ_p is a particular solution of $L(y) = b(x)$, and ψ is any other solution, then

$$L(\psi - \psi_p) = L(\psi) - L(\psi_p) = b - b = 0.$$

Thus $\psi - \psi_p$ is a solution of the homogeneous equation $L(y) = 0$, and this implies that any solution ψ of $L(y) = b(x)$ can be written in the form

$$\psi = \psi_p + c_1\phi_1 + c_2\phi_2 + \cdots + c_n\phi_n,$$

where ψ_p is a particular solution of $L(y) = b(x)$, the functions ϕ_1, ϕ_2, $\cdots$, ϕ_n are n linearly independent solutions of $L(y) = 0$, and c_1, $\cdots$, c_n are constants.

To find a particular solution ψ_p we proceed just as in the case $n = 2$, that is, we use the *variation of constants* method. We try to find n functions u_1, $\cdots$, u_n so that

$$\psi_p = u_1\phi_1 + \cdots + u_n\phi_n$$

is a solution. Taking our cue from Sec. 6 we see that if

$$u_1'\phi_1 + \cdots + u_n'\phi_n = 0,$$

then

$$\psi_p' = u_1\phi_1' + \cdots + u_n\phi_n',$$

and if

$$u_1'\phi_1' + \cdots + u_n'\phi_n' = 0,$$

we have

$$\psi_p'' = u_1\phi_1'' + \cdots + u_n\phi_n''.$$

Thus, if u_1', $\cdots$, u_n' satisfy

$$u_1'\phi_1 + \cdots + u_n'\phi_n = 0$$

$$u_1'\phi_1' + \cdots + u_n'\phi_n' = 0$$

$$\cdot$$
$$\cdot \qquad\qquad (10.1)$$
$$\cdot$$

$$u_1'\phi_1^{(n-2)} + \cdots + u_n'\phi_n^{(n-2)} = 0$$

$$u_1'\phi_1^{(n-1)} + \cdots + u_n'\phi_n^{(n-1)} = b$$

we see that

$$\psi_p = u_1\phi_1 + \cdots + u_n\phi_n$$

$$\psi_p' = u_1\phi_1' + \cdots + u_n\phi_n'$$

$$\cdot$$
$$\cdot \qquad\qquad (10.2)$$
$$\cdot$$

$$\psi_p^{(n-1)} = u_1\phi_1^{(n-1)} + \cdots + u_n\phi_n^{(n-1)}$$

$$\psi_p^{(n)} = u_1\phi_1^{(n)} + \cdots + u_n\phi_n^{(n)} + b.$$

Hence

$$L(\psi_p) = u_1 L(\phi_1) + \cdots + u_n L(\phi_n) + b = b,$$

and indeed ψ_p is a solution of $L(y) = b(x)$. The whole problem is now reduced to solving the linear system (10.1) for $u_1', \cdots, u_n'$. The determinant of the coefficients is just $W(\phi_1, \cdots, \phi_n)$, which is never zero when $\phi_1, \cdots, \phi_n$ are linearly independent solutions of $L(y) = 0$. Therefore there are unique functions $u_1', \cdots, u_n'$ satisfying (10.1). It is an easy exercise to see that solutions are given by

$$u_k'(x) = \frac{W_k(x) b(x)}{W(\phi_1, \cdots, \phi_n)(x)}, \qquad (k = 1, \cdots, n),$$

where W_k is the determinant obtained from $W(\phi_1, \cdots, \phi_n)$ by replacing the k-th column (that is, $\phi_k, \phi_k', \cdots, \phi_k^{(n-1)}$) by $0, 0, \cdots, 0, 1$.

If x_0 is any point in I, we may take for u_k the function given by

$$u_k(x) = \int_{x_0}^{x} \frac{W_k(t) b(t)}{W(\phi_1, \cdots, \phi_n)(t)} \, dt, \qquad (k = 1, \cdots, n).$$

The particular solution ψ_p now takes the form

$$\psi_p(x) = \sum_{k=1}^{n} \phi_k(x) \int_{x_0}^{x} \frac{W_k(t) b(t)}{W(\phi_1, \cdots, \phi_n)(t)} \, dt. \tag{10.3}$$

Theorem 20. *Let* b *be continuous on an interval* I, *and let* $\phi_1, \cdots, \phi_n$ *be* n *linearly independent solutions of* L(y) = 0 *on* I. *Every solution* ψ *of* L(y) = b(x) *can be written as*

$$\psi = \psi_p + c_1\phi_1 + \cdots + c_n\phi_n,$$

where ψ_p *is a particular solution of* L(y) = b(x), *and* $c_1, \cdots, c_n$ *are constants. Every such* ψ *is a solution of* L(y) = b(x). *A particular solution* ψ_p *is given by* (10.3).

It is left as an exercise for the student to show that the particular solution ψ_p given by (10.3) satisfies

$$\psi_p(x_0) = \psi_p'(x_0) = \cdots = \psi_p^{(n-1)}(x_0) = 0. \tag{10.4}$$

As an example let us compute the solution ψ of

$$y''' + y'' + y' + y = 1 \tag{10.5}$$

which satisfies

$$\psi(0) = 0, \qquad \psi'(0) = 1, \qquad \psi''(0) = 0. \tag{10.6}$$

The homogeneous equation is

$$y''' + y'' + y' + y = 0, \tag{10.7}$$

and the characteristic polynomial corresponding to it is

$$p(r) = r^3 + r^2 + r + 1.$$

The roots of p are i, $-i$, and -1. Since we are interested in a solution satisfying real initial conditions we take for independent solutions of (10.7)

$$\phi_1(x) = \cos x, \qquad \phi_2(x) = \sin x, \qquad \phi_3(x) = e^{-x}.$$

To obtain a particular solution of (10.5) of the form $u_1\phi_1 + u_2\phi_2 + u_3\phi_3$ we must solve the following equations for u_1', u_2', u_3':

$$u_1'\phi_1 + u_2'\phi_2 + u_3'\phi_3 = 0$$

$$u_1'\phi_1' + u_2'\phi_2' + u_3'\phi_3' = 0$$

$$u_1'\phi_1'' + u_2'\phi_2'' + u_3'\phi_3'' = 1,$$

which in this case reduce to

$$(\cos x)u_1' + (\sin x)u_2' + e^{-x}u_3' = 0$$

$$(-\sin x)u_1' + (\cos x)u_2' - e^{-x}u_3' = 0 \qquad (10.8)$$

$$(-\cos x)u_1' - (\sin x)u_2' + e^{-x}u_3' = 1.$$

The determinant of the coefficients is

$$W(\phi_1, \phi_2, \phi_3)(x) = \begin{vmatrix} \cos x & \sin x & e^{-x} \\ -\sin x & \cos x & -e^{-x} \\ -\cos x & -\sin x & e^{-x} \end{vmatrix}$$

Using (8.7) we have

$$W(\phi_1, \phi_2, \phi_3)(x) = e^{-x}W(\phi_1, \phi_2, \phi_3)(0),$$

since $a_1 = 1$ in this case. Now

$$W(\phi_1, \phi_2, \phi_3)(0) = \begin{vmatrix} 1 & 0 & 1 \\ 0 & 1 & -1 \\ -1 & 0 & 1 \end{vmatrix} = 2,$$

and thus

$$W(\phi_1, \phi_2, \phi_3)(x) = 2e^{-x}.$$

Solving (10.8) for u_1 we find that

$$u_1'(x) = \tfrac{1}{2}e^x \begin{vmatrix} 0 & \sin x & e^{-x} \\ 0 & \cos x & -e^{-x} \\ 1 & -\sin x & e^{-x} \end{vmatrix} = -\tfrac{1}{2}(\cos x + \sin x). \qquad (10.9)$$

Similarly we obtain

$$u_2'(x) = \tfrac{1}{2}(\cos x - \sin x), \tag{10.10}$$

$$u_3'(x) = \tfrac{1}{2}e^x. \tag{10.11}$$

Integrating (10.9)–(10.11), we obtain as choices for u_1, u_2, u_3:

$$u_1(x) = \tfrac{1}{2}(\cos x - \sin x),$$

$$u_2(x) = \tfrac{1}{2}(\sin x + \cos x),$$

$$u_3(x) = \tfrac{1}{2}e^x.$$

Therefore a particular solution of (10.5) is given by

$$u_1(x)\phi_1(x) + u_2(x)\phi_2(x) + u_3(x)\phi_3(x)$$

$$= \tfrac{1}{2}(\cos x - \sin x)\cos x + \tfrac{1}{2}(\sin x + \cos x)\sin x + \tfrac{1}{2}e^x e^{-x}$$

$$= 1.$$

(*Note*: There are simpler ways of discovering such a particular solution; see Sec. 11. We are interested in illustrating the general method here.) The most general solution ψ of (10.5) is of the form

$$\psi(x) = 1 + c_1 \cos x + c_2 \sin x + c_3 e^{-x},$$

where c_1, c_2, c_3 are constants. We must choose these constants so that the conditions (10.6) are valid. This leads to the following equations for c_1, c_2, c_3:

$$c_1 + c_3 = -1, \qquad c_2 - c_3 = 1, \qquad c_1 - c_3 = 0,$$

which have the unique solution

$$c_1 = -\tfrac{1}{2}, \qquad c_2 = \tfrac{1}{2}, \qquad c_3 = -\tfrac{1}{2}.$$

Therefore the solution of our problem is given by

$$\psi(x) = 1 + \tfrac{1}{2}(\sin x - \cos x - e^{-x}).$$

The solution corresponding to that given in (10.3), with $x_0 = 0$, is easily seen to be

$$\psi_p(x) = 1 - \tfrac{1}{2}(\cos x + \sin x + e^{-x}),$$

and this satisfies

$$\psi_p(0) = 0, \qquad \psi_p'(0) = 0, \qquad \psi_p''(0) = 0.$$

EXERCISES

1. Find all solutions of the following equations:
   (a) $y''' - y' = x$
   (b) $y''' - 8y = e^{ix}$
   (c) $y^{(4)} + 16y = \cos x$
   (d) $y^{(4)} - 4y^{(3)} + 6y'' - 4y' + y = e^x$
   (e) $y^{(4)} - y = \cos x$
   (f) $y'' - 2iy' - y = e^{ix} - 2e^{-ix}$

2. Let
$$L(y) = y^{(n)} + a_1 y^{(n-1)} + \cdots + a_n y,$$

and let p be the characteristic polynomial

$$p(r) = r^n + a_1 r^{n-1} + \cdots + a_n.$$

(a) If A and α are constants, and $p(\alpha) \neq 0$, show that there is a solution of $L(y) = Ae^{\alpha x}$ of the form $Be^{\alpha x}$, where B is a constant. What is B?
(b) Compute a solution of $L(y) = Ae^{\alpha x}$ in case α is a simple root of p (that is, a root of multiplicity one). (*Hint:* If B and r are any constants show that

$$BL(xe^{rx}) = B[p'(r) + xp(r)]e^{rx}.$$

Let $r = \alpha$.)
(c) Compute a solution of $L(y) = Ae^{\alpha x}$ in case α is a root of p of multiplicity k.

3. Prove that the solution ψ_p given by (10.3) satisfies the initial conditions (10.4). (*Hint:* Use (10.2).)

4. Let $g(x, t)$ be defined by

$$g(x, t) = \sum_{k=1}^{n} \frac{\phi_k(x) W_k(t)}{W(t)},$$

where $W(t) = W(\phi_1, \cdots, \phi_n)(t)$ is the Wronskian of n linearly independent solutions of $L(y) = 0$. For any continuous function b on any interval I containing x_0, let $G(b)$ be the function given by

$$G(b)(x) = \int_{x_0}^{x} g(x, t)b(t)\, dt.$$

Thus $G(b)$ is just the ψ_p of (10.3), and hence $L(G(b)) = b$.
(a) Show that g, as a function of x for each fixed t, is a solution of $L(y) = 0$ which satisfies

$$g(t, t) = 0, \qquad \frac{\partial g}{\partial x}(t, t) = 0, \qquad \cdots, \qquad \frac{\partial^{n-2} g}{\partial x^{n-2}}(t, t) = 0, \qquad \frac{\partial^{n-1} g}{\partial x^{n-1}}(t, t) = 1.$$

(*Note:* This shows that $g(x, t)$ is independent of the functions $\phi_1, \cdots, \phi_n$ used to define it.) (*Hint:* The functions $u_k' = W_k/W$, $k = 1, \cdots, n$, satisfy (10.1) with $b(x) = 1$ for all x in I.)

(b) Prove that $g(x, t) = g(x - t, 0)$. Thus g is a function of $x - t$ alone, and if $h(x) = g(x, 0)$, $g(x, t) = h(x - t)$. (*Hint*: Let for fixed t, $\phi_t(x) = g(x, t)$, $\psi_t(x) = g(x - t, 0) = \phi_0(x - t)$. Prove that $L(\phi_t) = L(\psi_t) = 0$, and that ϕ_t, ψ_t satisfy the same initial conditions at $x = t$.)

(c) Show that $g(x, t) = \sin(x - t)$ for the case $y'' + y = 0$.

(d) Compute $h(x)$ for the case

$$L(y) = y'' + 2ky' + \omega^2 y,$$

where k and ω are positive constants.

5. The formula (10.3) for a particular solution ψ_p of $L(y) = b(x)$ makes sense for some discontinuous functions b. Then ψ_p will be a solution of $L(y) = b(x)$ at the continuity points of b. Find a continuously differentiable solution of the equation

$$y'' + y = b(x),$$

where

$$b(x) = -1, \qquad (-\pi \leq x < 0),$$
$$= 1, \qquad (0 \leq x \leq \pi),$$
$$= 0, \qquad (|x| > \pi).$$

6. Consider the equation $L(y) = b(x)$, where b is continuous on an interval I. If $\alpha_1, \cdots, \alpha_n$ are any n constants, and x_0 is a point in I, show that there is exactly one solution ψ of $L(y) = b(x)$ on I satisfying

$$\psi(x_0) = \alpha_1, \quad \psi'(x_0) = \alpha_2, \quad \cdots, \quad \psi^{(n-1)}(x_0) = \alpha_n.$$

(*Hint*: Let ϕ be the solution of $L(y) = 0$ satisfying the same initial conditions. Let $\psi = \phi + \psi_p$, where ψ_p is given by (10.3). Show that ψ is unique.)

7. Consider the equation

$$y^{(n)} + a_1 y^{(n-1)} + \cdots + a_n y = b(x),$$

where $a_1, \cdots, a_n$ are real constants and b is a real-valued continuous function on some interval I. Show that any solution which satisfies real initial conditions is real-valued.

11. A special method for solving the non-homogeneous equation

Although the variation of constants method yields a solution of the non-homogeneous equation it sometimes requires more labor than necessary. We now give a method, which is often faster, of solving the non-homogeneous equation $L(y) = b(x)$ *when b is a solution of some homogeneous equation* $\mathrm{M}(\mathrm{y}) = 0$ *with constant coefficients*. Thus $b(x)$ must be a sum of terms of the type $P(x)e^{ax}$, where P is a polynomial and a is a constant.

Suppose L and M have constant coefficients, and have orders n and m respectively. If ψ is a solution of $L(y) = b(x)$, and $M(b) = 0$, then clearly

$$M(L(\psi)) = M(b) = 0.$$

This shows that ψ is a solution of a homogeneous equation $M(L(y)) = 0$ with constant coefficients of order $m + n$. Thus ψ can be written as a linear combination with constant coefficients of $m + n$ linearly independent solutions of $M(L(y)) = 0$. Not every linear combination will be a solution of $L(y) = b(x)$ however. Thus, to find out what conditions must be satisfied by the constants, we substitute back into $L(y) = b(x)$. This always leads to a determination of a set of coefficients; see Sec. 12, Theorem 22, for a justification.

We give an example to show the usefulness of this method. Suppose we consider

$$L(y) = y'' - 3y' + 2y = x^2.$$

Since x^2 is a solution of $M(y) = y''' = 0$, we see that every solution ψ of $L(y) = x^2$ is a solution of

$$M(L(y)) = y^{(5)} - 3y^{(4)} + 2y^{(3)} = 0.$$

The characteristic polynomial of this equation is $r^3(r^2 - 3r + 2)$, just the product of the characteristic polynomials for L and M. The roots are $0, 0, 0, 1, 2$, and hence ψ must have the form

$$\psi(x) = c_0 + c_1 x + c_2 x^2 + c_3 e^x + c_4 e^{2x}.$$

We notice immediately that $c_3 e^x + c_4 e^{2x}$ is just a solution of $L(y) = 0$. Since we are interested only in a particular solution ψ_p of $L(y) = x^2$, we can assume ψ_p has the form

$$\psi_p(x) = c_0 + c_1 x + c_2 x^2.$$

The problem is to determine the constants c_0, c_1, c_2 so that $L(\psi_p) = x^2$. Computing we find

$$\psi_p'(x) = c_1 + 2c_2 x, \qquad \psi_p''(x) = 2c_2,$$

and

$$L(\psi_p) = (2c_2 - 3c_1 + 2c_0) + (-6c_2 + 2c_1)x + 2c_2 x^2 = x^2.$$

Thus

$$2c_2 = 1, \quad \text{or} \quad c_2 = \tfrac{1}{2}, \quad \text{and} \quad -6c_2 + 2c_1 = 0,$$

or

$$c_1 = \tfrac{3}{2}, \quad \text{and} \quad 2c_2 - 3c_1 + 2c_0 = 0,$$

or

$$c_0 = \tfrac{7}{4}.$$

Therefore

$$\psi_p(x) = \tfrac{1}{4}(7 + 6x + 2x^2)$$

is a particular solution of $L(y) = x^2$.

We call this method the *annihilator method*, since to solve $L(y) = b(x)$, we find an M which makes $M(b) = 0$, that is, *annihilates b*. Once M has been found the problem becomes algebraic in nature, no integrations being necessary. Actually, as we have seen from the example, all we require is the characteristic polynomial q of M. The following is a table of some functions together with characteristic polynomials of annihilators. In this table a is constant, and k is a non-negative integer.

	Function	*Characteristic Polynomial of an Annihilator*
(a)	e^{ax}	$r - a$
(b)	$x^k e^{ax}$	$(r - a)^{k+1}$
(c)	$\sin ax, \cos ax$ (a real)	$r^2 + a^2$
(d)	$x^k \sin ax, x^k \cos ax$ (a real)	$(r^2 + a^2)^{k+1}$

The validity of this table is a consequence of Theorem 11.

Let us consider another example of the annihilator method. Consider the equation

$$L(y) = Ae^{ax}, \tag{11.1}$$

where L has characteristic polynomial p, and A, a are constants. We assume that a is *not* a root of p. The operator M given by $M(y) = y' - ay$, with characteristic polynomial $r - a$, annihilates Ae^{ax}. The characteristic polynomial of ML is $(r - a)p(r)$, and a is a simple root (multiplicity 1) of this. Thus any solution ψ of (11.1) has the form

$$\psi = Be^{ax} + \phi,$$

where $L(\phi) = 0$, and B is a constant. Placing ψ back into (11.1) we obtain

$$L(\psi) = BL(e^{ax}) + L(\phi) = Bp(a)e^{ax} = Ae^{ax}.$$

Since $p(a) \neq 0$ we see that $B = A/p(a)$. Therefore we have shown that, if a is not a root of the characteristic polynomial of L, there is a solution ψ of (11.1) of the form

$$\psi(x) = \frac{A}{p(a)}e^{ax}.$$

The example

$$y''' + y'' + y' + y = 1$$

considered in Sec. 10 illustrates this situation. The right side is of the form Ae^{ax} with $A = 1$, $a = 0$. The characteristic polynomial is $p(r) = r^3 +$

$r^2 + r + 1$, with roots i, $-i$, -1. Therefore a solution of this equation is given by

$$\psi(x) = \frac{1}{p(0)} e^{0x} = 1,$$

a result which we found with considerably more effort using the variation of constants method.

Lest the reader feel, after this example, that the variation of constants method is of little importance, we stress that the annihilator method depends very much on the fact that b is a solution of a homogeneous equation with constant coefficients. If $b(x) = \tan x$, for example, the method does not work, and we must use something like the variation of constants method. Moreover, as we shall see in Chap. 3, the variation of constants method is valid for linear equations with *variable* coefficients.

EXERCISES

1. Using the annihilator method find a particular solution of each of the following equations:

    (a) $y'' + 4y = \cos x$

    (b) $y'' + 4y = \sin 2x$

    (c) $y'' - 4y = 3e^{2x} + 4e^{-x}$

    (d) $y'' - y' - 2y = x^2 + \cos x$

    (e) $y'' + 9y = x^2 e^{3x}$

    (f) $y'' + y = xe^x \cos 2x$

    (g) $y'' + iy' + 2y = 2 \cosh 2x + e^{-2x}$ (*Note*: $\cosh u = (e^u + e^{-u})/2$.)

    (h) $y''' = x^2 + e^{-x} \sin x$

    (i) $y''' + 3y'' + 3y' + y = x^2 e^{-x}$

2. Let L be a constant coefficient operator, and suppose ψ_k is a solution of

$$L(y) = b_k(x), \quad k = 1, \cdots, m,$$

where the b_k are continuous functions on some interval I. Show that $\psi = \psi_1 + \cdots + \psi_m$ is a solution of

$$L(y) = b(x), \qquad b = b_1 + \cdots + b_m.$$

3. Suppose $b = b_1 + \cdots + b_m$, where b_k is annihilated by the constant coefficient operator M_k. Show that b is annihilated by $M = M_1 M_2 \cdots M_m$.

4. Consider the constant coefficient operator L with characteristic polynomial p. Consider the equation $L(y) = e^{ax}$, where a is a constant. If a is a root of p with multiplicity k, show by the annihilator method that a solution is given by

$$\psi(x) = \frac{x^k e^{ax}}{p^{(k)}(a)}.$$

5. (a) If $\cosh u = (e^u + e^{-u})/2$ and $\sinh u = (e^u - e^{-u})/2$, show that if a is a real constant $\cosh ax$ and $\sinh ax$ satisfy $y'' - a^2 y = 0$.

(b) Show that the constant coefficient operator M with characteristic polynomial $q(r) = (r^2 - a^2)^{k+1}$ annihilates both $x^k \sinh ax$ and $x^k \cosh ax$.

12. Algebra of constant coefficient operators

In order to justify the annihilator method we study the algebra of constant coefficient operators a little more carefully. For the type of equation we have in mind

$$a_0 y^{(n)} + a_1 y^{(n-1)} + \cdots + a_n y = b(x),$$

where $a_0 \neq 0, a_1, \cdots, a_n$ are constants, and b is a sum of products of polynomials and exponentials, every solution ψ has all derivatives on $-\infty < x < \infty$. This follows from the fact that ψ has n derivatives there, and

$$\psi^{(n)} = \frac{b}{a_0} - \frac{a_1}{a_0}\psi^{(n-1)} - \cdots - \frac{a_n}{a_0}\psi,$$

where b has all derivatives on $-\infty < x < \infty$.

All the operators we now define will be assumed to be defined on the set of all functions ϕ on $-\infty < x < \infty$ which have all derivatives there. Let L and M denote the operators given by

$$L(\phi) = a_0 \phi^{(n)} + a_1 \phi^{(n-1)} + \cdots + a_n \phi,$$

$$M(\phi) = b_0 \phi^{(m)} + b_1 \phi^{(m-1)} + \cdots + b_m \phi,$$

where $a_0, a_1, \cdots, a_n, b_0, b_1, \cdots, b_m$ are constants, with $a_0 \neq 0, b_0 \neq 0$. It will be convenient in what follows to consider a_0, b_0 which are not necessarily 1. The characteristic polynomials of L and M are thus

$$p(r) = a_0 r^n + a_1 r^{n-1} + \cdots + a_n,$$

and

$$q(r) = b_0 r^m + b_1 r^{m-1} + \cdots + b_m,$$

respectively.

We define the *sum* $L + M$ to be the operator given by

$$(L + M)(\phi) = L(\phi) + M(\phi),$$

and the *product* ML to be the operator given by

$$(ML)(\phi) = M(L(\phi)).$$

If α is a constant we define αL by

$$(\alpha L)(\phi) = \alpha(L(\phi)).$$

We note that $L + M$, ML and αL are all linear differential operators with constant coefficients.

Two operators L and M are said to be *equal* if

$$L(\phi) = M(\phi)$$

for all ϕ which have an infinite number of derivatives on $-\infty < x < \infty$. Suppose L, M have characteristic polynomials p, q respectively. Since e^{rx}, for any constant r, has an infinite number of derivatives on $-\infty < x < \infty$, we see that if $L = M$ then

$$L(e^{rx}) = p(r)e^{rx} = M(e^{rx}) = q(r)e^{rx},$$

and hence $p(r) = q(r)$ for all r. This implies that $m = n$, and $a_k = b_k$, $k = 0, 1, \cdots, n$. Thus $L = M$ if and only if L and M have the same order and the same coefficients, or, what is the same, if and only if $p = q$.

If D is the differentiation operator

$$D(\phi) = \phi',$$

we define $D^2 = DD$, and successively

$$D^k = DD^{k-1}, \qquad (k = 2, 3, \cdots).$$

For completeness we define D^0 by $D^0(\phi) = \phi$, but do not usually write it explicitly. If α is a constant we understand by α operating on a function ϕ just multiplication by α. Thus

$$\alpha(\phi) = (\alpha D^0)(\phi) = \alpha\phi.$$

Now, using our definitions, it is clear that

$$L = a_0 D^n + a_1 D^{n-1} + \cdots + a_n,$$

and

$$M = b_0 D^m + b_1 D^{m-1} + \cdots + b_m.$$

Theorem 21. *The correspondence which associates with each*

$$L = a_0 D^n + a_1 D^{n-1} + \cdots + a_n$$

its characteristic polynomial p *given by*

$$p(r) = a_0 r^n + a_1 r^{n-1} + \cdots + a_n$$

is a one-to-one correspondence between all linear differential operators with constant coefficients and all polynomials. If L, M *are associated with* p, q *respectively, then* L + M *is associated with* p + q, ML *is associated with* pq, *and* αL *is associated with* αp (α *a constant*).

Proof. We have already seen that the correspondence is one-to-one since $L = M$ if and only if $p = q$. The remainder of the theorem can be shown directly, or by noting that

$$(L + M)(e^{rx}) = L(e^{rx}) + M(e^{rx}) = [p(r) + q(r)]e^{rx},$$

$$(ML)(e^{rx}) = M(L(e^{rx})) = M(p(r)e^{rx}) = p(r)M(e^{rx}) = p(r)q(r)e^{rx},$$

$$(\alpha L)(e^{rx}) = \alpha(L(e^{rx})) = \alpha p(r)e^{rx}.$$

This result implies that the algebraic properties of the constant coefficient operators are the same as those of the polynomials. For example, since LM and ML both have the characteristic polynomial pq, we have $LM = ML$.* If the roots of p are $r_1, \cdots, r_n$, then

$$p(r) = a_0(r - r_1) \cdots (r - r_n),$$

and since the operator

$$a_0(D - r_1) \cdots (D - r_n)$$

has p as characteristic polynomial, we must have

$$L = a_0(D - r_1) \cdots (D - r_n).$$

This gives a factorization of L into a product of constant coefficient operators of the first order.

We apply Theorem 21 to give a justification of the annihilator method.

Theorem 22. *Consider the equation with constant coefficients*

$$L(y) = P(x)e^{ax}, \tag{12.1}$$

where P *is the polynomial given by*

$$P(x) = b_0 x^m + b_1 x^{m-1} + \cdots + b_m, \qquad (b_0 \neq 0). \tag{12.2}$$

Suppose a *is a root of the characteristic polynomial* p *of* L *of multiplicity* j. *Then there is a unique solution* ψ *of* (12.1) *of the form*

$$\psi(x) = x^j(c_0 x^m + c_1 x^{m-1} + \cdots + c_m)e^{ax},$$

where $c_0, c_1, \cdots, c_m$ *are constants determined by the annihilator method.*

---

* We remark that if L and M are not constant coefficient operators, then it may not be true that $LM = ML$. For example,

if $L(\phi)(x) = \phi'(x)$, $M(\phi)(x) = x\phi(x)$, then $(LM - ML)(\phi)(x) = \phi(x)$.

Proof. The proof makes use of the formula

$$L(x^k e^{rx}) = \left[p(r)x^k + kp'(r)x^{k-1} + \frac{k(k-1)}{2!}p''(r)x^{k-2} \right.$$

$$\left. + \cdots + kp^{(k-1)}(r)x + p^{(k)}(r) \right] e^{rx} \qquad (12.3)$$

which we proved in Sec. 7. The coefficient of $p^{(l)}(r)x^{k-l}$ in the bracket is the binomial coefficient

$$\binom{k}{l} = \frac{k!}{(k-l)!l!}.$$

Thus we may write

$$L(x^k e^{rx}) = \left[\sum_{l=0}^{k} \binom{k}{l} p^{(l)}(r)x^{k-l} \right] e^{rx},$$

where we understand $0! = 1$.

An annihilator of the right side of (12.1) is

$$M = (D - a)^{m+1},$$

with characteristic polynomial given by

$$q(r) = (r - a)^{m+1}.$$

Since a is a root of p with multiplicity j, it is a root of pq with multiplicity $j + m + 1$. Thus solutions of $ML(y) = 0$ are of the form

$$\psi(x) = (c_0 x^{j+m} + c_1 x^{j+m-1} + \cdots + c_{j+m})e^{ax} + \phi(x),$$

where $L(\phi) = 0$, and ϕ involves exponentials of the form e^{sx}, with s a root of p, $s \neq a$. Since a is a root of p with multiplicity j, we have that

$$(c_{m+1}x^{j-1} + c_{m+2}x^{j-2} + \cdots + c_{m+j})e^{ax}$$

is also a solution of $L(y) = 0$. Consequently we see that there is a solution ψ of (12.1) having the form

$$\psi(x) = x^j(c_0 x^m + c_1 x^{m-1} + \cdots + c_m)e^{ax} \qquad (12.4)$$

where $c_0, c_1, \cdots, c_m$ are constants.

We now show that these constants are uniquely determined by the requirement that ψ satisfy (12.1). Substituting (12.4) into L we obtain

$$L(\psi) = c_0 L(x^{j+m}e^{ax}) + c_1 L(x^{j+m-1}e^{ax}) + \cdots + c_m L(x^j e^{ax}). \qquad (12.5)$$

The terms in this sum can be computed using (12.3). We note that

$$p(a) = p'(a) = \cdots = p^{(j-1)}(a) = 0, \qquad p^{(j)}(a) \neq 0,$$

since a is a root of p with multiplicity j. Thus, if $k \geq j$,

$$L(x^k e^{ax}) = \left[\binom{k}{k-j} p^{(j)}(a) x^{k-j} + \binom{k}{k-j-1} p^{(j+1)}(a) x^{k-j-1} \right.$$

$$\left. + \cdots + p^{(k)}(a) \right] e^{ax}.$$

We then have

$$L(x^{j+m} e^{ax}) = \left[\binom{j+m}{m} p^{(j)}(a) x^m + \binom{j+m}{m-1} p^{(j+1)}(a) x^{m-1} \right.$$

$$\left. + \cdots + p^{(j+m)}(a) \right] e^{ax},$$

$$L(x^{j+m-1} e^{ax}) = \left[\binom{j+m-1}{m-1} p^{(j)}(a) x^{m-1} + \cdots + p^{(j+m-1)}(a) \right] e^{ax},$$

$$\vdots$$

$$L(x^j e^{ax}) = \binom{j}{0} p^{(j)}(a) e^{ax} = p^{(j)}(a) e^{ax}.$$

Using these computations in (12.5), and noting (12.2), we see that ψ satisfies (12.1) if and only if

$$c_0 \binom{j+m}{m} p^{(j)}(a) = b_0,$$

$$c_0 \binom{j+m}{m-1} p^{(j+1)}(a) + c_1 \binom{j+m-1}{m-1} p^{(j)}(a) = b_1,$$

$$\vdots$$

$$c_0 p^{(j+m)}(a) + c_1 p^{(j+m-1)}(a) + \cdots + c_m p^{(j)}(a) = b_m.$$

This is a set of $m+1$ linear equations for the constants $c_0, c_1, \cdots, c_m$. They have a unique solution, which can be obtained by solving the equations in succession since $p^{(j)}(a) \neq 0$. Alternately, we see that the determinant of the coefficients is just

$$\binom{j+m}{m} \binom{j+m-1}{m-1} \cdots 1 [p^{(j)}(a)]^{m+1} \neq 0.$$

This completes the proof of Theorem 22.

The justification of the annihilator method when the right side of

$$L(y) = b(x)$$

is a sum of terms of the form $P(x)e^{ax}$ can be reduced to Theorem 22, by noting that if ψ_1, ψ_2 satisfy

$$L(\psi_1) = b_1, \qquad L(\psi_2) = b_2,$$

respectively, then $\psi_1 + \psi_2$ satisfies

$$L(\psi_1 + \psi_2) = b_1 + b_2.$$

EXERCISES

1. (a) Show that if f, g are two functions with k derivatives then

$$D^k(fg) = \sum_{l=0}^{k} \binom{k}{l} D^l(f) D^{k-l}(g),$$

where

$$\binom{k}{l} = \frac{k!}{(k-l)!l!}.$$

(b) Show that if g has k derivatives, and r is a constant,

$$D^k(e^{rx}g) = e^{rx}(D+r)^k(g).$$

2. Let L be a linear differential operator with constant coefficients with characteristic polynomial $p(r) = (r - a)^k$, that is $L = (D - a)^k$. Using the result of Ex. 1 (b) show that any solution ϕ of $L(y) = 0$ has the form

$$\phi(x) = e^{ax}P(x),$$

where P is a polynomial such that $\deg P \leq k - 1$. Also show that any such ϕ is a solution of $L(y) = 0$.

3. Let b be a continuous function on an interval I, and let x_0 be a fixed point in I. Show that the ϕ given by

$$\phi(x) = e^{ax} \int_{x_0}^{x} \frac{(x-t)^{k-1}}{(k-1)!} e^{-at}b(t)\, dt$$

satisfies

$$(D - a)^k(\phi) = b,$$

and

$$\phi(x_0) = \phi'(x_0) = \cdots = \phi^{(k-1)}(x_0) = 0.$$

Here a is a constant. (*Hint*: From Ex. 1 (b),

$$(D - a)^k(\phi) = e^{ax}D^k(e^{-ax}\phi).$$

To differentiate a function F of the form

$$F(x) = \int_{\alpha(x)}^{\beta(x)} f(x, t) \, dt,$$

where α, β, f are "nice" functions, use the formula

$$F'(x) = f(x, \beta(x))\beta'(x) - f(x, \alpha(x))\alpha'(x) + \int_{\alpha(x)}^{\beta(x)} \frac{\partial f}{\partial x}(x, t) \, dt.$$

A proof of this formula for appropriate α, β, f may be found in most texts on advanced calculus.)

4. Let b be a continuous function on some interval I, and let x_0 be a fixed point in I. Define $(D - a)^{-k}(b)$ to be the function ϕ given in Ex. 3, that is

$$(D - a)^{-k}(b)(x) = e^{ax} \int_{x_0}^{x} \frac{(x - t)^{k-1}}{(k - 1)!} e^{-at} b(t) \, dt$$

for x in I. Thus $(D - a)^{-k}$ is an operator which is defined for continuous functions b on I. Show that

$$(D - a)^k[(D - a)^{-k}(b)] = b, \tag{*}$$

and

$$(D - a)^{-k}[(D - a)^k(\phi)] = \phi, \tag{**}$$

for any continuous b on I, and function ϕ on I which has k continuous derivatives there, and satisfies

$$\phi(x_0) = \phi'(x_0) = \cdots = \phi^{(k-1)}(x_0) = 0.$$

(*Hint*: The relation (*) follows from Ex. 3. For (**) let

$$b = (D - a)^k(\phi), \quad \text{and} \quad \psi = (D - a)^{-k}b.$$

Then from (*) $(D - a)^k(\psi) = (D - a)^k(\phi)$. Thus $(D - a)^k(\psi - \phi) = 0$. Show that $\psi = \phi$.) (*Note*: Let $\mathcal{C}$ denote the set of all continuous functions on I, and let $\mathcal{C}^k$ denote the set of all functions ϕ on I which have k continuous derivatives there and satisfy

$$\phi(x_0) = \phi'(x_0) = \cdots = \phi^{(k-1)}(x_0) = 0.$$

Then $(D - a)^k$ takes each ϕ in $\mathcal{C}^k$ into a function in $\mathcal{C}$, and $(D - a)^{-k}$ takes each b in $\mathcal{C}$ into a function in $\mathcal{C}^k$. The relations (*) and (**) show that $(D - a)^{-k}$ is both a right and a left reciprocal of $(D - a)^k$, and therefore the correspondence between $\mathcal{C}^k$ and $\mathcal{C}$ given by $(D - a)^k$ is one-to-one.)

5. (a) Let p be a polynomial with leading coefficient one with n distinct roots $r_1, \cdots, r_n$. Show that

$$\frac{1}{p(r)} = \frac{1}{p'(r_1)} \frac{1}{r - r_1} + \cdots + \frac{1}{p'(r_n)} \frac{1}{r - r_n}$$

if r is not any of the roots of p. This is the partial fraction decomposition of $1/p$.

(b) Let p be as in part (a), and let

$$L = (D - r_1)(D - r_2) \cdots (D - r_n).$$

If b is a continuous function on an interval I, show that a solution ϕ of

$$L(y) = b$$

is given by

$$\phi = \sum_{k=1}^{n} \frac{1}{p'(r_k)} (D - r_k)^{-1}(b),$$

where $(D - r_k)^{-1}$ is defined as in Ex. 4.

6. For any polynomial p with leading coefficient one, let $r_1, \cdots, r_s$ be the distinct roots, with r_k having multiplicity m_k. Then

$$p(r) = (r - r_1)^{m_1}(r - r_2)^{m_2} \cdots (r - r_s)^{m_s},$$

and there is a partial fraction decomposition

$$\frac{1}{p(r)} = \sum_{k=1}^{s} \sum_{j=1}^{m_k} \frac{c_{kj}}{(r - r_k)^j},$$

where the c_{kj} are certain constants. Let

$$L = (D - r_1)^{m_1}(D - r_2)^{m_2} \cdots (D - r_s)^{m_s},$$

and let b be a continuous function on an interval I. Corresponding to the partial fraction decomposition for $1/p$ show that a solution of $L(y) = b$ is given by

$$\phi = \sum_{k=1}^{s} \sum_{j=1}^{m_k} c_{kj}(D - r_k)^{-j}(b).$$

7. Let L_1, L_2 be two constant coefficient differential operators with characteristic polynomials p_1, p_2 respectively. Assume that p_1 and p_2 have no common roots. Let L be the operator with characteristic polynomial $p = p_1 p_2$, that is $L = L_1 L_2$. Prove that every solution ϕ of $L(y) = 0$ can be written uniquely as a sum

$$\phi = \phi_1 + \phi_2,$$

where $L_1(\phi_1) = 0$, $L_2(\phi_2) = 0$.

CHAPTER 3

Linear Equations with
Variable Coefficients

1. Introduction

A linear differential equation of order n with variable coefficients is an equation of the form

$$a_0(x)y^{(n)} + a_1(x)y^{(n-1)} + \cdots + a_n(x)y = b(x),$$

where $a_0, a_1, \cdots, a_n, b$ are complex-valued functions on some real interval I. Points where $a_0(x) = 0$ are called *singular points*, and often the equation requires special consideration at such points. Therefore in this chapter we assume that $a_0(x) \neq 0$ on I. By dividing by a_0 we can obtain an equation of the same form, but with a_0 replaced by the constant 1. Thus we consider the equation

$$y^{(n)} + a_1(x)y^{(n-1)} + \cdots + a_n(x)y = b(x). \tag{1.1}$$

As in the case when $a_1, \cdots, a_n$ are constants we designate the left side of (1.1) by $L(y)$. Thus

$$L(y) = y^{(n)} + a_1(x)y^{(n-1)} + \cdots + a_n(x)y, \tag{1.2}$$

and (1.1) becomes simply $L(y) = b(x)$. If $b(x) = 0$ for all x on I we say $L(y) = 0$ is a *homogeneous equation*, whereas if $b(x) \neq 0$ for some x in I, the equation $L(y) = b(x)$ is called a *non-homogeneous equation*.

We give a meaning to L itself as an operator which takes each function ϕ, which has n derivatives on I, into the function $L(\phi)$ on I whose value at x is given by

$$L(\phi)(x) = \phi^{(n)}(x) + a_1(x)\phi^{(n-1)}(x) + \cdots + a_n(x)\phi(x).$$

Thus a solution of (1.1) on I is a function ϕ on I which has n derivatives there, and which satisfies $L(\phi) = b$.

In this chapter we assume that the complex-valued functions $a_1, \cdots, a_n, b$ are *continuous* on some real interval I, and $L(y)$ will always denote the expression (1.2).

Most of the results we developed in Chap. 2 for the case when $a_1, \cdots, a_n$ are constants continue to be valid in the more general case we are now considering. Sections 2, 3, 4, and 6 of this chapter are devoted to showing this, and can thus be considered as a review. The major difficulty with linear equations with variable coefficients, from a practical point of view, is that it is rare that we can solve the equations in terms of elementary functions, such as the exponential and trigonometric functions. Thus there is no analogue of the rather powerful Theorem 11 of Chap. 2. However, in case $a_1, \cdots, a_n, b$ have convergent power series expansions the solutions will have this property also, and these series solutions can be obtained by a simple formal process.

2. Initial value problems for the homogeneous equation

Although in many cases it is not possible to express a solution of (1.1) in terms of elementary functions, it can be proved that solutions always exist. In fact we assume for now the following result, which includes Theorem 16 of Chap. 2 as a special case. A proof is given in Chap. 6, Theorem 8.

Theorem 1. (*Existence Theorem*) *Let* $a_1, \cdots, a_n$ *be continuous functions on an interval* I *containing the point* x_0. *If* $\alpha_1, \cdots, \alpha_n$ *are any* n *constants, there exists a solution* ϕ *of*

$$L(y) = y^{(n)} + a_1(x)y^{(n-1)} + \cdots + a_n(x)y = 0$$

on I *satisfying*

$$\phi(x_0) = \alpha_1, \quad \phi'(x_0) = \alpha_2, \quad \cdots, \quad \phi^{(n-1)}(x_0) = \alpha_n.$$

We stress two things about this theorem: (i) the solution exists on the *entire* interval I where $a_1, \cdots, a_n$ are continuous, and (ii) *every* initial value problem has a solution. Neither of these results may be true if the coefficient of $y^{(n)}$ vanishes somewhere in I. For example, consider the equation

$$xy' + y = 0,$$

whose coefficients are continuous for all real x. This equation and the initial condition $y(1) = 1$ has the solution ϕ_1, where

$$\phi_1(x) = \frac{1}{x}.$$

But this solution exists only for $0 < x < \infty$. Also, if ϕ is any solution, then

$$x\phi(x) = c,$$

where c is some constant. Thus only the trivial solution $(c = 0)$ exists at the origin, which implies that the only initial value problem

$$xy' + y = 0, \qquad y(0) = \alpha_1,$$

which has a solution is the one for which $\alpha_1 = 0$.

Just as in the case where the coefficients a_j $(j = 1, \cdots, n)$ are con-stants, the uniqueness of the solution ϕ given in Theorem 1 is demon-strated with the aid of an estimate for

$$\| \phi(x) \| = [\, | \phi(x) |^2 + | \phi'(x) |^2 + \cdots + | \phi^{(n-1)}(x) |^2]^{1/2}.$$

Theorem 2. *Let* $b_1, \cdots, b_n$ *be non-negative constants such that for all* x *in* I

$$| a_j(x) | \leqq b_j, \qquad (j = 1, \cdots, n),$$

and define k *by*

$$k = 1 + b_1 + \cdots + b_n.$$

If x_0 *is a point in* I, *and* ϕ *is a solution of* $L(y) = 0$ *on* I, *then*

$$\| \phi(x_0) \| e^{-k|x-x_0|} \leqq \| \phi(x) \| \leqq \| \phi(x_0) \| e^{k|x-x_0|} \tag{2.1}$$

for all x *in* I.

Proof. Since $L(\phi) = 0$ we have

$$\phi^{(n)}(x) = -a_1(x)\phi^{(n-1)}(x) - \cdots - a_n(x)\phi(x),$$

and therefore

$$| \phi^{(n)}(x) | \leqq | a_1(x) | \, | \phi^{(n-1)}(x) | + \cdots + | a_n(x) | \, | \phi(x) |$$

$$\leqq b_1 | \phi^{(n-1)}(x) | + \cdots + b_n | \phi(x) |.$$

The proof of Theorem 13, Chap. 2, now applies if we substitute b_j every-where in place of $| a_j |$.

We remark that if I is a *closed bounded* interval, that is, of the form $a \leqq x \leqq b$ with a, b real, and if the a_j are continuous on I, then there always exist finite constants b_j such that $| a_j(x) | \leqq b_j$ on I.

Theorem 3. (*Uniqueness Theorem*) *Let* x_0 *be in* I, *and let* $\alpha_1, \cdots, \alpha_n$ *be any* n *constants. There is at most one solution* ϕ *of* $L(y) = 0$ *on* I *satisfying*

$$\phi(x_0) = \alpha_1, \quad \phi'(x_0) = \alpha_2, \quad \cdots, \quad \phi^{(n-1)}(x_0) = \alpha_n. \tag{2.2}$$

Proof. Let ϕ, ψ be two solutions of $L(y) = 0$ on I satisfying the conditions (2.2) at x_0, and consider $\chi = \phi - \psi$. We wish to prove $\chi(x) = 0$ for all x on I. Even though the functions a_j are continuous on I they need not be bounded there.* Therefore we can not apply Theorem 2 directly. However, let x be any point on I other than x_0, and let J be any closed bounded interval in I which contains x_0 and x. On this interval the functions a_j are bounded, that is,

$$| a_j(x) | \leqq b_j, \qquad (j = 1, \cdots, n),$$

on J for some constants b_j, which may depend on J. Now we apply Theorem 2 to χ defined on J. We have $L(\chi) = 0$ on J, and $\| \chi(x_0) \| = 0$. Therefore (2.1) implies that $\| \chi(x) \| = 0$, and hence $\phi(x) = \psi(x)$. Since x was chosen to be *any* point in I other than x_0, we have proved $\phi(x) = \psi(x)$ for all x on I.

3. Solutions of the homogeneous equation

If $\phi_1, \cdots, \phi_m$ are any m solutions of the n-th order equation $L(y) = 0$ on an interval I, and $c_1, \cdots, c_m$ are any m constants, then

$$L(c_1\phi_1 + \cdots + c_m\phi_m) = c_1L(\phi_1) + \cdots + c_mL(\phi_m),$$

which implies that $c_1\phi_1 + \cdots c_m\phi_m$ is also a solution. In words, any linear combination of solutions is again a solution. The *trivial solution* is the function which is identically zero on I.

As in the case of an L with constant coefficients, every solution of $L(y) = 0$ is a linear combination of any n linearly independent solutions. Recall that n functions $\phi_1, \cdots, \phi_n$ defined on an interval I are said to be *linearly independent* if the only constants $c_1, \cdots, c_n$ such that

$$c_1\phi_1(x) + \cdots + c_n\phi_n(x) = 0$$

for all x in I are the constants

$$c_1 = c_2 = \cdots = c_n = 0.$$

Using Theorem 1 we construct n linearly independent solutions, and show that every solution is a linear combination of these. In Sec. 4 we show that every solution is a linear combination of *any* n linearly independent solutions.

Theorem 4. *There exist* n *linearly independent solutions of* $L(y) = 0$ *on* I.

---

* For example, $a_1(x) = x$ is not bounded on $0 \leqq x < \infty$; and $a_1(x) = 1/x$ is not bounded on $0 < x \leqq 1$.

Proof. Let x_0 be a point in I. According to Theorem 1 there is a solution ϕ_1 of $L(y) = 0$ satisfying

$$\phi_1(x_0) = 1, \quad \phi_1'(x_0) = 0, \quad \cdots, \quad \phi_1^{(n-1)}(x_0) = 0.$$

In general for each $i = 1, 2, \cdots, n$ there is a solution ϕ_i satisfying

$$\phi_i^{(i-1)}(x_0) = 1, \quad \phi_i^{(j-1)}(x_0) = 0, \quad j \neq i. \tag{3.1}$$

The solutions $\phi_1, \cdots, \phi_n$ are linearly independent on I, for suppose there are constants $c_1, c_2, \cdots, c_n$ such that

$$c_1\phi_1(x) + c_2\phi_2(x) + \cdots + c_n\phi_n(x) = 0 \tag{3.2}$$

for all x in I. Differentiating we see that

$$c_1\phi_1'(x) + c_2\phi_2'(x) + \cdots + c_n\phi_n'(x) = 0$$

$$c_1\phi_1''(x) + c_2\phi_2''(x) + \cdots + c_n\phi_n''(x) = 0$$

$$\vdots \tag{3.3}$$

$$c_1\phi_1^{(n-1)}(x) + c_2\phi_2^{(n-1)}(x) + \cdots + c_n\phi_n^{(n-1)}(x) = 0$$

for all x in I. In particular, the equations (3.2), (3.3) must hold at x_0. Putting $x = x_0$ in (3.2) we find, using (3.1), that $c_1 1 + 0 + \cdots + 0 = 0$, or $c_1 = 0$. Putting $x = x_0$ in the equations (3.3) we obtain $c_2 = c_3 = \cdots = c_n = 0$, and thus the solutions $\phi_1, \cdots, \phi_n$ are linearly independent.

Theorem 5. *Let $\phi_1, \cdots, \phi_n$ be the n solutions of* $L(y) = 0$ *on* I *satisfying* (3.1). *If ϕ is any solution of* $L(y) = 0$ *on* I, *there are n constants* $c_1, \cdots, c_n$ *such that*

$$\phi = c_1\phi_1 + \cdots + c_n\phi_n.$$

Proof. Let

$$\phi(x_0) = \alpha_1, \quad \phi'(x_0) = \alpha_2, \quad \cdots, \quad \phi^{(n-1)}(x_0) = \alpha_n,$$

and consider the function

$$\psi = \alpha_1\phi_1 + \alpha_2\phi_2 + \cdots + \alpha_n\phi_n.$$

It is a solution of $L(y) = 0$, and clearly

$$\psi(x_0) = \alpha_1\phi_1(x_0) + \alpha_2\phi_2(x_0) + \cdots + \alpha_n\phi_n(x_0) = \alpha_1,$$

since

$$\phi_1(x_0) = 1, \quad \phi_2(x_0) = 0, \quad \cdots, \quad \phi_n(x_0) = 0.$$

Using the other relations in (3.1) we see that

$$\psi(x_0) = \alpha_1, \quad \psi'(x_0) = \alpha_2, \quad \cdots, \quad \psi^{(n-1)}(x_0) = \alpha_n.$$

Thus ψ is a solution of $L(y) = 0$ having the same initial conditions at x_0 as ϕ. By Theorem 3, we must have $\phi = \psi$, that is

$$\phi = \alpha_1\phi_1 + \alpha_2\phi_2 + \cdots + \alpha_n\phi_n.$$

We have proved the theorem with the constants

$$c_1 = \alpha_1, \quad c_2 = \alpha_2, \quad \cdots, \quad c_n = \alpha_n.$$

A set of functions which has the property that, if ϕ_1, ϕ_2 belong to the set, and c_1, c_2 are any two constants, then $c_1\phi_1 + c_2\phi_2$ belongs to the set also, is called a *linear space* of functions. We have just seen that the set of all solutions of $L(y) = 0$ on an interval I is a linear space of functions. If a linear space of functions contains n functions $\phi_1, \cdots, \phi_n$ which are linearly independent and such that every function in the space can be represented as a linear combination of these, then $\phi_1, \cdots, \phi_n$ is called a *basis* for the linear space*, and the *dimension* of the linear space is the integer n. The content of Theorem 5 is that the functions $\phi_1, \cdots, \phi_n$ satisfying the initial conditions (3.1) form a basis for the solutions of $L(y) = 0$ on I, and this linear space of functions has dimension n.

EXERCISES

1. Consider the equation

$$y'' + \frac{1}{x}y' - \frac{1}{x^2}y = 0$$

for $x > 0$.
   (a) Show that there is a solution of the form x^r, where r is a constant.
   (b) Find two linearly independent solutions for $x > 0$, and prove that they are linearly independent.
   (c) Find the two solutions ϕ_1, ϕ_2 satisfying

$$\phi_1(1) = 1, \quad \phi_2(1) = 0,$$
$$\phi_1'(1) = 0, \quad \phi_2'(1) = 1.$$

2. Find two linearly independent solutions of the equation

$$(3x - 1)^2y'' + (9x - 3)y' - 9y = 0$$

for $x > \frac{1}{3}$. (*Hint*: See Ex. 1(a), with x replaced by $3x - 1$.)

3. Consider the equation

$$L(y) = y'' + a_1(x)y' + a_2(x)y = 0,$$

---

* A basis is sometimes called a *fundamental set*, and a linear space is often called a *vector space*. The dimension of a linear space does not depend on a choice of basis.

where a_1, a_2 are continuous on some interval I, and a_1 has a continuous derivative there.

(a) If ϕ is a solution of $L(y) = 0$ let $\phi = u\psi$, and determine a differential equation for u which will make ψ the solution of an equation in which the first derivative term is absent.

(b) Solve this differential equation for u.

(c) Show that ψ will then satisfy the equation

$$y'' + \alpha(x)y = 0,$$

where

$$\alpha = a_2 - \frac{a_1^2}{4} - \frac{a_1'}{2}.$$

4. The equation $y' + a(x)y = 0$ has for a solution

$$\phi(x) = \exp\left[-\int_{x_0}^{x} a(t)\, dt \right]$$

(Here let a be continuous on an interval I containing x_0.) This suggests trying to find a solution of

$$L(y) = y'' + a_1(x)y' + a_2(x)y = 0$$

of the form

$$\phi(x) = \exp\left[\int_{x_0}^{x} p(t)\, dt \right],$$

where p is a function to be determined. Show that ϕ is a solution of $L(y) = 0$ if, and only if, p satisfies the first order non-linear equation

$$y' = -y^2 - a_1(x)y - a_2(x).$$

(*Remark*: This last equation is called a *Riccati equation*.)

5. Let

$$L(y) = y^{(n)} + a_1(x)y^{(n-1)} + \cdots + a_n(x)y,$$

where $a_1, \cdots, a_n$ are continuous *real-valued* functions on an interval I.

(a) Show that if ϕ is a solution of $L(y) = 0$, then so are Re ϕ and Im ϕ.

(b) Let ϕ be a solution of $L(y) = 0$ satisfying

$$\phi(x_0) = \alpha_1, \quad \phi'(x_0) = \alpha_2, \quad \cdots, \quad \phi^{(n-1)}(x_0) = \alpha_n,$$

where x_0 is some point in I, and $\alpha_1, \cdots, \alpha_n$ are real constants. Prove that ϕ is real-valued.

(c) Show that there is a basis for the solutions of $L(y) = 0$ consisting of real-valued functions. (*Hint*: Consider the basis $\phi_1, \cdots, \phi_n$ satisfying

$$\phi_i^{(j-1)}(x_0) = \delta_{ij}, \qquad (i, j = 1, \cdots, n),$$

where

$$\delta_{ij} = 1 \text{ if } i = j, \qquad \delta_{ij} = 0 \text{ if } i \neq j.)$$

6. Consider the equation

$$y'' + a_1(x)y' + a_2(x)y = 0,$$

where a_1, a_2 are continuous functions on $-\infty < x < \infty$ of period $\xi > 0$, that is,

$$a_1(x + \xi) = a_1(x), \qquad a_2(x + \xi) = a_2(x),$$

for all x.

(a) Let ϕ be a non-trivial solution, and let $\psi(x) = \phi(x + \xi)$. Prove that ψ is also a solution.

(b) Show that ϕ is a periodic solution of period ξ if, and only if,

$$\phi(0) = \phi(\xi), \qquad \phi'(0) = \phi'(\xi).$$

(c) Let ϕ_1, ϕ_2 be the two solutions satisfying

$$\phi_1(0) = 1, \qquad \phi_2(0) = 0,$$

$$\phi_1'(0) = 0, \qquad \phi_2'(0) = 1.$$

Show that there are constants a, b, c, d such that

$$\phi_1(x + \xi) = a\phi_1(x) + b\phi_2(x),$$

$$\phi_2(x + \xi) = c\phi_1(x) + d\phi_2(x),$$

for all x. (*Hint*: See (a).)

(d) Compute the constants a, b, c, d in (b) by considering the point $x = 0$.

7. Let $\phi_1, \cdots, \phi_n$ be n continuous functions on the interval $a \leq x \leq b$. Let

$$\alpha_{ij} = \int_a^b \overline{\phi_i(x)}\, \phi_j(x)\, dx, \qquad (i, j = 1, \cdots, n),$$

and let Δ denote the determinant

$$\Delta = \begin{vmatrix} \alpha_{11} & \alpha_{12} & \cdots & \alpha_{1n} \\ \alpha_{21} & \alpha_{22} & \cdots & \alpha_{2n} \\ \cdot & & & \cdot \\ \cdot & & & \cdot \\ \cdot & & & \cdot \\ \alpha_{n1} & \alpha_{n2} & \cdots & \alpha_{nn} \end{vmatrix}$$

Prove that $\phi_1, \cdots, \phi_n$ are linearly independent on $a \leq x \leq b$ if, and only if, $\Delta \neq 0$. (*Hint*: Suppose

$$\Delta \neq 0, \quad \text{and} \quad c_1\phi_1 + \cdots + c_n\phi_n = 0.$$

Multiply this equation in turn by $\overline{\phi_1}, \overline{\phi_2}, \cdots, \overline{\phi_n}$ and integrate to obtain

$$c_1\alpha_{11} + c_2\alpha_{12} + \cdots + c_n\alpha_{1n} = 0,$$

$$\vdots$$

$$\quad(*)$$

$$c_1\alpha_{n1} + c_2\alpha_{n2} + \cdots + c_n\alpha_{nn} = 0.$$

The only solution of these is $c_1 = c_2 = \cdots = c_n = 0$. Conversely, if $\phi_1, \cdots, \phi_n$ are linearly independent and $\Delta = 0$, then there are $c_1, \cdots, c_n$ satisfying (*) not all zero. Multiply the first equation by $\overline{c_1}$, the second by $\overline{c_2}$, etc., to obtain

$$0 = \sum_{j=1}^{n} \sum_{i=1}^{n} \overline{c_i}\, \alpha_{ij} c_j = \int_a^b \left| \sum_{i=1}^{n} c_i \phi_i(x) \right|^2 dx.$$

Show that this implies $c_1 \phi_1 + \cdots + c_n \phi_n = 0$.

The determinant Δ is called the *Gramian* of $\phi_1, \cdots, \phi_n$. Note that $\alpha_{ij} = \overline{\alpha_{ji}}$.)

4. The Wronskian and linear independence

In order to show that any set of n linearly independent solutions of $L(y) = 0$ can serve as a basis for the solutions of $L(y) = 0$, we consider the Wronskian $W(\phi_1, \cdots, \phi_n)$ of any n solutions $\phi_1, \cdots, \phi_n$. Recall that this is defined to be the determinant

$$W(\phi_1, \cdots, \phi_n) = \begin{vmatrix} \phi_1 & \phi_2 & \cdots & \phi_n \\ \phi_1' & \phi_2' & \cdots & \phi_n' \\ \cdot & \cdot & & \cdot \\ \cdot & \cdot & & \cdot \\ \cdot & \cdot & & \cdot \\ \phi_1^{(n-1)} & \phi_2^{(n-1)} & \cdots & \phi_n^{(n-1)} \end{vmatrix}$$

Theorem 6. *If* $\phi_1, \cdots, \phi_n$ *are* n *solutions of* L(y) $= 0$ *on an interval* I, *they are linearly independent there if, and only if,*

$$W(\phi_1, \cdots, \phi_n)(x) \neq 0 \quad \text{for all } x \text{ in } I.$$

Proof. First suppose $W(\phi_1, \cdots, \phi_n)(x) \neq 0$ for all x in I. If there are constants $c_1, \cdots, c_n$ such that

$$c_1 \phi_1(x) + \cdots + c_n \phi_n(x) = 0 \tag{4.1}$$

for all x in I, then clearly

$$c_1 \phi_1'(x) + \cdots + c_n \phi_n'(x) = 0$$
$$\cdot$$
$$\cdot \tag{4.2}$$
$$\cdot$$
$$c_1 \phi_1^{(n-1)}(x) + \cdots + c_n \phi_n^{(n-1)}(x) = 0$$

for all x in I. For a fixed x in I the equations (4.1), (4.2) are n linear homogeneous equations satisfied by $c_1, \cdots, c_n$. The determinant of the coefficients is just $W(\phi_1, \cdots, \phi_n)(x)$, which is not zero. Hence there is only one solution to this system, namely

$$c_1 = c_2 = \cdots = c_n = 0.$$

Therefore $\phi_1, \cdots, \phi_n$ are linearly independent on I.

Conversely, suppose $\phi_1, \cdots, \phi_n$ are linearly independent on I. Suppose there is an x_0 in I such that

$$W(\phi_1, \cdots, \phi_n)(x_0) = 0.$$

Then this implies that the system of n linear equations

$$
\begin{aligned}
c_1\phi_1(x_0) + \cdots + c_n\phi_n(x_0) &= 0 \\
c_1\phi_1'(x_0) + \cdots + c_n\phi_n'(x_0) &= 0 \\
&\vdots \\
c_1\phi_1^{(n-1)}(x_0) + \cdots + c_n\phi_n^{(n-1)}(x_0) &= 0
\end{aligned}
\tag{4.3}
$$

has a solution $c_1, \cdots, c_n$, where not all the constants $c_1, \cdots, c_n$ are zero. Let $c_1, \cdots, c_n$ be such a solution, and consider the function

$$\psi = c_1\phi_1 + \cdots + c_n\phi_n.$$

Now $L(\psi) = 0$, and from (4.3) we see that

$$\psi(x_0) = 0, \quad \psi'(x_0) = 0, \quad \cdots, \quad \psi^{(n-1)}(x_0) = 0.$$

From Theorem 3 it follows that $\psi(x) = 0$ for *all* x in I, and thus

$$c_1\phi_1(x) + \cdots + c_n\phi_n(x) = 0$$

for all x in I. But this contradicts the fact that $\phi_1, \cdots, \phi_n$ are linearly independent on I. Thus the supposition that there was a point x_0 in I such that

$$W(\phi_1, \cdots, \phi_n)(x_0) = 0$$

must be false. We have consequently proved that

$$W(\phi_1, \cdots, \phi_n)(x) \neq 0 \quad \text{for all } x \text{ in } I.$$

Theorem 7. *Let $\phi_1, \cdots, \phi_n$ be* n *linearly independent solutions of* $L(y) = 0$ *on an interval* I. *If ϕ is any solution of* $L(y) = 0$ *on* I, *it can be represented in the form*

$$\phi = c_1\phi_1 + \cdots + c_n\phi_n,$$

where $c_1, \cdots, c_n$ *are constants. Thus any set of* n *linearly independent solutions of* $L(y) = 0$ *on* I *is a basis for the solutions of* $L(y) = 0$ *on* I.

Proof. Let x_0 be a point in I, and suppose

$$\phi(x_0) = \alpha_1, \quad \phi'(x_0) = \alpha_2, \quad \cdots, \quad \phi^{(n-1)}(x_0) = \alpha_n.$$

We show that there exist unique constants $c_1, \cdots, c_n$ such that

$$\psi = c_1\phi_1 + \cdots + c_n\phi_n$$

is a solution of $L(y) = 0$ satisfying

$$\psi(x_0) = \alpha_1, \quad \psi'(x_0) = \alpha_2, \quad \cdots, \quad \psi^{(n-1)}(x_0) = \alpha_n.$$

By the uniqueness result Theorem 3 we then have $\phi = \psi$, or

$$\phi = c_1\phi_1 + \cdots + c_n\phi_n.$$

The initial conditions for ψ are equivalent to the following equations for $c_1, \cdots, c_n$:

$$c_1\phi_1(x_0) + \cdots + c_n\phi_n(x_0) = \alpha_1$$
$$c_1\phi_1'(x_0) + \cdots + c_n\phi_n'(x_0) = \alpha_2$$
$$\vdots \qquad\qquad\qquad\qquad (4.4)$$
$$c_1\phi_1^{(n-1)}(x_0) + \cdots + c_n\phi_n^{(n-1)}(x_0) = \alpha_n.$$

This is a set of n linear equations for $c_1, \cdots, c_n$. The determinant of the coefficients is $W(\phi_1, \cdots, \phi_n)(x_0)$, which is not zero since $\phi_1, \cdots, \phi_n$ are linearly independent (Theorem 6). Therefore there is a unique solution $c_1, \cdots, c_n$ of the equations (4.4), and this completes the proof.

The analogue of Theorem 18, Chap. 2, is the following result.

Theorem 8. *Let* $\phi_1, \cdots, \phi_n$ *be* n *solutions of* $L(y) = 0$ *on an interval* I, *and let* x_0 *be any point in* I. *Then*

$$W(\phi_1, \cdots, \phi_n)(x) = \exp\left[-\int_{x_0}^{x} a_1(t)\, dt\right] W(\phi_1, \cdots, \phi_n)(x_0). \qquad (4.5)$$

Proof. We first prove this result for the simple case $n = 2$, and then give a proof which is valid for general n. The latter proof makes use of some general properties of determinants.

Proof for the case n $= 2$. In this case

$$W(\phi_1, \phi_2) = \phi_1\phi_2' - \phi_1'\phi_2,$$

and therefore

$$W'(\phi_1, \phi_2) = \phi_1'\phi_2' + \phi_1\phi_2'' - \phi_1''\phi_2 - \phi_1'\phi_2'$$
$$= \phi_1\phi_2'' - \phi_1''\phi_2.$$

Since ϕ_1, ϕ_2 satisfy $y'' + a_1(x)y' + a_2(x)y = 0$, we obtain

$$\phi_1'' = -a_1\phi_1' - a_2\phi_1,$$

$$\phi_2'' = -a_1\phi_2' - a_2\phi_2.$$

Thus

$$W'(\phi_1, \phi_2) = \phi_1(-a_1\phi_2' - a_2\phi_2) - (-a_1\phi_1' - a_2\phi_1)\phi_2$$

$$= -a_1(\phi_1\phi_2' - \phi_1'\phi_2) = -a_1 W(\phi_1, \phi_2).$$

We see that $W(\phi_1, \phi_2)$ satisfies the linear first order equation

$$y' + a_1(x)y = 0,$$

and hence

$$W(\phi_1, \phi_2)(x) = c \exp\left[-\int_{x_0}^x a_1(t)\, dt\right]$$

where c is a constant. By putting $x = x_0$, we obtain

$$c = W(\phi_1, \phi_2)(x_0),$$

thus proving (4.5) in case $n = 2$.

Proof for a general n. We let $W = W(\phi_1, \cdots, \phi_n)$ for brevity. From the definition of W as a determinant it follows that its derivative W' is a sum of n determinants

$$W' = V_1 + \cdots + V_n,$$

where V_k differs from W only in its k-th row, and the k-th row of V_k is obtained by differentiating the k-th row of W. Thus

$$W' = \begin{vmatrix} \phi_1' & \cdots & \phi_n' \\ \phi_1' & \cdots & \phi_n' \\ \phi_1'' & \cdots & \phi_n'' \\ \cdot & & \cdot \\ \cdot & & \cdot \\ \cdot & & \cdot \\ \phi_1^{(n-1)} & \cdots & \phi_n^{(n-1)} \end{vmatrix} + \begin{vmatrix} \phi_1 & \cdots & \phi_n \\ \phi_1'' & \cdots & \phi_n'' \\ \phi_1'' & \cdots & \phi_n'' \\ \cdot & & \cdot \\ \cdot & & \cdot \\ \cdot & & \cdot \\ \phi_1^{(n-1)} & \cdots & \phi_n^{(n-1)} \end{vmatrix}$$

$$+ \cdots + \begin{vmatrix} \phi_1 & \cdots & \phi_n \\ \phi_1' & \cdots & \phi_n' \\ \phi_1'' & \cdots & \phi_n'' \\ \cdot & & \cdot \\ \cdot & & \cdot \\ \cdot & & \cdot \\ \phi_j^{(n)} & \cdots & \phi_n^{(n)} \end{vmatrix}$$

The first $n - 1$ determinants $V_1, \cdots, V_{n-1}$ are all zero, since they each have two identical rows. Since $\phi_1, \cdots, \phi_n$ are solutions of $L(y) = 0$ we have

$$\phi_i^{(n)} = -a_1 \phi_i^{(n-1)} - \cdots - a_n \phi_i, \qquad (i = 1, \cdots, n),$$

and therefore

$$W' = \begin{vmatrix} \phi_1 & \cdots & \phi_n \\ \phi_1' & \cdots & \phi_n' \\ \cdot & & \cdot \\ \cdot & & \cdot \\ \phi_1^{(n-2)} & \cdots & \phi_n^{(n-2)} \\ -\sum_{j=0}^{n-1} a_{n-j} \phi_1^{(j)} & \cdots & -\sum_{j=0}^{n-1} a_{n-j} \phi_n^{(j)} \end{vmatrix}$$

The value of this determinant is unchanged if we multiply any row by a number and add to the last row. We multiply the first row by a_n, the second by $a_{n-1}, \cdots$, the $(n - 1)$-st row by a_2, and add these to the last row, obtaining

$$W' = \begin{vmatrix} \phi_1 & \cdots & \phi_n \\ \phi_1' & \cdots & \phi_n' \\ \cdot & & \cdot \\ \cdot & & \cdot \\ \phi_1^{(n-2)} & \cdots & \phi_n^{(n-2)} \\ -a_1 \phi_1^{(n-1)} & \cdots & -a_1 \phi_n^{(n-1)} \end{vmatrix} = -a_1 W$$

Therefore W satisfies the linear first order equation $y' + a_1(x)y = 0$, and thus

$$W(x) = \exp\left[-\int_{x_0}^{x} a_1(t) \, dt \right] W(x_0).$$

Corollary. *If the coefficients a_k of L are constants, then*

$$W(\phi_1, \cdots, \phi_n)(x) = e^{-a_1(x - x_0)} W(\phi, \cdots, \phi_n)(x_0).$$

Note that this corollary is just Theorem 18, Chap. 2.

A consequence of Theorem 8 is that n solutions $\phi_1, \cdots, \phi_n$ **of**

$$L(y) = 0$$

on an interval I are linearly independent there if and only if

$$W(\phi_1, \cdots, \phi_n)(x_0) \neq 0$$

for any particular x_0 in I.

EXERCISES

1. Consider the equation

$$L(y) = y'' + a_1(x)y' + a_2(x)y = 0,$$

where a_1, a_2 are continuous on some interval I. Let ϕ_1, ϕ_2 and ψ_1, ψ_2 be two bases for the solutions of $L(y) = 0$. Show that there is a non-zero constant k such that

$$W(\psi_1, \psi_2)(x) = kW(\phi_1, \phi_2)(x).$$

2. Consider the same equation as in Ex. 1. Show that a_1 and a_2 are uniquely determined by any basis ϕ_1, ϕ_2 for the solutions of $L(y) = 0$. (*Hint:* Try solving for a_1, a_2 from the equations

$$L(\phi_1) = 0, \qquad L(\phi_2) = 0.$$

Show that

$$a_1 = -\frac{\begin{vmatrix} \phi_1 & \phi_2 \\ \phi_1'' & \phi_2'' \end{vmatrix}}{W(\phi_1, \phi_2)}, \qquad a_2 = \frac{\begin{vmatrix} \phi_1' & \phi_2' \\ \phi_1'' & \phi_2'' \end{vmatrix}}{W(\phi_1, \phi_2)}.)$$

3. Consider the equation

$$y'' + \alpha(x)y = 0,$$

where α is a continuous function on $-\infty < x < \infty$ which is of period $\xi > 0$. Let ϕ_1, ϕ_2 be the basis for the solutions satisfying

$$\phi_1(0) = 1, \qquad \phi_2(0) = 0,$$

$$\phi_1'(0) = 0, \qquad \phi_2'(0) = 1.$$

(a) Show that $W(\phi_1, \phi_2)(x) = 1$ for all x.
(b) Show that there is at least one non-trivial solution ϕ of period ξ if, and only if,

$$\phi_1(\xi) + \phi_2'(\xi) = 2.$$

(*Hint:* Ex. 6, Sec. 3.)
(c) Show that there exists a non-trivial solution ϕ satisfying

$$\phi(x + \xi) = -\phi(x)$$

if, and only if,

$$\phi_1(\xi) + \phi_2'(\xi) = -2.$$

(*Hint*: Show that such a ϕ exists if, and only if,

$$\phi(\xi) = -\phi(0) \quad \text{and} \quad \phi'(\xi) = -\phi'(0).$$

See Ex. 6, Sec. 3.)

(d) If $\phi_1(\xi) + \phi_2'(\xi) = -2$ show that there exists a non-trivial solution of period 2ξ. (*Hint*: Use (c). Alternately, use (b) with ξ replaced by 2ξ.)

4. (a) Let ϕ be a real-valued non-trivial solution of

$$y'' + \alpha(x)y = 0$$

on $a < x < b$, and let ψ be a real-valued non-trivial solution of

$$y'' + \beta(x)y = 0$$

on $a < x < b$. Here α, β are real-valued continuous functions. Suppose that

$$\beta(x) > \alpha(x), \qquad (a < x < b).$$

Show that if x_1 and x_2 are successive zeros of ϕ on $a < x < b$, then ψ must vanish at some point ξ, $x_1 < \xi < x_2$. (*Hint*: Suppose $\psi(x) > 0$ for $x_1 \leqslant x < x_2$, and assume $\phi(x) > 0$ for $x_1 < x < x_2$. Then

$$(\psi\phi' - \phi\psi')' = \psi\phi'' - \phi\psi'' = (\beta - \alpha)\phi\psi,$$

and an integration yields

$$\psi(x_2)\phi'(x_2) - \psi(x_1)\phi'(x_1) > 0,$$

since $\phi(x_1) = \phi(x_2) = 0$. Show that $\phi'(x_2) < 0$ and $\phi'(x_1) > 0$.)

(b) Show that any real-valued solution ψ of

$$y'' + xy = 0$$

on $0 < x < \infty$ has an infinity of zeros there. (*Hint*: Consider the equation $y'' + y = 0$, and use (a) with

$$\alpha(x) = 1, \qquad \beta(x) = x, \qquad \phi(x) = \cos x.)$$

5. Let ϕ and ψ be two real-valued linearly independent solutions of

$$y'' + \alpha(x)y = 0$$

on $a < x < b$, where α is real-valued. Show that between any two successive zeros of ϕ there is a zero of ψ. (*Hint*: Use the method of Ex. 4(a). Alternately, suppose

$$\phi(x_1) = \phi(x_2) = 0, \quad \text{and} \quad \psi(x) > 0 \quad \text{for} \quad x_1 \leqq x \leqq x_2.$$

Let $\chi = \phi/\psi$, and show that

$$\chi' = \frac{-W(\phi, \psi)}{\psi^2}, \qquad (x_1 \leqq x \leqq x_2).$$

Apply Rolle's theorem to χ on $x_1 \leqq x \leqq x_2$. Note that ϕ and ψ cannot vanish simultaneously for $W(\phi, \psi)(x) \neq 0$.)

6. One solution of

$$L(y) = y'' + \frac{1}{4x^2} y = 0$$

for $x > 0$ is $\phi(x) = x^{1/2}$. Show that there is another solution ψ of the form $\psi = u\phi$, where u is some function. (*Hint*: Try to find u so that $L(u\phi) = 0$. This is a variation of the variation of constants idea.)

7. Consider the equation

$$y'' + \alpha(x)y = 0,$$

where α is a real-valued continuous function on $0 < x < \infty$.

(a) If $\alpha(x) \geqq \epsilon$ for $0 < x < \infty$, where ϵ is a positive constant, show that every real-valued solution has an infinity of zeros on $0 < x < \infty$. (*Hint*: Ex. 4.)

(b) Show that this conclusion is not valid if α just satisfies $\alpha(x) > 0$ for $0 < x < \infty$. (*Hint*: Ex. 6.)

8. Consider the equation

$$y'' + \alpha(x)y = 0,$$

where α is a real-valued continuous function for $a < x < b$.

(a) If ϕ is a non-trivial solution which has a zero at x_0, show that $\phi'(x_0) \neq 0$. (*Remark*: Such a zero is called a *simple* zero.)

(b) Show that the zeros of a non-trivial solution ϕ are *isolated*, that is, if $\phi(x_0) = 0$, there is no sequence of distinct $x_n \to x_0$, $(n \to \infty)$, such that $\phi(x_n) = 0$. (*Hint*: If $\phi(x_n) = 0$, $x_n \to x_0$, show that $\phi'(x_0) = 0$.)

5. Reduction of the order of a homogeneous equation

Suppose we have found by some means one solution ϕ_1 of the equation

$$L(y) = y^{(n)} + a_1(x)y^{(n-1)} + \cdots + a_n(x)y = 0.$$

It is then possible to take advantage of this information to reduce the order of the equation to be solved by one. The idea is the same one employed in the variation of constants method. We try to find solutions ϕ of $L(y) = 0$ of the form $\phi = u\phi_1$, where u is some function. If $\phi = u\phi_1$ is to be a solution we must have

$$0 = (u\phi_1)^{(n)} + a_1(u\phi_1)^{(n-1)} + \cdots + a_{n-1}(u\phi_1)' + a_n(u\phi_1)$$

$$= u^{(n)}\phi_1 + \cdots + u\phi_1^{(n)} + a_1u^{(n-1)}\phi_1 + \cdots + a_1u\phi_1^{(n-1)}$$

$$+ \cdots$$

$$+ a_{n-1}u'\phi_1 + a_{n-1}u\phi_1'$$

$$+ a_nu\phi_1.$$

The coefficient of u in this equation is just $L(\phi_1) = 0$. Therefore, if $v = u'$, this is a linear equation of order $n - 1$ in v,

$$\phi_1 v^{(n-1)} + \cdots + \left[n\phi_1^{(n-1)} + a_1 (n - 1) \phi_1^{(n-2)} + \cdots + a_{n-1}\phi_1 \right] v = 0. \quad (5.1)$$

The coefficient of $v^{(n-1)}$ is ϕ_1, and hence if $\phi_1(x) \neq 0$ on an interval I this equation has $n - 1$ linearly independent solutions $v_2, \cdots, v_n$ on I. If x_0 is some point in I, and

$$u_k(x) = \int_{x_0}^{x} v_k(t)\, dt, \qquad (k = 2, \cdots, n),$$

then we have $u_k' = v_k$, and the functions

$$\phi_1, u_2\phi_1, \cdots, u_n\phi_1 \quad (5.2)$$

are solutions of $L(y) = 0$. Moreover these functions form a basis for the solutions of $L(y) = 0$ on I. For suppose we have constants $c_1, \cdots, c_n$ such that

$$c_1\phi_1 + c_2 u_2\phi_1 + \cdots + c_n u_n\phi_1 = 0.$$

Since $\phi_1(x) \neq 0$ on I this implies

$$c_1 + c_2 u_2 + \cdots + c_n u_n = 0, \quad (5.3)$$

and differentiating we obtain

$$c_2 u_2' + \cdots + c_n u_n' = 0,$$

or

$$c_2 v_2 + \cdots + c_n v_n = 0.$$

Since $v_2, \cdots, v_n$ are linearly independent on I we have

$$c_2 = c_3 = \cdots = c_n = 0,$$

and from (5.3) we obtain $c_1 = 0$ also. Thus the functions in (5.2) form a basis for the solutions of $L(y) = 0$ on I.

Theorem 9. *Let ϕ_1 be a solution of* $L(y) = 0$ *on an interval* I, *and suppose $\phi_1(x) \neq 0$ on* I. *If $v_2, \cdots, v_n$ is any basis on* I *for the solutions of the linear equation* (5.1) *of order* $n - 1$, *and if*

$$v_k = u_k', \qquad (k = 2, \cdots, n),$$

then $\phi_1, u_2\phi_1, \cdots, u_n\phi_1$ is a basis for the solutions of $L(y) = 0$ *on* I.

The case $n = 2$ of Theorem 9 merits further discussion, since in this case the equation for v is linear of the *first* order, and therefore can be solved explicitly (Chap. 1). Here we have

$$L(y) = y'' + a_1(x)y' + a_2(x)y = 0,$$

and if ϕ_1 is a solution on I we have

$$L(u\phi_1) = (u\phi_1)'' + a_1(u\phi_1)' + a_2(u\phi_1)$$
$$= u''\phi_1 + 2u'\phi_1' + u\phi_1'' + a_1u'\phi_1 + a_1u\phi_1' + a_2u\phi_1$$
$$= u''\phi_1 + u'(2\phi_1' + a_1\phi_1).$$

Thus, if $v = u'$, and u is such that $L(u\phi_1) = 0$,

$$\phi_1 v' + (2\phi_1' + a_1\phi_1)v = 0. \tag{5.4}$$

But (5.4) is a linear equation of order one, and can always be solved explicitly provided $\phi_1(x) \neq 0$ on I. Indeed v satisfies

$$\phi_1^2 v' + (2\phi_1\phi_1' + a_1\phi_1^2)v = 0, \tag{5.5}$$

which is just (5.4) multiplied by ϕ_1. Thus

$$(\phi_1^2 v)' + a_1(\phi_1^2 v) = 0,$$

which implies that

$$\phi_1^2(x)v(x) = c \exp\left[-\int_{x_0}^x a_1(t)\ dt\right],$$

where x_0 is a point in I, and c is a constant. Since any constant multiple of a solution of (5.5) is again a solution, we see that

$$v(x) = \frac{1}{[\phi_1(x)]^2} \exp\left[-\int_{x_0}^x a_1(t)\ dt\right]$$

is a solution of (5.5), and also of (5.4). Therefore two independent solutions of

$$L(y) = y'' + a_1(x)y' + a_2(x)y = 0 \tag{5.6}$$

on I are ϕ_1 and ϕ_2, where

$$\phi_2(x) = \phi_1(x) \int_{x_0}^x \frac{1}{[\phi_1(s)]^2} \exp\left[-\int_{x_0}^s a_1(t)\ dt\right] ds. \tag{5.7}$$

Theorem 10. *If ϕ_1 is a solution of (5.6) on an interval* I, *and $\phi_1(x) \neq 0$ on* I, *a second solution ϕ_2 of (5.6) on* I *is given by (5.7). The functions ϕ_1, ϕ_2 form a basis for the solutions of (5.6) on* I.

As a simple example consider the equation

$$y'' - \frac{2}{x^2} y = 0, \qquad (0 < x < \infty).$$

It is easy to verify that the ϕ_1 given by $\phi_1(x) = x^2$ is a solution on $0 < x < \infty$, and since this function does not vanish on this interval there is another independent solution ϕ_2 of the form $\phi_2 = u\phi_1$. If $v = u'$ we find that v satisfies

$$x^2 v' + 4xv = 0, \quad \text{or} \quad xv' + 4v = 0.$$

A solution for this is given by

$$v(x) = x^{-4}, \quad (0 < x < \infty),$$

and therefore a choice for u is

$$u(x) = -\frac{1}{3x^3}, \quad (0 < x < \infty).$$

This leads to

$$\phi_2(x) = -\frac{1}{3x}, \quad (0 < x < \infty),$$

but since any constant times a solution is a solution, we may as well choose for a second solution $\phi_2(x) = x^{-1}$. Thus x^2, x^{-1} form a basis for the solutions on $0 < x < \infty$.

EXERCISES

1. A differential equation and a function ϕ_1 are given in each of the following. Verify that the function ϕ_1 satisfies the equation, and find a second independent solution.

(a) $x^2 y'' - 7xy' + 15y = 0, \phi_1(x) = x^3, (x > 0).$

(b) $x^2 y'' - xy' + y = 0, \phi_1(x) = x, (x > 0).$

(c) $y'' - 4xy' + (4x^2 - 2)y = 0, \phi_1(x) = e^{x^2}.$

(d) $xy'' - (x + 1)y' + y = 0, \phi_1(x) = e^x, (x > 0).$

(e) $(1 - x^2)y'' - 2xy' + 2y = 0, \phi_1(x) = x, (0 < x < 1).$

(f) $y'' - 2xy' + 2y = 0, \phi_1(x) = x, (x > 0).$

2. One solution of

$$x^3 y''' - 3x^2 y'' + 6xy' - 6y = 0$$

for $x > 0$ is $\phi_1(x) = x$. Find a basis for the solutions for $x > 0$.

3. Consider the equation

$$L(y) = y''' + a_1(x)y'' + a_2(x)y' + a_3(x)y = 0.$$

Suppose ϕ_1, ϕ_2 are given linearly independent solutions of $L(y) = 0$.

(a) Let $\phi = u\phi_1$, and compute the equation of order two satisfied by u' in order that $L(\phi) = 0$. Show that $(\phi_2/\phi_1)'$ is a solution of this equation of order two.

(b) Use the fact that $(\phi_2/\phi_1)'$ satisfies the equation of order two to reduce the order of this equation by one.

4. Two solutions of

$$x^3 y''' - 3xy' + 3y = 0, \qquad (x > 0),$$

are $\phi_1(x) = x$, $\phi_2(x) = x^3$. Use this information to find a third independent solution. (*Hint*: See Ex. 3.)

5. Consider the equation

$$y'' + a_1(x)y' + a_2(x)y = 0,$$

where a_1, a_2 are continuous on some interval I containing x_0. Suppose ϕ_1 is a solution such that $\phi_1(x) \neq 0$ for all x in I.

(a) Show that there is a second solution ϕ_2 on I such that

$$W(\phi_1, \phi_2)(x_0) = 1.$$

(b) Compute such a ϕ_2 in terms of ϕ_1, by solving the first order equation

$$\phi_1(x)\phi_2'(x) - \phi_1'(x)\phi_2(x) = \exp\left[-\int_{x_0}^{x} a_1(t)\, dt \right],$$

for ϕ_2.

6. The non-homogeneous equation

Let $a_1, \cdots, a_n, b$ be continuous functions on an interval I, and consider the equation

$$L(y) = y^{(n)} + a_1(x)y^{(n-1)} + \cdots + a_n(x)y = b(x). \qquad (6.1)$$

We have already seen that, in the case where the a_k are all constants, this equation may be solved using the variation of constants method (Sec. 10, Chap. 2.). The method does not depend on the fact that the a_k are constants, and is therefore valid for the equation (6.1). We outline briefly the results.

If ψ_p is a particular solution of (6.1), any other solution ψ has the form

$$\psi = \psi_p + c_1\phi_1 + \cdots + c_n\phi_n,$$

where $c_1, \cdots, c_n$ are constants, and $\phi_1, \cdots, \phi_n$ is a basis for the solutions of $L(y) = 0$. Every such ψ is a solution of $L(y) = b(x)$. A particular solution ψ_p can be found which has the form

$$\psi_p = u_1\phi_1 + \cdots + u_n\phi_n,$$

where $u_1, \cdots, u_n$ are functions satisfying

$$u_1'\phi_1 + \cdots + u_n'\phi_n = 0$$

$$u_1'\phi_1' + \cdots + u_n'\phi_n' = 0$$

$$\cdot$$
$$\cdot$$
$$\cdot$$

$$u_1'\phi_1^{(n-2)} + \cdots + u_n'\phi_n^{(n-2)} = 0$$

$$u_1'\phi_1^{(n-1)} + \cdots + u_n'\phi_n^{(n-1)} = b.$$

If x_0 is any point of I we may take for u_k the function given by

$$u_k(x) = \int_{x_0}^{x} \frac{W_k(t)b(t)}{W(\phi_1, \cdots, \phi_n)(t)} \, dt, \qquad (k = 1, \cdots, n),$$

and then ψ_p has the form

$$\psi_p(x) = \sum_{k=1}^{n} \phi_k(x) \int_{x_0}^{x} \frac{W_k(t)b(t)}{W(\phi_1, \cdots, \phi_n)(t)} \, dt. \qquad (6.2)$$

Here $W(\phi_1, \cdots, \phi_n)$ is the Wronskian of the basis $\phi_1, \cdots, \phi_n$, and W_k is the determinant obtained from $W(\phi_1, \cdots, \phi_n)$ by replacing the k-th column $(\phi_k, \phi_k', \cdots, \phi_k^{(n-1)})$ by $(0, 0, \cdots, 0, 1)$.

Theorem 11. *Let* b *be continuous on an interval* I, *and let* $\phi_1, \cdots, \phi_n$ *be a basis for the solutions of* $L(y) = 0$ *on* I. *Every solution* ψ *of* $L(y) = b(x)$ *can be written as*

$$\psi = \psi_p + c_1\phi_1 + \cdots + c_n\phi_n,$$

where ψ_p *is a particular solution of* $L(y) = b(x)$, *and* $c_1, \cdots, c_n$ *are constants. Every such* ψ *is a solution of* $L(y) = b(x)$. *A particular solution* ψ_p *is given by* (6.2).

As an illustration let us find all solutions of the equation

$$y'' - \frac{2}{x^2} y = x, \qquad (0 < x < \infty). \qquad (6.3)$$

We have already seen in Sec. 5 that a basis for the solutions of the homogeneous equation is given by

$$\phi_1(x) = x^2, \qquad \phi_2(x) = x^{-1}$$

A solution ψ_p of the non-homogeneous equation has the form

$$\psi_p = u_1 x^2 + u_2 x^{-1}, \tag{6.4}$$

where u_1', u_2' satisfy

$$x^2 u_1' + x^{-1} u_2' = 0$$

$$2x u_1' - x^{-2} u_2' = x.$$

Now $W(\phi_1, \phi_2)(x) = -3$, and we find that

$$u_1'(x) = \frac{1}{3}, \qquad u_2'(x) = -\frac{x^3}{3}.$$

For u_1, u_2 we may take

$$u_1(x) = \frac{x}{3}, \qquad u_2(x) = -\frac{x^4}{12},$$

and from (6.4) we see that

$$\psi_p(x) = \frac{x^3}{3} - \frac{x^3}{12} = \frac{x^3}{4}.$$

Every solution ϕ of (6.3) then has the form

$$\phi(x) = \frac{x^3}{4} + c_1 x^2 + c_2 x^{-1},$$

where c_1, c_2 are constants.

Since we can always solve the non-homogeneous equation $L(y) = b(x)$ by using algebraic methods and an integration, we now concentrate our attention on methods for solving the homogeneous equation.

EXERCISES

1. One solution of

$$x^2 y'' - 2y = 0$$

on $0 < x < \infty$ is $\phi_1(x) = x^2$. Find all solutions of

$$x^2 y'' - 2y = 2x - 1$$

on $0 < x < \infty$.

2. One solution of

$$x^2 y'' - xy' + y = 0, \qquad (x > 0),$$

is $\phi_1(x) = x$. Find the solution ψ of

$$x^2 y'' - xy' + y = x^2$$

satisfying $\psi(1) = 1, \psi'(1) = 0$.

3. (a) Show that there is a basis ϕ_1, ϕ_2 for the solutions of

$$x^2 y'' + 4xy' + (2 + x^2)y = 0, \qquad (x > 0),$$

of the form

$$\phi_1(x) = \frac{\psi_1(x)}{x^2}, \qquad \phi_2(x) = \frac{\psi_2(x)}{x^2}.$$

(*Hint*: If ϕ is a solution, let $\phi = v/x^2$.)

(b) Find all solutions of

$$x^2 y'' + 4xy' + (2 + x^2)y = x^2$$

for $x > 0$.

4. (a) Consider the equation

$$L(y) = y'' + a_1(x)y' + a_2(x)y = b(x),$$

where a_1, a_2, b are continuous on some interval I. Suppose ϕ_1 is a solution of $L(y) = 0$ such that $\phi_1(x) \neq 0$ for all x in I. Show that there is a particular solution ψ_p of $L(y) = b(x)$ of the form $\psi_p = u_p \phi_1$, where $v_p = u_p'$ is a particular solution of the first order equation

$$\phi_1(x)v' + [2\phi_1'(x) + a_1(x)\phi_1(x)]v = b(x).$$

(b) Use the idea in (a) to find all solutions of

$$x^2 y'' - xy' + y = x^2$$

for $x > 0$. (*Hint*: From Ex. 2 one solution of $x^2 y'' - xy' + y = 0$ is given by $\phi_1(x) = x$.)

5. Show that the function ψ_p given by (6.2) satisfies

$$\psi_p(x_0) = \psi_p'(x_0) = \cdots = \psi_p^{(n-1)}(x_0) = 0.$$

6. Let $g(x, t)$ be defined by

$$g(x, t) = \sum_{k=1}^{n} \frac{\phi_k(x)W_k(t)}{W(t)},$$

where $W = W(\phi_1, \cdots, \phi_n)$ is the Wronskian of n linearly independent solutions of $L(y) = 0$. Then the ψ_p of (6.2) can be written as

$$\psi_p(x) = \int_{x_0}^{x} g(x, t)b(t) \, dt.$$

(a) Prove that

$$g(x, t) = \frac{k(x, t)}{W(t)},$$

where

$$
k(x, t) = \begin{vmatrix}
\phi_1(t) & \phi_2(t) & \cdots & \phi_n(t) \\
\phi_1'(t) & \phi_2'(t) & \cdots & \phi_n'(t) \\
\cdot & \cdot & & \cdot \\
\cdot & \cdot & & \cdot \\
\cdot & \cdot & & \cdot \\
\phi_1^{(n-2)}(t) & \phi_2^{(n-2)}(t) & \cdots & \phi_n^{(n-2)}(t) \\
\phi_1(x) & \phi_2(x) & \cdots & \phi_n(x)
\end{vmatrix}
$$

(b) Show that

$$
g(t, t) = 0, \quad \frac{\partial g}{\partial x}(t, t) = 0, \quad \cdots, \quad \frac{\partial^{n-2} g}{\partial x^{n-2}}(t, t) = 0, \quad \frac{\partial^{n-1} g}{\partial x^{n-1}}(t, t) = 1.
$$

7. Consider the equation

$$
y'' + y = b(x),
$$

where b is a continuous function on $1 \leqq x < \infty$ satisfying

$$
\int_1^\infty |b(t)| \, dt < \infty.
$$

(a) Show that a particular solution ψ_p is given by

$$
\psi_p(x) = \int_1^x \sin(x - t) b(t) \, dt.
$$

(b) Show that any solution is bounded on $1 \leqq x < \infty$.

7. Homogeneous equations with analytic coefficients

If g is a function defined on an interval I containing a point x_0, we say that g is *analytic* at x_0 if g can be expanded in a power series about x_0 which has a positive radius of convergence. Thus g is analytic at x_0 if it can be represented in the form

$$
g(x) = \sum_{k=0}^\infty c_k (x - x_0)^k, \tag{7.1}
$$

where the c_k are constants, and the series converges for $|x - x_0| < r_0$, $r_0 > 0$. Recall (Sec. 5, Chap. 0) that one of the important properties of a function g which has the form (7.1), where the series converges for

$|x - x_0| < r_0$, is that all of its derivatives exist on $|x - x_0| < r_0$, and they may be computed by differentiating the series term by term. Thus, for example

$$g'(x) = \sum_{k=1}^{\infty} k c_k (x - x_0)^{k-1},$$

and

$$g''(x) = \sum_{k=2}^{\infty} k(k - 1) c_k (x - x_0)^{k-2},$$

and the differentiated series converge on $|x - x_0| < r_0$ also.

If the coefficients $a_1, \cdots, a_n$ of L are analytic at x_0 it turns out that the solutions are also. In fact solutions can be computed by a formal algebraic process. We illustrate by considering the example

$$L(y) = y'' - xy = 0.$$

Here $a_1(x) = 0$, $a_2(x) = -x$, and hence a_1, a_2 are analytic for all real x_0. We try for a solution the series

$$\phi(x) = c_0 + c_1 x + c_2 x^2 + \cdots.$$

Then

$$\phi''(x) = 2c_2 + 3 \cdot 2 c_3 x + 4 \cdot 3 c_4 x^2 + \cdots$$

$$= \sum_{k=0}^{\infty} (k + 2)(k + 1) c_{k+2} x^k.$$

Also

$$x\phi(x) = c_0 x + c_1 x^2 + c_2 x^3 + \cdots = \sum_{k=1}^{\infty} c_{k-1} x^k,$$

$$\phi''(x) - x\phi(x) = 2c_2 + \sum_{k=1}^{\infty} [(k + 2)(k + 1) c_{k+2} - c_{k-1}] x^k.$$

In order for ϕ to be a solution of $L(y) = 0$ we must have

$$\phi''(x) - x\phi(x) = 0,$$

or

$$2c_2 + \sum_{k=1}^{\infty} [(k + 2)(k + 1) c_{k+2} - c_{k-1}] x^k = 0,$$

and this is true only if all the coefficients of the powers of x are zero. Thus

$$2c_2 = 0, \quad (k + 2)(k + 1) c_{k+2} - c_{k-1} = 0, \quad (k = 1, 2, \cdots).$$

This gives an infinite set of equations, which can be solved for the c_k. Thus, for $k = 1$, we have

$$3 \cdot 2 c_3 = c_0, \quad \text{or} \quad c_3 = \frac{c_0}{3 \cdot 2}.$$

Putting $k = 2$ we find

$$c_4 := \frac{c_1}{4 \cdot 3}.$$

Continuing in this way we see that

$$c_5 = \frac{c_2}{5 \cdot 4} = 0, \quad c_6 = \frac{c_3}{6 \cdot 5} = \frac{c_0}{6 \cdot 5 \cdot 3 \cdot 2}, \quad c_7 = \frac{c_4}{7 \cdot 6} = \frac{c_1}{7 \cdot 6 \cdot 4 \cdot 3}.$$

It can be shown by induction that

$$c_{3m} = \frac{c_0}{2 \cdot 3 \cdot 5 \cdot 6 \cdots (3m - 1) 3m}, \quad (m = 1, 2, \cdots),$$

$$c_{3m+1} = \frac{c_1}{3 \cdot 4 \cdot 6 \cdot 7 \cdots 3m(3m + 1)}, \quad (m = 1, 2, \cdots),$$

$$c_{3m+2} = 0, \quad (m = 0, 1, 2, \cdots).$$

Thus all the constants are determined in terms of c_0 and c_1. Collecting together terms with c_0 and c_1 as a factor we have

$$\phi(x) = c_0 \left[1 + \frac{x^3}{3 \cdot 2} + \frac{x^6}{6 \cdot 5 \cdot 3 \cdot 2} + \cdots \right] + c_1 \left[x + \frac{x^4}{4 \cdot 3} + \frac{x^7}{7 \cdot 6 \cdot 4 \cdot 3} + \cdots \right].$$

Let ϕ_1, ϕ_2 represent the two series in the brackets. Thus

$$\phi_1(x) = 1 + \sum_{m=1}^{\infty} \frac{x^{3m}}{2 \cdot 3 \cdot 5 \cdot 6 \cdots (3m - 1) 3m},$$

$$\phi_2(x) = x + \sum_{m=1}^{\infty} \frac{x^{3m+1}}{3 \cdot 4 \cdot 6 \cdot 7 \cdots 3m(3m + 1)}. \tag{7.2}$$

We have shown, in a formal way, that ϕ satisfies $y'' - xy = 0$ for any two constants c_0, c_1. In particular the choice $c_0 = 1$, $c_1 = 0$ shows that ϕ_1 satisfies this equation, and the choice $c_0 = 0$, $c_1 = 1$ implies ϕ_2 also satisfies the equation.

The only question that remains concerns the convergence of the series defining $\phi_1(x)$ and $\phi_2(x)$. It is readily checked by the ratio test that both series converge for all finite x. For example, let us consider the series for $\phi_1(x)$. Writing it as

$$\phi_1(x) = 1 + \sum_{m=1}^{\infty} d_m(x),$$

we see that

$$\frac{d_{m+1}(x)}{d_m(x)} = \frac{x^{3m+3}}{2 \cdot 3 \cdot 5 \cdot 6 \cdots (3m - 1)(3m)(3m + 2)(3m + 3)}$$

$$\times \frac{2 \cdot 3 \cdot 5 \cdot 6 \cdots (3m - 1)(3m)}{x^{3m}},$$

and therefore

$$\left| \frac{d_{m+1}(x)}{d_m(x)} \right| = \frac{|x|^3}{(3m+2)(3m+3)},$$

which tends to zero, as $m \to \infty$, provided only that $|x| < \infty$.

Summarizing, we have found in a purely formal way two series, which are convergent for all finite x, and thus represent two functions ϕ_1, ϕ_2, and from the way we obtained ϕ_1, ϕ_2 it is apparent that they are solutions of the equation $y'' - xy = 0$ on $-\infty < x < \infty$. They are linearly independent solutions for it is clear from the series (7.2) defining ϕ_1 and ϕ_2 that

$$\phi_1(0) = 1, \qquad \phi_2(0) = 0,$$
$$\phi_1'(0) = 0, \qquad \phi_2'(0) = 1,$$

and therefore

$$W(\phi_1, \phi_2)(0) = 1 \neq 0.$$

The method illustrated by this example works in general when the coefficients are analytic, and always yields a convergent power series solution for any initial value problem. We state this result formally, and devote Section 9 to its justification.

Theorem 12. (*Existence Theorem for Analytic Coefficients*) *Let* x_0 *be a real number, and suppose that the coefficients* $a_1, \cdots, a_n$ *in*

$$L(y) = y^{(n)} + a_1(x)y^{(n-1)} + \cdots + a_n(x)y$$

have convergent power series expansions in powers of $x - x_0$ *on an interval*

$$|x - x_0| < r_0, \qquad r_0 > 0.$$

If $\alpha_1, \cdots, \alpha_n$ *are any* n *constants, there exists a solution* ϕ *of the problem*

$$L(y) = 0, \quad y(x_0) = \alpha_1, \quad \cdots, \quad y^{(n-1)}(x_0) = \alpha_n,$$

with a power series expansion

$$\phi(x) = \sum_{k=0}^{\infty} c_k(x - x_0)^k \tag{7.3}$$

convergent for $|x - x_0| < r_0$. *We have*

$$k! c_k = \alpha_{k+1}, \qquad (k = 0, 1, \cdots, n-1),$$

and c_k *for* k $\geq$ n *may be computed in terms of* $c_0, c_1, \cdots, c_{n-1}$ *by substituting the series* (7.3) *into* L(y) = 0.

It follows from Theorem 12, and the Uniqueness Theorem 3, that *any* solution ϕ of $L(y) = 0$ on $|x - x_0| < r_0$ has a convergent power series expansion there of the form (7.3).

EXERCISES

1. Find two linearly independent power series solutions (in powers of x) of the following equations:

(a) $y'' - xy' + y = 0$ (b) $y'' + 3x^2y' - xy = 0$

(c) $y'' - x^2y = 0$ (d) $y'' + x^3y' + x^2y = 0$

(e) $y'' + y = 0$

For what values of x do the series converge?

2. Find the solution ϕ of

$$y'' + (x - 1)^2y' - (x - 1)y = 0$$

in the form

$$\phi(x) = \sum_{k=0}^{\infty} c_k(x - 1)^k,$$

which satisfies $\phi(1) = 1, \phi'(1) = 0$. (*Hint*: Let $x - 1 = \xi$.)

3. Find the solution ϕ of

$$(1 + x^2)y'' + y = 0$$

of the form

$$\phi(x) = \sum_{k=0}^{\infty} c_k x^k,$$

which satisfies $\phi(0) = 0, \phi'(0) = 1$. (*Note*: When the equation is written in the form

$$y'' + \frac{1}{1 + x^2} y = 0,$$

it is one with analytic coefficients at $x = 0$, since

$$\frac{1}{1 + x^2} = 1 - x^2 + x^4 - x^6 + \cdots = \sum_{k=0}^{\infty} (-1)^k x^{2k},$$

which converges for $|x| < 1$. However to compute ϕ it is best to substitute the series for ϕ directly into the given equation.) What is the largest $r > 0$ such that the series for ϕ converges for $|x| < r$?

4. The equation

$$y'' + e^x y = 0$$

has a solution ϕ of the form

$$\phi(x) = \sum_{k=0}^{\infty} c_k x^k$$

which satisfies $\phi(0) = 1, \phi'(0) = 0$. Compute $c_0, c_1, c_2, c_3, c_4, c_5$. (*Hint*: $c_k = \phi^{(k)}(0)/k!$ and $\phi''(x) = -e^x\phi(x)$.)

5. Compute the solution ϕ of

$$y''' - xy = 0$$

which satisfies $\phi(0) = 1, \phi'(0) = 0, \phi''(0) = 0$.

6. The equation

$$(1 - x^2)y'' - 2xy' + \alpha(\alpha + 1)y = 0, \tag{*}$$

where α is a constant, is called the *Legendre equation*.

(a) Show that if it is written in the form

$$y'' + a_1(x)y' + a_2(x)y = 0,$$

then a_1, a_2 have convergent power series expansions (in powers of x) on $|x| < 1$.

(b) Compute two linearly independent solutions for $|x| < 1$. (*Hint*: Leave the equation in the form (*).)

(c) Show that if α is a non-negative integer n there is a polynomial solution of degree n.

7. The equation

$$(1 - x^2)y'' - xy' + \alpha^2 y = 0,$$

where α is a constant, is called the *Chebyshev equation*.

(a) Compute two linearly independent series solutions for $|x| < 1$.

(b) Show that for every non-negative integer $\alpha = n$ there is a polynomial solution of degree n. When appropriately normalized these are called the *Chebyshev polynomials*.

8. The equation

$$y'' - 2xy' + 2\alpha y = 0,$$

where α is a constant, is called the *Hermite equation*.

(a) Find two linearly independent solutions on $-\infty < x < \infty$.

(b) Show that there is a polynomial solution of degree n, in case $\alpha = n$ is a non-negative integer.

(c) Show that the polynomial H_n defined by

$$H_n(x) = (-1)^n e^{x^2} \frac{d^n}{dx^n} e^{-x^2}$$

is a solution of the Hermite equation in case $\alpha = n$ is a non-negative integer. This solution H_n is called the *n-th Hermite polynomial*. (*Hint*: If $u(x) = e^{-x^2}$ show that $u'(x) + 2xu(x) = 0$. Differentiate this equation n times to obtain

$$H_{n+1}(x) - 2xH_n(x) + 2nH_{n-1}(x) = 0 \tag{*}$$

for $n \geqq 1$. Differentiate H_n to obtain

$$H_n'(x) = 2xH_n(x) - H_{n+1}(x) \tag{**}$$

for $n \geqq 0$. Use (*) and (**) to show H_n is a solution of the Hermite equation.)

(d) Compute H_0, H_1, H_2, H_3.

8. The Legendre equation

Some of the important differential equations met in physical problems are second order linear equations with analytic coefficients. One of these is the *Legendre equation*

$$L(y) = (1 - x^2)y'' - 2xy' + \alpha(\alpha + 1)y = 0, \qquad (8.1)$$

where α is a constant. If we write this equation as

$$y'' - \frac{2x}{1 - x^2} y' + \frac{\alpha(\alpha + 1)}{1 - x^2} y = 0,$$

we see that the functions a_1, a_2 given by

$$a_1(x) = \frac{-2x}{1 - x^2}, \qquad a_2(x) = \frac{\alpha(\alpha + 1)}{1 - x^2},$$

are analytic at $x = 0$. Indeed,

$$\frac{1}{1 - x^2} = 1 + x^2 + x^4 + \cdots = \sum_{k=0}^{\infty} x^{2k},$$

and this series converges for $|x| < 1$. Thus a_1 and a_2 have the series expansions

$$a_1(x) = \sum_{k=0}^{\infty} (-2)x^{2k+1}, \qquad a_2(x) = \sum_{k=0}^{\infty} \alpha(\alpha + 1)x^{2k},$$

which converge for $|x| < 1$. From Theorem 12 it follows that the solutions of $L(y) = 0$ on $|x| < 1$ have convergent power series expansions there. We proceed to find a basis for these solutions.

Let ϕ be any solution of the Legendre equation on $|x| < 1$, and suppose

$$\phi(x) = c_0 + c_1x + c_2x^2 + \cdots = \sum_{k=0}^{\infty} c_kx^k. \qquad (8.2)$$

We have

$$\phi'(x) = c_1 + 2c_2x + 3c_3x^2 + \cdots = \sum_{k=0}^{\infty} kc_kx^{k-1},$$

$$-2x\phi'(x) = \sum_{k=0}^{\infty} -2kc_kx^k, \qquad (8.3)$$

$$\phi''(x) = 2c_2 + 3\cdot2c_3x + \cdots = \sum_{k=0}^{\infty} k(k - 1)c_kx^{k-2},$$

$$-x^2\phi''(x) = \sum_{k=0}^{\infty} - k(k - 1)c_kx^k. \qquad (8.4)$$

Note that $\phi''(x)$ may also be written as

$$\phi''(x) = \sum_{k=0}^{\infty} (k+2)(k+1)c_{k+2}x^k. \tag{8.5}$$

From (8.2)–(8.5) we obtain

$$L(\phi)(x) = (1 - x^2)\phi''(x) - 2x\phi'(x) + \alpha(\alpha + 1)\phi(x)$$

$$= \sum_{k=0}^{\infty} [(k+2)(k+1)c_{k+2} - k(k-1)c_k - 2kc_k + \alpha(\alpha+1)c_k]x^k$$

$$= \sum_{k=0}^{\infty} [(k+2)(k+1)c_{k+2} + (\alpha + k + 1)(\alpha - k)c_k]x^k.$$

For ϕ to satisfy $L(\phi) = 0$ we must have all the coefficients of the powers of x equal to zero. Hence

$$(k+2)(k+1)c_{k+2} + (\alpha + k + 1)(\alpha - k)c_k = 0,$$
$$(k = 0, 1, 2, \cdots). \tag{8.6}$$

This is the recursion relation which gives c_{k+2} in terms of c_k. For $k = 0$ we obtain

$$c_2 = -\frac{(\alpha + 1)\alpha}{2}c_0,$$

and for $k = 1$ we get

$$c_3 = -\frac{(\alpha + 2)(\alpha - 1)}{3 \cdot 2}c_1.$$

Similarly, letting $k = 2, 3$ in (8.6) we obtain

$$c_4 = -\frac{(\alpha + 3)(\alpha - 2)}{4 \cdot 3}c_2 = \frac{(\alpha + 3)(\alpha + 1)\alpha(\alpha - 2)}{4 \cdot 3 \cdot 2}c_0,$$

$$c_5 = -\frac{(\alpha + 4)(\alpha - 3)}{5 \cdot 4}c_3 = \frac{(\alpha + 4)(\alpha + 2)(\alpha - 1)(\alpha - 3)}{5 \cdot 4 \cdot 3 \cdot 2}c_1.$$

The pattern now becomes clear, and it follows by induction that for $m = 1, 2, \cdots$,

$$c_{2m} = (-1)^m \frac{(\alpha + 2m - 1)(\alpha + 2m - 3)\cdots(\alpha + 1)\alpha(\alpha - 2)\cdots(\alpha - 2m + 2)}{(2m)!}c_0,$$

$$c_{2m+1}$$
$$= (-1)^m \frac{(\alpha + 2m)(\alpha + 2m - 2)\cdots(\alpha + 2)(\alpha - 1)(\alpha - 3)\cdots(\alpha - 2m + 1)}{(2m+1)!}c_1.$$

All coefficients are determined in terms of c_0 and c_1, and we must have

$$\phi(x) = c_0\phi_1(x) + c_1\phi_2(x),$$

where

$$\phi_1(x) = 1 - \frac{(\alpha + 1)\alpha}{2!} x^2 + \frac{(\alpha + 3)(\alpha + 1)\alpha(\alpha - 2)}{4!} x^4 - \cdots,$$

or

$$\phi_1(x) = 1 + \sum_{m=1}^{\infty} (-1)^m$$

$$\times \frac{(\alpha+2m-1)(\alpha+2m-3)\cdots(\alpha+1)\alpha(\alpha-2)\cdots(\alpha-2m+2)}{(2m)!} x^{2m},$$

(8.7)

and

$$\phi_2(x)$$

$$= x - \frac{(\alpha + 2)(\alpha - 1)}{3!} x^3 + \frac{(\alpha + 4)(\alpha + 2)(\alpha - 1)(\alpha - 3)}{5!} x^5 - \cdots,$$

or

$$\phi_2(x) = x + \sum_{m=1}^{\infty} (-1)^m$$

$$\times \frac{(\alpha+2m)(\alpha+2m-2)\cdots(\alpha+2)(\alpha-1)(\alpha-3)\cdots(\alpha-2m+1)}{(2m + 1)!} x^{2m+1}. \quad (8.8)$$

Both ϕ_1 and ϕ_2 are solutions of the Legendre equation, those corresponding to the choices

$$c_0 = 1, \quad c_1 = 0, \qquad \text{and} \qquad c_0 = 0, \quad c_1 = 1,$$

respectively. They form a basis for the solutions, since

$$\phi_1(0) = 1, \qquad \phi_2(0) = 0,$$
$$\phi_1'(0) = 0, \qquad \phi_2'(0) = 1.$$

We notice that if α is a non-negative even integer

$$n = 2m, \qquad (m = 0, 1, 2, \cdots),$$

then ϕ_1 has only a finite number of non-zero terms. Indeed, in this case ϕ_1 is a polynomial of degree n containing only even powers of x. For example,

$$\phi_1(x) = 1, \qquad\qquad (\alpha = 0),$$
$$\phi_1(x) = 1 - 3x^2, \qquad\qquad (\alpha = 2),$$
$$\phi_1(x) = 1 - 10x^2 + \tfrac{3 \cdot 5}{3} x^4, \qquad (\alpha = 4).$$

The solution ϕ_2 is *not* a polynomial in this case since none of the coefficients in the series (8.8) vanish.

A similar situation occurs when α is a positive odd integer n. Then ϕ_2 is a polynomial of degree n having only odd powers of x, and ϕ_1 is not a polynomial. For example,

$$\phi_2(x) = x, \qquad\qquad (\alpha = 1),$$

$$\phi_2(x) = x - \tfrac{5}{3} x^3, \qquad\qquad (\alpha = 3),$$

$$\phi_2(x) = x - \tfrac{14}{3} x^3 + \tfrac{21}{5} x^5, \qquad (\alpha = 5).$$

We consider in more detail these polynomial solutions when $\alpha = n$, a non-negative integer. The polynomial solution P_n of degree n of

$$(1 - x^2)y'' - 2xy' + n(n + 1)y = 0, \qquad (8.9)$$

satisfying $P_n(1) = 1$ is called the n-th *Legendre polynomial*. In order to justify this definition we must show that there is just one such solution for each non-negative integer n. This will be established by way of a slight detour, which is of interest in itself.

Let ϕ be the polynomial of degree n defined by

$$\phi(x) = \frac{d^n}{dx^n} (x^2 - 1)^n.$$

This ϕ satisfies the Legendre equation (8.9). Indeed, let

$$u(x) = (x^2 - 1)^n.$$

Then we obtain by differentiating

$$(x^2 - 1)u' - 2nxu = 0.$$

Differentiating this expression $n + 1$ times yields

$$(x^2 - 1)u^{(n+2)} + 2x(n + 1)u^{(n+1)} + (n + 1)nu^{(n)}$$
$$- 2nxu^{(n+1)} - 2n(n + 1)u^{(n)} = 0.$$

Since $\phi = u^{(n)}$ we obtain

$$(1 - x^2)\phi''(x) - 2x\phi'(x) + n(n + 1)\phi(x) = 0,$$

and we have shown that ϕ satisfies (8.9).

This polynomial ϕ satisfies

$$\phi(1) = 2^n n \; !.$$

This can be seen by noting that

$$\phi(x) = [(x^2 - 1)^n]^{(n)} = [(x-1)^n(x+1)^n]^{(n)}$$

$$= [(x-1)^n]^{(n)}(x+1)^n + \text{terms with } (x-1) \text{ as a factor}$$

$$= n!\,(x+1)^n + \text{terms with } (x-1) \text{ as a factor.}$$

Hence $\phi(1) = n!\,2^n$, as stated.

It is now clear that the function P_n given by

$$P_n(x) = \frac{1}{2^n n!} \frac{d^n}{dx^n} (x^2 - 1)^n \tag{8.10}$$

is the n-th Legendre polynomial, provided we can show that there is no other polynomial solution of (8.9) which is 1 at $x = 1$.

Suppose ψ is any polynomial solution of (8.9). Then for some constant c we must have $\psi = c\phi_1$ or $\psi = c\phi_2$, according as n is even or odd. Here ϕ_1, ϕ_2 are the solutions (8.7), (8.8). Suppose n is even, for example. Then, for $|x| < 1$,

$$\psi = c\phi_1 + d\phi_2$$

for some constants c, d, since ϕ_1, ϕ_2 form a basis for the solutions on $|x| < 1$. But then $\psi - c\phi_1$ is a polynomial, whereas $d\phi_2$ is *not* a polynomial in case $d \neq 0$. Hence $d = 0$. In particular the function P_n given by (8.10) satisfies $P_n = c\phi_1$ for some constant c, if n is even. Since

$$1 = P_n(1) = c\phi_1(1),$$

we see that $\phi_1(1) \neq 0$. A similar result is valid if n is odd. Thus no nontrivial polynomial solution of the Legendre equation can be zero at $x = 1$. From this it follows that there is only one polynomial P_n satisfying (8.9) and $P_n(1) = 1$, for if $\tilde{P}_n$ was another, then $P_n - \tilde{P}_n$ would be a polynomial solution, and

$$P_n(1) - \tilde{P}_n(1) = 0.$$

The first few Legendre polynomials are

$$P_0(x) = 1, \qquad P_1(x) = x, \qquad P_2(x) = \tfrac{3}{2} x^2 - \tfrac{1}{2},$$

$$P_3(x) = \tfrac{5}{2} x^3 - \tfrac{3}{2} x, \qquad P_4(x) = \tfrac{35}{8} x^4 - \tfrac{15}{4} x^2 + \tfrac{3}{8}.$$

EXERCISES

1. Show that the series defining the functions ϕ_1, ϕ_2 in (8.7), (8.8) converge for $|x| < 1$. (*Hint*: Use the ratio test.)

2. Show that $P_n(-x) = (-1)^n P_n(x)$, and hence that $P_n(-1) = (-1)^n$.

3. Show that the coefficient of x^n in $P_n(x)$ is

$$\frac{(2n)!}{2^n(n!)^2}.$$

(*Hint*: Use (8.10).)

4. Show that there are constants $\alpha_0, \alpha_1, \cdots, \alpha_n$ such that

$$x^n = \alpha_0 P_0(x) + \alpha_1 P_1(x) + \cdots + \alpha_n P_n(x).$$

(*Hint*: For $n = 0, 1 = P_0(x)$. For $n = 1, x = P_1(x)$. Use induction.)

5. Show that any polynomial of degree n is a linear combination of $P_0, P_1, \cdots, P_n$. (*Hint*: Ex. 4.)

6. Show that

$$\int_{-1}^{1} P_n(x)P_m(x)\, dx = 0, \qquad (n \neq m).$$

(*Hint*: Note that

$$[(1 - x^2)P_n']' = -n(n + 1)P_n,$$
$$[(1 - x^2)P_m']' = -m(m + 1)P_m.$$

Hence

$$P_m[(1 - x^2)P_n']' - P_n[(1 - x^2)P_m']' = \{(1 - x^2)[P_mP_n' - P_m'P_n]\}'$$
$$= [m(m + 1) - n(n + 1)]P_mP_n.$$

Integrate from -1 to 1.)

7. Show that

$$\int_{-1}^{1} P_n^2(x)\, dx = \frac{2}{2n + 1}.$$

(*Hint*: Let $u(x) = (x^2 - 1)^n$. Then from (8.10)

$$P_n(x) = \frac{1}{2^n n!}\, u^{(n)}(x).$$

Show that $u^{(k)}(1) = u^{(k)}(-1) = 0$ if $0 \leq k < n$. Then, integrating by parts,

$$\int_{-1}^{1} u^{(n)}(x)u^{(n)}(x)\, dx = u^n(x)u^{(n-1)}(x)\Big|_{-1}^{1} - \int_{-1}^{1} u^{(n+1)}(x)u^{(n-1)}(x)\, dx$$

$$= -\int_{-1}^{1} u^{(n+1)}(x)u^{(n-1)}(x)\, dx$$

$$= \cdots = (-1)^n \int_{-1}^{1} u^{(2n)}(x)u(x)\, dx.$$

$$= (2n)! \int_{-1}^{1} (1 - x^2)^n\, dx.$$

To compute the latter integral let $x = \sin\theta$, and obtain

$$\int_{-1}^{1}(1-x^2)^n\,dx = 2\int_{0}^{\pi/2}\cos^{2n+1}\theta\,d\theta = \frac{2(2^n n!)^2}{(2n+1)!}.)$$

8. Let P be any polynomial of degree n, and let

$$P = c_0 P_0 + c_1 P_1 + \cdots + c_n P_n, \qquad (*)$$

where $c_0, c_1, \cdots, c_n$ are constants. (Such constants exist by Ex. 5.) Show that

$$c_k = \frac{2k+1}{2}\int_{-1}^{1}P(x)P_k(x)\,dx, \qquad (k = 0, 1, \cdots, n).$$

(*Hint*: Multiply (*) by P_k and integrate from -1 to 1. Use the results of Exs. 6 and 7.)

9. Using the fact that $P_0(x) = 1$ is a solution of

$$(1 - x^2)y'' - 2xy' = 0,$$

find a second independent solution by the method of Sec. 5.

10. (a) Verify that the function Q_1 defined by

$$Q_1(x) = \frac{x}{2}\log\left(\frac{1+x}{1-x}\right) - 1, \qquad (|x| < 1),$$

is a solution of the Legendre equation when $\alpha = 1$.
(b) Express Q_1 as a linear combination of the solutions ϕ_1, ϕ_2 given by (8.7), (8.8) with $\alpha = 1$. (*Hint*: Compute $Q_1(0)$ and $Q_1'(0)$.)

*9. Justification of the power series method

We now consider the proof of Theorem 12. In order not to complicate matters too much we shall give a proof for the case when $n = 2$ and $x_0 = 0$. All the essential ideas appear in this case. We shall make use of two results concerning power series. The first is that if we have two power series

$$\sum_{k=0}^{\infty} c_k x^k, \qquad \sum_{k=0}^{\infty} C_k x^k,$$

and we know that

$$|c_k| \leqq C_k, \qquad C_k \geqq 0, \qquad (k = 0, 1, 2, \cdots),$$

and that the series

$$\sum_{k=0}^{\infty} C_k x^k$$

converges for $|x| < r$, for some $r > 0$, then the series

$$\sum_{k=0}^{\infty} c_k x^k$$

also converges for $|x| < r$. This is usually called the *comparison test* for convergence. The second result we require is that if a series

$$\sum_{k=0}^{\infty} \alpha_k x^k \qquad (9.1)$$

is convergent for $|x| < r_0$, then for any x, $|x| = r < r_0$, there is a constant $M > 0$ such that

$$r^k |\alpha_k| \leqq M, \qquad (k = 0, 1, 2, \cdots). \qquad (9.2)$$

This is not difficult to show. Since the series (9.1) is convergent for $|x| = r$ its terms must tend to zero,

$$|\alpha_k x^k| = |\alpha_k| r^k \to 0, \qquad (k \to \infty).$$

In particular there is an integer $N > 0$ such that

$$|\alpha_k| r^k \leqq 1, \qquad (k > N).$$

Let M be the largest number among

$$|\alpha_0|, \quad |\alpha_1| r, \quad \cdots, \quad |\alpha_N| r^N, \quad 1.$$

Then clearly (9.2) is valid for this M.

We now consider the equation

$$L(y) = y'' + a(x)y' + b(x)y = 0, \qquad (9.3)$$

where a, b are functions having expansions

$$a(x) = \sum_{k=0}^{\infty} \alpha_k x^k, \qquad b(x) = \sum_{k=0}^{\infty} \beta_k x^k, \qquad (9.4)$$

which converge for $|x| < r_0$ for some $r_0 > 0$. Given any constants a_1, a_2 we want to produce a solution ϕ of (9.3) satisfying

$$\phi(0) = a_1, \qquad \phi'(0) = a_2,$$

and which can be written in the form

$$\phi(x) = \sum_{k=0}^{\infty} c_k x^k, \tag{9.5}$$

where the series converges for $|x| < r_0$. If this series is convergent we must have

$$c_0 = a_1, \qquad c_1 = a_2,$$

and the constants $c_k (k \geq 2)$ must satisfy a recursion relation, which we now compute. We have

$$\phi'(x) = \sum_{k=0}^{\infty} (k+1) c_{k+1} x^k,$$

and

$$\phi''(x) = \sum_{k=0}^{\infty} (k+2)(k+1) c_{k+2} x^k \tag{9.6}$$

Now from (9.4) we obtain

$$a(x)\phi'(x) = \left(\sum_{k=0}^{\infty} \alpha_k x^k \right) \left(\sum_{k=0}^{\infty} (k+1) c_{k+1} x^k \right)$$

$$= \sum_{k=0}^{\infty} \left(\sum_{j=0}^{k} \alpha_{k-j}(j+1) c_{j+1} \right) x^k, \tag{9.7}$$

and

$$b(x)\phi(x) = \left(\sum_{k=0}^{\infty} \beta_k x^k \right) \left(\sum_{k=0}^{\infty} c_k x^k \right)$$

$$= \sum_{k=0}^{\infty} \left(\sum_{j=0}^{k} \beta_{k-j} c_j \right) x^k. \tag{9.8}$$

Adding (9.6), (9.7), and (9.8) we get

$$L(\phi)(x) = \sum_{k=0}^{\infty} \left[(k+2)(k+1) c_{k+2} + \sum_{j=0}^{k} \alpha_{k-j}(j+1) c_{j+1} \right.$$

$$\left. + \sum_{j=0}^{k} \beta_{k-j} c_j \right] x^k = 0.$$

Thus the c_k must satisfy

$$(k+2)(k+1) c_{k+2} = - \sum_{j=0}^{k} [\alpha_{k-j}(j+1) c_{j+1} + \beta_{k-j} c_j]. \tag{9.9}$$

$$(k = 0, 1, 2, \cdots).$$

Our job now is to show that *if* the c_k, for $k \geqq 2$, are *defined* by (9.9), then the series

$$\sum_{k=0}^{\infty} c_k x^k \qquad (9.10)$$

is convergent for $|x| < r_0$. To do this we make use of the two results concerning power series we mentioned earlier. Let r be any number satisfying $0 < r < r_0$. Since the series in (9.4) are convergent for $|x| = r$ we have a constant $M > 0$ such that

$$|\alpha_j| r^j \leqq M, \qquad |\beta_j| r^j \leqq M, \qquad (j = 0, 1, 2, \cdots).$$

Using this in (9.9) we find that

$$(k+2)(k+1)|c_{k+2}| \leqq \frac{M}{r^k} \sum_{j=0}^{k} [(j+1)|c_{j+1}| + |c_j|]r^j$$

$$\leqq \frac{M}{r^k} \sum_{j=0}^{k} [(j+1)|c_{j+1}| + |c_j|]r^j + M|c_{k+1}|r. \qquad (9.11)$$

Now let us define

$$C_0 = |c_0|, \qquad C_1 = |c_1|,$$

and C_k for $k \geqq 2$ by

$$(k+2)(k+1)C_{k+2} = \frac{M}{r^k} \sum_{j=0}^{k} [(j+1)C_{j+1} + C_j]r^j + MC_{k+1}r, \qquad (9.12)$$

$$(k = 0, 1, 2, \cdots).$$

Comparing (9.12) with (9.11) we see that an induction yields

$$|c_k| \leqq C_k, \qquad C_k \geqq 0, \qquad (k = 0, 1, 2, \cdots). \qquad (9.13)$$

We now investigate for what x the series

$$\sum_{k=0}^{\infty} C_k x^k \qquad (9.14)$$

is convergent. From (9.12) we find that

$$(k+1)kC_{k+1} = \frac{M}{r^{k-1}} \sum_{j=0}^{k-1} [(j+1)C_{j+1} + C_j]r^j + MC_k r,$$

and

$$k(k-1)C_k = \frac{M}{r^{k-2}} \sum_{j=0}^{k-2} [(j+1)C_{j+1} + C_j]r^j + MC_{k-1}r,$$

for large k. From these expressions we obtain

$$r(k+1)kC_{k+1} = \frac{M}{r^{k-2}} \sum_{j=0}^{k-2} [(j+1)C_{j+1} + C_j]r^j$$
$$+ M[kC_k + C_{k-1}]r + MC_k r^2$$
$$= k(k-1)C_k - MC_{k-1}r$$
$$+ MkC_k r + MC_{k-1}r + MC_k r^2$$
$$= [k(k-1) + Mkr + Mr^2]C_k.$$

Hence

$$\left| \frac{C_{k+1}x^{k+1}}{C_k x^k} \right| = \frac{[k(k-1) + Mkr + Mr^2]}{r(k+1)k} |x|,$$

which tends to $|x|/r$ as $k \to \infty$. Thus, by the ratio test, the series (9.14) converges for $|x| < r$. This implies that the series (9.10) converges for $|x| < r$, and since r was any number satisfying $0 < r < r_0$, we have shown at last that the series (9.10) converges for $|x| < r_0$.

This completes our justification of Theorem 12.

CHAPTER 4

Linear Equations with Regular Singular Points

1. Introduction

In this chapter we continue our investigation of linear equations with variable coefficients

$$a_0(x)y^{(n)} + a_1(x)y^{(n-1)} + \cdots + a_n(x)y = 0. \qquad (1.1)$$

We shall assume that the coefficients $a_0, a_1, \cdots, a_n$ are analytic at some point x_0, and we shall be interested in an important case when $a_0(x_0) = 0$. A point x_0 such that $a_0(x_0) = 0$ is called a *singular point* of the equation (1.1). In this case we can not apply directly the existence result Theorem 1, Chap. 3, concerning initial value problems at x_0. Indeed, it is usually rather difficult to determine the nature of the solutions in the vicinity of such singular points. However there is a large class of equations for which the singularity is rather "weak," in the sense that slight modifications of the methods used for solving equations with analytic coefficients in Chap. 3 serve to yield solutions near the singularities.

We say that x_0 is a *regular singular point* for (1.1) if the equation can be written in the form

$$(x - x_0)^n y^{(n)} + b_1(x)(x - x_0)^{n-1}y^{(n-1)} + \cdots + b_n(x)y = 0 \qquad (1.2)$$

near x_0, where the functions $b_1, \cdots, b_n$ are analytic at x_0. If the functions $b_1, \cdots, b_n$ can be written in the form

$$b_k(x) = (x - x_0)^k \beta_k(x), \qquad (k = 1, \cdots, n),$$

where $\beta_1, \cdots, \beta_n$ are analytic at x_0, we see that (1.2) becomes

$$y^{(n)} + \beta_1(x)y^{(n-1)} + \cdots + \beta_n(x)y = 0 \qquad (1.3)$$

143

upon dividing out $(x - x_0)^n$. Thus (1.2) is a generalization of the equation with analytic coefficients considered in Chap. 3, Secs. 7–9.

An equation of the form

$$c_0(x)(x - x_0)^n y^{(n)} + c_1(x)(x - x_0)^{n-1} y^{(n-1)} + \cdots + c_n(x)y = 0$$

has a regular singular point at x_0 if $c_0, c_1, \cdots, c_n$ are analytic at x_0, and $c_0(x_0) \neq 0$. This is because we may divide by $c_0(x)$, for x near x_0, to obtain an equation of the form (1.2) with $b_k(x) = c_k(x)/c_0(x)$, and it can be shown that these b_k are analytic at x_0.

We first consider the simplest case of an equation, not of the type (1.3), having a regular singular point. This is the Euler equation, which is the case of (1.2) with $b_1, \cdots, b_n$ all constants. Next we investigate the general equation of the second order with a regular singular point, and indicate how solutions may be obtained near the singular point. For $x > x_0$ such solutions ϕ turn out to be of the form

$$\phi(x) = (x - x_0)^r \sigma(x) + (x - x_0)^s \rho(x) \log (x - x_0),$$

where r, s are constants, and σ, ρ are analytic at x_0. As an example the solutions of the important Bessel equation are computed in detail. Regular singular points at infinity are briefly discussed.

The method used is to show that the coefficients of the series for the analytic functions σ, ρ can be computed in a recursive fashion, and then to indicate that the series obtained actually converge near the singular point. Fortunately many of the equations with singular points which arise in physical problems have regular singular points.

To indicate how lucky we are in this situation consider the equation

$$x^2 y'' - y' - \tfrac{3}{4} y = 0. \tag{1.4}$$

The origin $x_0 = 0$ is a singular point, but not a regular singular point since the coefficient -1 of y' is not of the form $x\, b_1(x)$, where b_1 is analytic at 0. Nevertheless we may *formally* solve this equation by a series

$$\sum_{k=0}^{\infty} c_k x^k, \tag{1.5}$$

where the coefficients c_k satisfy the recursion formula

$$(k + 1)c_{k+1} = (k^2 - k - \tfrac{3}{4})c_k, \qquad (k = 0, 1, 2, \cdots). \tag{1.6}$$

If $c_0 \neq 0$, the ratio test applied to (1.5), (1.6), shows that

$$\left| \frac{c_{k+1}x^{k+1}}{c_k x^k} \right| = \left| \frac{k^2 - k - \tfrac{3}{4}}{k + 1} \right| |x| \to \infty,$$

as $k \to \infty$, provided $|x| \neq 0$. Thus the series (1.5) will only converge for $x = 0$, and therefore does not represent a function near $x = 0$, much less a solution of (1.4).

2. The Euler equation

The simplest example of a second order equation, not of the type considered in Chap. 3, having a regular singular point at the origin is the *Euler equation*

$$L(y) = x^2 y'' + axy' + by = 0, \tag{2.1}$$

where a, b are constants. We first consider this equation for $x > 0$, and observe that the coefficient of $y^{(k)}$ in $L(y)$ is a constant times x^k. If r is any constant, x^r has the property that its k-th derivative times x^k is a constant times x^r. For example

$$x(x^r)' = rx^r, \qquad x^2(x^r)'' = r(r-1)x^r.$$

This suggests trying for a solution of $L(y) = 0$ a power of x. We find that

$$L(x^r) = [r(r-1) + ar + b]x^r.$$

If q is the polynomial defined by

$$q(r) = r(r-1) + ar + b,$$

we may write

$$L(x^r) = q(r)x^r, \tag{2.2}$$

and it is clear that if r_1 is a root of q then

$$L(x^{r_1}) = 0.$$

Thus the function ϕ_1 given by $\phi_1(x) = x^{r_1}$ is a solution of (2.1) for $x > 0$. If r_2 is the other root of q, and $r_2 \neq r_1$, we obtain another solution ϕ_2 given by $\phi_2(x) = x^{r_2}$.

In case the roots r_1, r_2 of q are equal we know that

$$q(r_1) = 0, \qquad q'(r_1) = 0,$$

and this suggests differentiating (2.2) with respect to r. Indeed

$$\frac{\partial}{\partial r} L(x^r) = L\left(\frac{\partial}{\partial r}x^r\right) = L(x^r \log x)$$

$$= [q'(r) + q(r) \log x]x^r,$$

and if $r = r_1$ we see that

$$L(x^{r_1} \log x) = 0.$$

Therefore $\phi_2(x) = x^{r_1} \log x$ is a second solution associated with the root r_1 in this case.

In either case the solutions ϕ_1, ϕ_2 are linearly independent for $x > 0$. The proof is easy. If $r_1 \neq r_2$ and c_1, c_2 are constants such that

$$c_1 x^{r_1} + c_2 x^{r_2} = 0, \qquad (x > 0),$$

then

$$c_1 + c_2 x^{r_2 - r_1} = 0, \qquad (x > 0). \tag{2.3}$$

Differentiating we see that

$$c_2 (r_2 - r_1) x^{r_2 - r_1 - 1} = 0,$$

which implies $c_2 = 0$, and from (2.3) we obtain $c_1 = 0$ also. In case $r_1 = r_2$, and c_1, c_2 are constants such that

$$c_1 x^{r_1} + c_2 x^{r_1} \log x = 0, \qquad (x > 0),$$

then

$$c_1 + c_2 \log x = 0, \qquad (x > 0), \tag{2.4}$$

and differentiating we obtain

$$\frac{c_2}{x} = 0, \qquad (x > 0),$$

or $c_2 = 0$. From (2.4) we see that $c_1 = 0$.

We have glossed over one point in the above calculations, and that is the definition of x^r in case r is *complex*. This possibility must be taken into account since the roots of q could be complex. We define x^r for r complex by

$$x^r = e^{r \log x}, \qquad (x > 0).$$

Then we have

$$(x^r)' = r(\log x)' e^{r \log x} = r\, x^{-1} x^r = r\, x^{r-1},$$

and

$$\frac{\partial}{\partial r}(x^r) = \frac{\partial}{\partial r}(e^{r \log x}) = (\log x) e^{r \log x} = x^r \log x,$$

which are the formulas we used in the calculations.

Solutions for (2.1) can be found for $x < 0$ also. In this case consider $(-x)^r$, where r is a constant. Then we have for $x < 0$

$$[(-x)^r]' = -r(-x)^{r-1}, \qquad [(-x)^r]'' = r(r-1)(-x)^{r-2},$$

and hence

$$x[(-x)^r]' = r(-x)^r, \qquad x^2[(-x)^r]'' = r(r-1)(-x)^r.$$

Thus

$$L((-x)^r) = q(r)(-x)^r, \qquad (x < 0).$$

Also

$$\frac{\partial}{\partial r}[(-x)^r] = (-x)^r \log(-x), \qquad (x < 0),$$

as can be easily checked. Therefore we see that if the roots r_1, r_2 of q are distinct two independent solutions ϕ_1, ϕ_2 of (2.1) for $x < 0$ are given by

$$\phi_1(x) = (-x)^{r_1}, \qquad \phi_2(x) = (-x)^{r_2}, \qquad (x < 0),$$

and if $r_1 = r_2$ two solutions are given by

$$\phi_1(x) = (-x)^{r_1}, \qquad \phi_2(x) = (-x)^{r_1} \log(-x), \qquad (x < 0).$$

These are just the formulas for the solutions obtained for $x > 0$, with x replaced by $-x$ everywhere. Since $|x| = x$ for $x > 0$, and $|x| = -x$ for $x < 0$, we can write the solutions for any $x \neq 0$ in the following way:

$$\phi_1(x) = |x|^{r_1}, \qquad \phi_2(x) = |x|^{r_2}, \qquad (x \neq 0),$$

in case $r_1 \neq r_2$, and

$$\phi_1(x) = |x|^{r_1}, \qquad \phi_2(x) = |x|^{r_1} \log|x|, \qquad (x \neq 0),$$

in case $r_1 = r_2$.

Theorem 1. *Consider the second order Euler equation*

$$x^2 y'' + axy' + by = 0, \qquad (a, b \text{ constants}),$$

and the polynomial q *given by*

$$q(r) = r(r - 1) + ar + b.$$

A basis for the solutions of the Euler equation on any interval not containing x $= 0$ *is given by*

$$\phi_1(x) = |x|^{r_1}, \qquad \phi_2(x) = |x|^{r_2},$$

in case r_1, r_2 *are distinct roots of* q, *and by*

$$\phi_1(x) = |x|^{r_1}, \qquad \phi_2(x) = |x|^{r_1} \log|x|,$$

if r_1 *is a root of* q *of multiplicity two.*

As an example let us consider the equation

$$x^2 y'' + xy' + y = 0$$

for $x \neq 0$. The polynomial q is given by

$$q(r) = r(r - 1) + r + 1 = r^2 + 1,$$

and its roots are $r_1 = i$, $r_2 = -i$. Thus a basis for the solutions is furnished by

$$\phi_1(x) = |x|^i, \qquad \phi_2(x) = |x|^{-i}, \qquad (x \neq 0),$$

where we have

$$|x|^i = e^{i \log |x|}.$$

Note that in this case another basis ψ_1, ψ_2 is given by

$$\psi_1(x) = \cos(\log |x|), \qquad \psi_2(x) = \sin(\log |x|), \qquad (x \neq 0).$$

The extension of the result of Theorem 1 to the Euler equation of the n-th order

$$L(y) = x^n y^{(n)} + a_1 x^{n-1} y^{(n-1)} + \cdots + a_n y = 0, \qquad (2.5)$$

where $a_1, \cdots, a_n$ are constants, is straightforward. We have for any constant r

$$x^k [\,|x|^r\,]^{(k)} = r(r-1) \cdots (r-k+1) |x|^r, \qquad (x \neq 0),$$

and hence

$$L(|x|^r) = q(r) |x|^r, \qquad (2.6)$$

where q is now the polynomial of degree n

$$q(r) = r(r-1) \quad \cdots \quad (r-n+1) + a_1 r(r-1) \quad \cdots \quad (r-n+2)$$
$$+ \cdots + a_n.$$

This polynomial is called the *indicial polynomial* for the Euler equation (2.5). Differentiating (2.6) k times with respect to r we obtain

$$\frac{\partial^k}{\partial r^k} L(|x|^r) = L\left(\frac{\partial^k}{\partial r^k} |x|^r\right) = L(|x|^r \log^k |x|)$$

$$= \left[q^{(k)}(r) + k q^{(k-1)}(r) \log |x| + \frac{k(k-1)}{2!} q^{(k-2)}(r) \log^2 |x| \right. \qquad (2.7)$$

$$\left. + \cdots + q(r) \log^k |x| \right] |x|^r.$$

If r_1 is a root of q of multiplicity m_1 then

$$q(r_1) = 0, \quad q'(r_1) = 0, \quad \cdots, \quad q^{(m_1-1)}(r_1) = 0,$$

and we see from (2.7) that

$$|x|^{r_1}, \quad |x|^{r_1} \log |x|, \quad \cdots, \quad |x|^{r_1} \log^{m_1-1} |x|$$

are solutions of $L(y) = 0$. Repeating this process for each root of q we obtain the following result.

Theorem 2. *Let* $r_1, \cdots, r_s$ *be the distinct roots of the indicial polynomial* q *for* (2.5), *and suppose* r_i *has multiplicity* m_i. *Then the* n *functions*

$$|x|^{r_1}, |x|^{r_1}\log|x|, \cdots, |x|^{r_1}\log^{m_1-1}|x|; |x|^{r_2}, |x|^{r_2}\log|x|,$$

$$\cdots, |x|^{r_2}\log^{m_2-1}|x|; \cdots; |x|^{r_s}, |x|^{r_s}\log|x|, \cdots, |x|^{r_s}\log^{m_s-1}|x|$$

form a basis for the solutions of the n-*th order Euler equation* (2.5) *on any interval not containing* $x = 0$.

A proof of the linear independence of the above solutions can be given along the lines of the proof of Theorem 12 of Chap. 2, and hence will be omitted. Note the similarity between Theorem 2 above and Theorem 11, Chap. 2.

EXERCISES

1. Find all solutions of the following equations for $x > 0$:

(a) $x^2 y'' + 2xy' - 6y = 0$ (b) $2x^2 y'' + xy' - y = 0$

(c) $x^2 y'' + xy' - 4y = x$ (d) $x^2 y'' - 5xy' + 9y = x^3$

(e) $x^3 y''' + 2x^2 y'' - xy' + y = 0$

2. Find all solutions of the following equations for $|x| > 0$:

(a) $x^2 y'' + xy' + 4y = 1$ (b) $x^2 y'' - 3xy' + 5y = 0$

(c) $x^2 y'' - (2 + i)xy' + 3iy = 0$ (d) $x^2 y'' + xy' - 4\pi y = x$

3. Let ϕ be a solution for $x > 0$ of the Euler equation

$$x^2 y'' + axy' + by = 0,$$

where a, b are constants. Let $\psi(t) = \phi(e^t)$.

(a) Show that ψ satisfies the equation

$$\psi''(t) + (a - 1)\psi'(t) + b\psi(t) = 0.$$

(b) Compute the characteristic polynomial of the equation satisfied by ψ, and compare it with the indicial polynomial of the given Euler equation.

(c) Show that $\phi(x) = \psi(\log x)$.

(d) Using (a), (b), (c), and similar facts for $x < 0$, prove Theorem 1.

4. Suppose the constants a, b in the Euler equation

$$x^2 y'' + axy' + by = 0$$

are real. Let r_1, r_2 denote the roots of the indicial polynomial q.

(a) If $r_1 = \sigma + i\tau$ with $\tau \neq 0$, show that $r_2 = \bar{r}_1 = \sigma - i\tau$.

(b) If $r_1 = \sigma + i\tau$ with $\tau \neq 0$, show that the functions ψ_1, ψ_2 given by

$$\psi_1(x) = |x|^\sigma \cos(\tau \log|x|),$$

$$\psi_2(x) = |x|^\sigma \sin(\tau \log|x|),$$

form a basis for the solutions of the Euler equation on any interval not containing $x = 0$.

5. The logarithm of a negative number can be defined in the following way. If $x < 0$, then $-x > 0$, and we have

$$x = (-x)(-1) = (-x)e^{i\pi}.$$

We define

$$\log x = \log [(-x)e^{i\pi}] = \log (-x) + \log e^{i\pi}$$

$$= \log (-x) + i\pi, \qquad (x < 0).$$

Thus $\log x$, for $x < 0$, is a complex number. Using this definition, let

$$x^r = e^{r \log x},$$

for $x > 0$ *and* for $x < 0$.

(a) Show that

$$x^r = e^{i\pi r} |x|^r, \qquad (x < 0).$$

(b) Let r_1, r_2 be the roots of the indicial polynomial for the Euler equation

$$x^2 y'' + axy' + by = 0.$$

Show that two independent solutions for $|x| > 0$ are given by

$$x^{r_1}, \qquad x^{r_2}$$

if $r_1 \neq r_2$, and by

$$x^{r_1}, \qquad x^{r_1} \log x$$

if $r_1 = r_2$.

(c) Obtain the linearly independent solutions of Theorem 1 from the linearly independent solutions of (b).

6. Let

$$L(y) = x^2 y'' + axy' + by$$

where a, b are constants, and let q be the indicial polynomial

$$q(r) = r(r - 1) + ar + b.$$

(a) Show that the equation $L(y) = x^k$ has a solution ψ of the form $\psi(x) = cx^k$ if $q(k) \neq 0$. Compute c. (*Hint:* $L(cx^k) = cL(x^k) = cq(k)x^k$.)

(b) Suppose k is a root of q of multiplicity one. Show that there is a solution ψ of $L(y) = x^k$ of the form

$$\psi(x) = cx^k \log x.$$

Compute c.

(c) Find a solution of $L(y) = x^k$ in case k is a double root of q.

(d) Do Exercises 1(c), 1(d), 2(a), 2(d), using the results of (a), (b), (c) above.

3. Second order equations with regular singular points—an example

A second order equation with a regular singular point at x_0 has the form

$$(x - x_0)^2 y'' + a(x)(x - x_0)y' + b(x)y = 0, \tag{3.1}$$

where a, b are analytic at x_0. Thus a, b have power series expansions

$$a(x) = \sum_{k=0}^{\infty} \alpha_k (x - x_0)^k, \qquad b(x) = \sum_{k=0}^{\infty} \beta_k (x - x_0)^k,$$

which are convergent on some interval $|x - x_0| < r_0$, for some $r_0 > 0$. We shall be interested in finding solutions of (3.1) near x_0. In order to simplify our notation we shall assume $x_0 = 0$.

If $x_0 \neq 0$ it is easy to change (3.1) into an equivalent equation with a regular singular point at the origin. We let $t = x - x_0$, and

$$\tilde{a}(t) = a(x_0 + t) = \sum_{k=0}^{\infty} \alpha_k t^k, \qquad \tilde{b}(t) = b(x_0 + t) = \sum_{k=0}^{\infty} \beta_k t^k.$$

The power series for $\tilde{a}$, $\tilde{b}$ converge on the interval $|t| < r_0$ about $t = 0$. Let ϕ be any solution of (3.1), and define $\tilde{\phi}$ by

$$\tilde{\phi}(t) = \tilde{\phi}(x_0 + t).$$

Then

$$\frac{d\tilde{\phi}}{dt}(t) = \frac{d\phi}{dx}(x_0 + t), \qquad \frac{d^2\tilde{\phi}}{dt^2}(t) = \frac{d^2\phi}{dx^2}(x_0 + t),$$

and we see that $\tilde{\phi}$ satisfies

$$t^2 u'' + \tilde{a}(t) t u' + \tilde{b}(t) u = 0, \tag{3.2}$$

where now $u' = du/dt$. This is an equation with a regular singular point at $t = 0$. Conversely, if $\tilde{\phi}$ satisfies (3.2) the function ϕ given by

$$\phi(x) = \tilde{\phi}(x - x_0)$$

satisfies (3.1). In this sense (3.2) is equivalent to (3.1).

With $x_0 = 0$ in (3.1) we may write (3.1) as

$$L(y) = x^2 y'' + a(x)xy' + b(x)y = 0, \tag{3.3}$$

where a, b are analytic at the origin, and have power series expansions

$$a(x) = \sum_{k=0}^{\infty} \alpha_k x^k, \qquad b(x) = \sum_{k=0}^{\infty} \beta_k x^k, \tag{3.4}$$

which are convergent on an interval $|x| < r_0$, $r_0 > 0$. The Euler equation is the special case of (3.3) with a, b constant. The effect of the higher order terms (terms with x as a factor) in the series (3.4) is to introduce series into the solutions of (3.3). We illustrate by an example.

Consider the equation

$$L(y) = x^2 y'' + \tfrac{3}{2} xy' + xy = 0, \tag{3.5}$$

which has a regular singular point at the origin. Let us restrict our attention to $x > 0$. Since it is not an Euler equation we can not expect it to have a solution of the form x^r there. However we try for a solution

$$\phi(x) = x^r \sum_{k=0}^{\infty} c_k x^k = c_0 x^r + c_1 x^{r+1} + \cdots, \qquad (c_0 \neq 0), \tag{3.6}$$

that is, x^r times a power series. This simple idea works. We operate formally and see what conditions must be satisfied by r and $c_0, c_1, c_2, \cdots$ in order that this ϕ be a solution of (3.5). Computing we find that

$$\phi'(x) = c_0 r x^{r-1} + c_1 (r+1) x^r + c_2 (r+2) x^{r+1} + \cdots,$$

$$\phi''(x) = c_0 r (r-1) x^{r-2} + c_1 (r+1) r x^{r-1} + c_2 (r+2)(r+1) x^r + \cdots,$$

and hence

$$x^2 \phi''(x) = c_0 r (r-1) x^r + c_1 (r+1) r x^{r+1} + c_2 (r+2)(r+1) x^{r+2} + \cdots,$$

$$\tfrac{3}{2} x \phi'(x) = \tfrac{3}{2} c_0 r x^r \qquad + \tfrac{3}{2} c_1 (r+1) x^{r+1} + \tfrac{3}{2} c_2 (r+2) x^{r+2} \qquad + \cdots,$$

$$x \phi(x) = \qquad\qquad c_0 x^{r+1} + \qquad\qquad c_1 x^{r+2} + \cdots.$$

Adding we obtain

$$L(\phi)(x) = [r(r-1) + \tfrac{3}{2} r] c_0 x^r + \{[(r+1)r + \tfrac{3}{2}(r+1)] c_1 + c_0\} x^{r+1}$$
$$+ \{[(r+2)(r+1) + \tfrac{3}{2}(r+2)] c_2 + c_1\} x^{r+2} + \cdots.$$

If we let

$$q(r) = r(r-1) + \tfrac{3}{2} r = r(r + \tfrac{1}{2}),$$

this may be written as

$$L(\phi)(x) = q(r) c_0 x^r + [q(r+1) c_1 + c_0] x^{r+1} + [q(r+2) c_2 + c_1] x^{r+2}$$
$$+ \cdots$$

$$= q(r) c_0 x^r + x^r \sum_{k=1}^{\infty} [q(r+k) c_k + c_{k-1}] x^k.$$

If ϕ is to satisfy $L(\phi)(x) = 0$ all coefficients of the powers of x must vanish. Since we assumed $c_0 \neq 0$ this implies

$$q(r) = 0,$$
$$q(r+k) c_k + c_{k-1} = 0, \qquad (k = 1, 2, \cdots). \tag{3.7}$$

The polynomial q is called the *indicial polynomial* for (3.5). It is the coefficient of the lowest power of x appearing in $L(\phi)(x)$, and from (3.7) we see that its roots are the only permissible values of r for which there are solutions of the form (3.6). In our example these roots are

$$r_1 = 0, \qquad r_2 = -\tfrac{1}{2}.$$

The second set of equations in (3.7) delimits $c_1, c_2, \cdots$ in terms of c_0 and r. If $q(r + k) \neq 0$ for $k = 1, 2, \cdots$, then

$$c_k = -\frac{c_{k-1}}{q(r + k)}, \qquad (k = 1, 2, \cdots).$$

Thus

$$c_k = \frac{(-1)^k c_0}{q(r + k)q(r + k - 1) \cdots q(r + 1)}, \qquad (k = 1, 2, \cdots).$$

If $r_1 = 0$,

$$q(r_1 + k) = q(k) \neq 0 \quad \text{for} \quad k = 1, 2, \cdots;$$

since the other root of q is $r_2 = -\tfrac{1}{2}$. Similarly if $r_2 = -\tfrac{1}{2}$,

$$q(r_2 + k) = q(-\tfrac{1}{2} + k) \neq 0 \quad \text{for} \quad k = 1, 2, \cdots.$$

Letting $c_0 = 1$ and $r = r_1 = 0$ we obtain, at least formally, a solution ϕ_1 given by

$$\phi_1(x) = 1 + \sum_{k=1}^{\infty} \frac{(-1)^k x^k}{q(k)q(k - 1) \cdots q(1)},$$

and letting $c_0 = 1$ and $r = r_2 = -\tfrac{1}{2}$ we obtain another solution

$$\phi_2(x) = x^{-1/2} + x^{-1/2} \sum_{k=1}^{\infty} \frac{(-1)^k x^k}{q(k - \tfrac{1}{2})q(k - \tfrac{3}{2}) \cdots q(\tfrac{1}{2})}.$$

These functions ϕ_1, ϕ_2 will be solutions provided the series converge on some interval containing $x = 0$. Let us write the series for ϕ_1 in the form

$$\phi_1(x) = \sum_{k=0}^{\infty} d_k(x).$$

Using the ratio test we obtain

$$\left| \frac{d_{k+1}(x)}{d_k(x)} \right| = \frac{|x|}{|q(k + 1)|} = \frac{|x|}{(k + 1)(k + \tfrac{3}{2})} \to 0$$

as $k \to \infty$, provided $|x| < \infty$. Thus the series defining ϕ_1 is convergent for all finite x. The same can be shown to hold for the series multiplying $x^{-1/2}$ in the expression for ϕ_2. Thus ϕ_1, ϕ_2 are solutions of (3.5) for all $x > 0$.

To obtain solutions for $x < 0$ we note that all the above computations go through if x^r is replaced everywhere by $|x|^r$, where

$$|x|^r = e^{r \log |x|}. \tag{3.8}$$

Thus two solutions of (3.5) which are valid for all $x \neq 0$ are given by

$$\phi_1(x) = 1 + \sum_{k=1}^{\infty} \frac{(-1)^k x^k}{q(k)q(k-1) \cdots q(1)},$$

and

$$\phi_2(x) = |x|^{-1/2}\left[1 + \sum_{k=1}^{\infty} \frac{(-1)^k x^k}{q(k-\frac{1}{2})q(k-\frac{3}{2}) \cdots q(\frac{1}{2})}\right].$$

Note that the definition (3.8) implies that $|x|^{1/2}$ is the *positive* square root of $|x|$. It is left as an exercise for the student to show that ϕ_1, ϕ_2 are linearly independent on any interval not containing $x = 0$.

The above example illustrates the general fact that an equation (3.3) with a regular singular point at the origin always has a solution ϕ of the form

$$\phi(x) = |x|^r \sum_{k=0}^{\infty} c_k x^k, \tag{3.9}$$

where r is a constant, and the series converges on the interval $|x| < r_0$. Moreover r, and the constants c_k, may be computed by substituting (3.9) into the differential equation.

EXERCISES

1. Find the singular points of the following equations, and determine those which are regular singular points:

(a) $x^2y'' + (x + x^2)y' - y = 0$

(b) $3x^2y'' + x^6y' + 2xy = 0$

(c) $x^2y'' - 5y' + 3x^2y = 0$

(d) $xy'' + 4y = 0$

(e) $(1 - x^2)y'' - 2xy' + 2y = 0$

(f) $(x^2 + x - 2)^2y'' + 3(x + 2)y' + (x - 1)y = 0$

(g) $x^2y'' + (\sin x)y' + (\cos x)y = 0$

2. Compute the indicial polynomials, and their roots, for the following equations:

(a) $x^2y'' + (x + x^2)y' - y = 0$

(b) $x^2y'' + xy' + (x^2 - \frac{1}{4})y = 0$

(c) $4x^2y'' + (4x^4 - 5x)y' + (x^2 + 2)y = 0$

(d) $x^2y'' + (x - 3x^2)y' + e^x y = 0$

(e) $x^2y'' + (\sin x)y' + (\cos x)y = 0$

3. (a) Show that -1 and 1 are regular singular points for the Legendre equation

$$(1 - x^2)y'' - 2xy' + \alpha(\alpha + 1)y = 0.$$

(b) Find the indicial polynomial, and its roots, corresponding to the point $x = 1$.

4. Find a solution ϕ of the form

$$\phi(x) = x^r \sum_{k=0}^{\infty} c_k x^k, \qquad (x > 0),$$

for the following equations:

(a) $2x^2 y'' + (x^2 - x)y' + y = 0$ (b) $x^2 y'' + (x - x^2)y' + y = 0$

4. Second order equations with regular singular points—the general case

Let us verify the last statement in Sec. 3 for $x > 0$. Suppose we have a solution ϕ of the form

$$\phi(x) = x^r \sum_{k=0}^{\infty} c_k x^k, \qquad (c_0 \neq 0), \tag{4.1}$$

for the equation

$$x^2 y'' + a(x)xy' + b(x)y = 0, \tag{4.2}$$

where

$$a(x) = \sum_{k=0}^{\infty} \alpha_k x^k, \qquad b(x) = \sum_{k=0}^{\infty} \beta_k x^k, \tag{4.3}$$

for $|x| < r_0$. Then

$$\phi'(x) = x^{r-1} \sum_{k=0}^{\infty} (k + r) c_k x^k,$$

$$\phi''(x) = x^{r-2} \sum_{k=0}^{\infty} (k + r)(k + r - 1) c_k x^k,$$

and hence

$$b(x)\phi(x) = x^r \left(\sum_{k=0}^{\infty} c_k x^k \right)\left(\sum_{k=0}^{\infty} \beta_k x^k \right)$$

$$= x^r \sum_{k=0}^{\infty} \tilde{\beta}_k x^k, \qquad \left(\tilde{\beta}_k = \sum_{j=0}^{k} c_j \beta_{k-j} \right)$$

$$xa(x)\phi'(x) = x^r \left(\sum_{k=0}^{\infty} (k + r) c_k x^k \right)\left(\sum_{k=0}^{\infty} \alpha_k x^k \right)$$

$$= x^r \sum_{k=0}^{\infty} \tilde{\alpha}_k x^k, \qquad \left(\tilde{\alpha}_k = \sum_{j=0}^{k} (j + r) c_j \alpha_{k-j} \right),$$

$$x^2 \phi''(x) = x^r \sum_{k=0}^{\infty} (k + r)(k + r - 1) c_k x^k.$$

Thus

$$L(\phi)(x) = x^r \sum_{k=0}^{\infty} [(k + r)(k + r - 1) c_k + \tilde{\alpha}_k + \tilde{\beta}_k] x^k,$$

and we must have

$$[\quad]_k = [(k+r)(k+r-1)c_k + \bar{\alpha}_k + \bar{\beta}_k] = 0, \qquad (k = 0, 1, 2, \cdots).$$

Using the definitions of $\bar{\alpha}_k$, $\bar{\beta}_k$ we can write the bracket $[\quad]_k$ as

$$[\quad]_k = (k+r)(k+r-1)c_k + \sum_{j=0}^{k} (j+r)c_j\alpha_{k-j} + \sum_{j=0}^{k} c_j\beta_{k-j}$$

$$= [(k+r)(k+r-1) + (k+r)\alpha_0 + \beta_0]c_k$$

$$+ \sum_{j=0}^{k-1} [(j+r)\alpha_{k-j} + \beta_{k-j}]c_j.$$

For $k = 0$ we must have

$$r(r-1) + r\alpha_0 + \beta_0 = 0, \qquad (4.4)$$

since $c_0 \neq 0$. The second degree polynomial q given by

$$q(r) = r(r-1) + r\alpha_0 + \beta_0$$

is called the *indicial polynomial* for (4.2), and the only admissible values of r are the roots of q. We see that

$$[\quad]_k = q(r+k)c_k + d_k = 0, \qquad (k = 1, 2, \cdots), \qquad (4.5)$$

where

$$d_k = \sum_{j=0}^{k-1} [(j+r)\alpha_{k-j} + \beta_{k-j}]c_j, \qquad (k = 1, 2, \cdots). \qquad (4.6)$$

Note that d_k is a linear combination of $c_0, c_1, \cdots, c_{k-1}$ with coefficients involving the known functions a, b, and r. Leaving r and c_0 indeterminate for the moment we solve the equations (4.5), (4.6) successively in terms of c_0 and r. The solutions we denote by $C_k(r)$, and the corresponding d_k by $D_k(r)$. Thus

$$D_1(r) = (r\alpha_1 + \beta_1)c_0, \qquad C_1(r) = -\frac{D_1(r)}{q(r+1)},$$

and in general

$$D_k(r) = \sum_{j=0}^{k-1} [(j+r)\alpha_{k-j} + \beta_{k-j}]C_j(r), \qquad (4.7)$$

$$C_k(r) = -\frac{D_k(r)}{q(r+k)}, \qquad (k = 1, 2, \cdots). \qquad (4.8)$$

The C_k thus determined are rational functions of r (quotients of polynomials), and the only points where they cease to exist are the points r

for which $q(r + k) = 0$ for some $k = 1, 2, \cdots$. Only two such possible points exist. Let us define Φ by

$$\Phi(x, r) = c_0 x^r + x^r \sum_{k=1}^{\infty} C_k(r) x^k. \tag{4.9}$$

If the series in (4.9) converges for $0 < x < r_0$, then clearly

$$L(\Phi)(x, r) = c_0 q(r) x^r. \tag{4.10}$$

We have now arrived at the following situation. If the ϕ given by (4.1) is a solution of (4.2) then r must be a root of the indicial polynomial q, and the c_k ($k \geq 1$) are determined uniquely in terms of r and c_0 to be the $C_k(r)$ of (4.8), provided $q(r + k) \neq 0$, $k = 1, 2, \cdots$. Conversely, if r is a root of q, and if the $C_k(r)$ can be determined (that is, $q(r + k) \neq 0$ for $k = 1, 2, \cdots$) then the function ϕ given by $\phi(x) = \Phi(x, r)$ is a solution of (4.2) for any choice of c_0, provided the series in (4.9) can be shown to be convergent.

Let r_1, r_2 be the two roots of q, and suppose we have labeled them so that $\operatorname{Re} r_1 \geq \operatorname{Re} r_2$. Then $q(r_1 + k) \neq 0$ for any $k = 1, 2, \cdots$. Thus $C_k(r_1)$ exists for all $k = 1, 2, \cdots$, and letting $c_0 = C_0(r_1) = 1$ we see that the function ϕ_1 given by

$$\phi_1(x) = x^{r_1} \sum_{k=0}^{\infty} C_k(r_1) x^k, \qquad (C_0(r_1) = 1), \tag{4.11}$$

is a solution of (4.2), provided the series is convergent. This will be proved in Sec. 5.

If r_2 is a root of q distinct from r_1, and $q(r_2 + k) \neq 0$ for $k = 1, 2, \cdots$, then clearly $C_k(r_2)$ is defined for $k = 1, 2, \cdots$, and the function ϕ_2 given by

$$\phi_2(x) = x^{r_2} \sum_{k=0}^{\infty} C_k(r_2) x^k, \qquad (C_0(r_2) = 1), \tag{4.12}$$

is another solution of (4.2), provided the series is convergent. The condition

$$q(r_2 + k) \neq 0 \quad \text{for} \quad k = 1, 2, \cdots$$

is the same as

$$r_1 \neq r_2 + k \quad \text{for} \quad k = 1, 2, \cdots,$$

or $r_1 - r_2$ is not a *positive integer*.

Notice that since $\alpha_0 = a(0)$, $\beta_0 = b(0)$, the indicial polynomial q can be written as

$$q(r) = r(r - 1) + a(0)r + b(0).$$

Theorem 3. *Consider the equation*

$$x^2 y'' + a(x) x y' + b(x) y = 0,$$

where a, b *have convergent power series expansions for*

$$|x| < r_0, \qquad r_0 > 0.$$

Let r_1, r_2 *(Re* $r_1 \geqq$ *Re* r_2) *be the roots of the indicial polynomial*

$$q(r) = r(r-1) + a(0)r + b(0).$$

For $0 < |x| < r_0$ *there is a solution* ϕ_1 *of the form*

$$\phi_1(x) = |x|^{r_1} \sum_{k=0}^{\infty} c_k x^k, \qquad (c_0 = 1),$$

where the series converges for $|x| < r_0$. *If* $r_1 - r_2$ *is not zero or a positive integer, there is a second solution* ϕ_2 *for* $0 < |x| < r_0$ *of the form*

$$\phi_2(x) = |x|^{r_2} \sum_{k=0}^{\infty} \tilde{c}_k x^k, \qquad (\tilde{c}_0 = 1),$$

where the series converges for $|x| < r_0$.

The coefficients c_k, $\tilde{c}_k$ *can be obtained by substitution of the solutions into the differential equation.*

As we have seen in (4.11), (4.12), the coefficients c_k, $\tilde{c}_k$ appearing in the solutions ϕ_1, ϕ_2 of Theorem 3 are given by

$$c_k = C_k(r_1), \qquad \tilde{c}_k = C_k(r_2), \qquad (k = 0, 1, 2, \cdots),$$

where the $C_k(r)$, $(k = 1, 2, \cdots)$, are the solutions of the equations (4.7), (4.8), with $C_0(r) = 1$.

It is easy to check, as in the case of the Euler equation, that the calculations made for $x > 0$ remain valid for $x < 0$ provided x^r is replaced everywhere by $|x|^r$. Thus all that remains to be proved in Theorem 3 is the convergence of the series involved in ϕ_1 and ϕ_2. This will be done in Sec. 5.

If $r_1 - r_2$ is either zero or a positive integer we shall say that we have an *exceptional case*. The Euler equation shows that if $r_1 = r_2$ we must expect solutions involving log x. It turns out that even in the case when $r_1 - r_2$ is a positive integer log x may appear. In Sec. 6 we show how to obtain a solution associated with r_2 in the exceptional cases.

EXERCISES

1. Find all solutions ϕ of the form

$$\phi(x) = |x|^r \sum_{k=0}^{\infty} c_k x^k, \qquad (|x| > 0),$$

for the following equations:

(a) $3x^2 y'' + 5xy' + 3xy = 0$ (b) $x^2 y'' + xy' + x^2 y = 0$

(c) $x^2 y'' + xy' + (x^2 - \frac{1}{4})y = 0$

Test each of the series involved for convergence.

2. Consider the equation

$$x^2 y'' + xe^x y' + y = 0.$$

(a) Compute the indicial polynomial, and show that its roots are $-i$ and i.

(b) Compute the coefficients c_1, c_2, c_3 in the solution

$$\phi(x) = x^i \sum_{k=0}^{\infty} c_k x^k, \qquad (c_0 = 1).$$

3. (a) Find a solution ϕ of the form

$$\phi(x) = |x - 1|^r \sum_{k=0}^{\infty} c_k (x - 1)^k$$

for the Legendre equation

$$(1 - x^2) y'' - 2xy' + \alpha(\alpha + 1)y = 0.$$

For what values of x does the series converge? (*Hint*: Do not divide by $x + 1$ and multiply by $x - 1$, but note that $x = (x - 1) + 1$. Express the coefficients in terms of powers of $x - 1$.)

(b) Show that there is a polynomial solution if α is a non-negative integer.

4. The equation

$$xy'' + (1 - x)y' + \alpha y = 0,$$

where α is a constant, is called the *Laguerre equation*.

(a) Show that this equation has a regular singular point at $x = 0$.

(b) Compute the indicial polynomial and its roots.

(c) Find a solution ϕ of the form

$$\phi(x) = x^r \sum_{k=0}^{\infty} c_k x^k.$$

(d) Show that if $\alpha = n$, a non-negative integer, there is a polynomial solution of degree n.

5. (a) Let L_n denote the polynomial

$$L_n(x) = e^x \frac{d^n}{dx^n} (x^n e^{-x}).$$

Show that L_n satisfies the Laguerre equation if $\alpha = n$. This polynomial is called the n-th *Laguerre polynomial*. (*Hint*: See the treatment of the Legendre polynomials on p. 135.)

(b) Compute L_0, L_1, L_2.

*5. A convergence proof

The proof that the series involved in Theorem 3 converge for $|x| < r_0$ is similar to the proof of Theorem 12, Chap. 3 (Sec. 9, Chap. 3). Under consideration is the equation

$$x^2 y'' + a(x) xy' + b(x) y = 0,$$

with

$$a(x) = \sum_{k=0}^{\infty} \alpha_k x^k, \qquad b(x) = \sum_{k=0}^{\infty} \beta_k x^k, \tag{5.1}$$

where these series converge for $|x| < r_0$ for some $r_0 > 0$. The indicial polynomial q is given by

$$q(r) = r(r-1) + \alpha_0 r + \beta_0, \tag{5.2}$$

and its two roots are r_1, r_2 with $\operatorname{Re} r_1 \geq \operatorname{Re} r_2$.

The series we must show to be convergent are determined from

$$\sum_{k=0}^{\infty} C_k(r) x^k, \tag{5.3}$$

where the $C_k(r)$ are given recursively by

$$C_0(r) = 1,$$

$$q(r+k) C_k(r) = - \sum_{j=0}^{k-1} [(j+r)\alpha_{k-j} + \beta_{k-j}] C_j(r), \tag{5.4}$$

$$(k = 1, 2, \cdots);$$

see (4.7), (4.8). We must prove that the series (5.3) converges for $|x| < r_0$ if $r = r_1$, and if $r = r_2$, provided $r_1 - r_2$ is not a positive integer.

We note that

$$q(r) = (r - r_1)(r - r_2),$$

and hence that

$$q(r_1 + k) = k(k + r_1 - r_2),$$

$$q(r_2 + k) = k(k + r_2 - r_1).$$

Therefore

$$|q(r_1 + k)| \geqq k(k - |r_1 - r_2|),$$
$$|q(r_2 + k)| \geqq k(k - |r_2 - r_1|). \tag{5.5}$$

Now let ρ be any number satisfying $0 < \rho < r_0$. Since the series in (5.1) are convergent for $|x| = \rho$ there is a constant $M > 0$ such that

$$|\alpha_j|\rho^j \leqq M, \qquad |\beta_j|\rho^j \leqq M, \qquad (j = 0, 1, 2, \cdots). \tag{5.6}$$

Using (5.5) and (5.6) in (5.4) we obtain

$$k(k - |r_1 - r_2|)|C_k(r_1)| \leqq M \sum_{j=0}^{k-1} (j + 1 + |r_1|)\rho^{j-k}|C_j(r_1)|, \tag{5.7}$$
$$(k = 1, 2, \cdots).$$

Let N be that integer satisfying

$$N - 1 \leqq |r_1 - r_2| < N,$$

and let us define $\gamma_0, \gamma_1, \cdots$ by

$$\gamma_0 = C_0(r_1) = 1, \qquad \gamma_k = |C_k(r_1)|, \qquad (k = 1, 2, \cdots, N - 1),$$

and

$$k(k - |r_1 - r_2|)\gamma_k = M \sum_{j=0}^{k-1} (j + 1 + |r_1|)\rho^{j-k}\gamma_j, \tag{5.8}$$
$$(k = N, N + 1, \cdots).$$

Then comparing the definition of the γ_k with (5.7) we see that

$$|C_k(r_1)| \leqq \gamma_k, \qquad (k = 0, 1, 2, \cdots). \tag{5.9}$$

We show that the series

$$\sum_{k=0}^{\infty} \gamma_k x^k \tag{5.10}$$

is convergent for $|x| < \rho$. Replacing k by $k + 1$ in (5.8) we obtain

$$\rho(k + 1)(k + 1 - |r_1 - r_2|)\gamma_{k+1} = [k(k - |r_1 - r_2|)$$
$$+ M(k + 1 + |r_1|)]\gamma_k$$

for $k \geqq N$. Thus

$$\left|\frac{\gamma_{k+1}x^{k+1}}{\gamma_k x^k}\right| = \frac{[k(k - |r_1 - r_2|) + M(k + 1 + |r_1|)]}{\rho(k + 1)(k + 1 - |r_1 - r_2|)}|x|,$$

which tends to $|x|/\rho$ as $k \to \infty$. Thus, by the ratio test, the series (5.10) converges for $|x| < \rho$.

Using (5.9) and the comparison test we see that the series

$$\sum_{k=0}^{\infty} C_k(r_1)x^k, \qquad (C_0(r_1) = 1),$$

converges for $|x| < \rho$. But since ρ is any number satisfying $0 < \rho < r_0$ we have shown that this series converges for $|x| < r_0$.

The same computations with r_1 replaced by r_2 everywhere show that

$$\sum_{k=0}^{\infty} C_k(r_2)x^k, \qquad (C_0(r_2) = 1),$$

converges for $|x| < r_0$, provided $r_1 - r_2$ is not a positive integer.

6. The exceptional cases

We divide the exceptional cases into two groups according as the roots r_1, r_2 (Re $r_1 \geqq$ Re r_2) of the indicial polynomial satisfy

     (i) $r_1 = r_2$,

     (ii) $r_1 - r_2$ is a positive integer.

We try to find solutions for $0 < x < r_0$. We are going to work in a purely formal way in order to discover the form that the solutions should take. For such x we have from (4.9), (4.10)

$$L(\Phi)(x, r) = c_0 q(r)x^r, \tag{6.1}$$

where Φ is given by

$$\Phi(x, r) = c_0 x^r + x^r \sum_{k=1}^{\infty} C_k(r)x^k. \tag{6.2}$$

The $C_k(r)$ are determined recursively by the formulas

$$C_0(r) = c_0 \neq 0,$$
$$q(r + k)C_k(r) = -D_k(r), \tag{6.3}$$
$$D_k(r) = \sum_{j=0}^{k-1} [(j + r)\alpha_{k-j} + \beta_{k-j}]C_j(r), \qquad (k = 1, 2, \cdots);$$

see (4.7), (4.8).

In case (i) we have

$$q(r_1) = 0, \qquad q'(r_1) = 0,$$

and this suggests formally differentiating (6.1) with respect to r. We obtain

$$\frac{\partial}{\partial r} L(\Phi)(x, r) = L\left(\frac{\partial \Phi}{\partial r}\right)(x, r)$$
$$= c_0[q'(r) + (\log x)q(r)]x^r,$$

and we see that if $r = r_1 = r_2$, $c_0 = 1$, then

$$\phi_2(x) = \frac{\partial \Phi}{\partial r}(x, r_1)$$

will yield a solution of our equation, provided the series involved **converge**. Computing formally from (6.2) we find

$$\phi_2(x) = x^{r_1} \sum_{k=0}^{\infty} C_k'(r_1) x^k + (\log x) x^{r_1} \sum_{k=0}^{\infty} C_k(r_1) x^k$$

$$= x^{r_1} \sum_{k=0}^{\infty} C_k'(r_1) x^k + (\log x) \phi_1(x),$$

where ϕ_1 is the solution already obtained:

$$\phi_1(x) = x^{r_1} \sum_{k=0}^{\infty} C_k(r_1) x^k, \qquad (C_0(r_1) = 1).$$

Note that $C_k'(r_1)$ exists for all $k = 0, 1, 2, \cdots$, since C_k is a rational function of r whose denominator is not zero at $r = r_1$. Also $C_0(r) = 1$ implies that $C_0'(r_1) = 0$, and thus the series multiplying x^{r_1} in ϕ_2 starts with the first power of x.

Let us now turn to the case (ii), and suppose that $r_1 = r_2 + m$, where m is a positive integer. If c_0 is given,

$$C_1(r_2), \cdots, C_{m-1}(r_2)$$

all exist as finite numbers, but since

$$q(r + m) C_m(r) = - D_m(r), \tag{6.4}$$

we run into trouble in trying to compute $C_m(r_2)$. Now

$$q(r) = (r - r_1)(r - r_2),$$

and hence

$$q(r + m) = (r - r_2)(r + m - r_2).$$

If $D_m(r)$ also has $r - r_2$ as a factor (i.e., $D_m(r_2) = 0$) this would cancel the same factor in $q(r + m)$, and (6.4) would give $C_m(r_2)$ as a finite number. Then

$$C_{m+1}(r_2), \quad C_{m+2}(r_2), \quad \cdots$$

all exist. In this rather special situation we will have a solution ϕ_2 of **the** form

$$\phi_2(x) = x^{r_2} \sum_{k=0}^{\infty} C_k(r_2) x^k, \qquad (C_0(r_2) = 1).$$

We can always arrange it so that $D_m(r_2) = 0$ by choosing

$$C_0(r) = r - r_2.$$

From (6.3) we see that $D_k(r)$ is linear homogeneous in

$$C_0(r), \cdots, C_{k-1}(r),$$

and hence $D_k(r)$ has $C_0(r) = r - r_2$ as a factor. Thus $C_m(r_2)$ will exist as a finite number. Letting

$$\Psi(x, r) = x^r \sum_{k=0}^{\infty} C_k(r)x^k, \qquad (C_0(r) = r - r_2), \qquad (6.5)$$

we find formally that

$$L(\Psi)(x, r) = (r - r_2)q(r)x^r. \qquad (6.6)$$

Putting $r = r_2$ we obtain formally a solution ψ given by

$$\psi(x) = \Psi(x, r_2).$$

However $C_0(r_2) = C_1(r_2) = \cdots = C_{m-1}(r_2) = 0$. Thus the series for ψ actually starts with the m-th power of x, and hence ψ has the form

$$\psi(x) = x^{r_2+m}\sigma(x) = x^{r_1}\sigma(x),$$

where σ is some power series. It is not difficult to see that ψ is just a constant multiple of the solution ϕ_1 already obtained.

To get a solution really associated with r_2 we differentiate (6.6) with respect to r, obtaining

$$\frac{\partial}{\partial r} L(\Psi)(x, r) = L\left(\frac{\partial \Psi}{\partial r}\right)(x, r)$$

$$= q(r)x^r + (r - r_2)[q'(r) + (\log x)q(r)]x^r.$$

Now letting $r = r_2$ we find that the ϕ_2 given by

$$\phi_2(x) = \frac{\partial \Psi}{\partial r}(x, r_2)$$

is a solution, provided the series involved are convergent. It has the form

$$\phi_2(x) = x^{r_2} \sum_{k=0}^{\infty} C_k'(r_2)x^k + (\log x)x^{r_2} \sum_{k=0}^{\infty} C_k(r_2)x^k,$$

where $C_0(r) = r - r_2$. Since

$$C_0(r_2) = \cdots = C_{m-1}(r_2) = 0,$$

we may write this as

$$\phi_2(x) = x^{r_2} \sum_{k=0}^{\infty} C_k'(r_2)x^k + c (\log x) \phi_1(x),$$

where $c = C_m(r_2)$.

The method used in this section to obtain solutions is called the *Frobenius method.* All the series obtained converge for $|x| < r_0$, and the ϕ_2 computed formally will be a solution in both the cases (i) and (ii). This requires justifying the differentiating of the various series term by term with respect to r, and this can be done.

Another approach which leads to a justification of the results is the following. Once we have discovered what form a second solution ϕ_2 should take, we can substitute this back into the equation and compute the coefficients of the various series involved. Then a proof of the convergence of these series can be patterned after the convergence proof in Sec. 5. We omit this proof.

Solutions for $x < 0$ can be obtained by replacing

$$x^{r_1}, \quad x^{r_2}, \quad \log x$$

everywhere by

$$|x|^{r_1}, \quad |x|^{r_2}, \quad \log|x|$$

respectively. We summarize our results in the following theorem.

Theorem 4. *Consider the equation*

$$x^2 y'' + a(x)xy' + b(x)y = 0,$$

where a, b *have power series expansions which are convergent for* $|x| < r_0$, $r_0 > 0$. *Let* r_1, r_2 *(Re* $r_1 \geqq$ *Re* r_2*) be the roots of the indicial polynomial*

$$q(r) = r(r-1) + a(0)r + b(0).$$

If $r_1 = r_2$ *there are two linearly independent solutions* ϕ_1, ϕ_2 *for* $0 < |x| < r_0$ *of the form*

$$\phi_1(x) = |x|^{r_1}\sigma_1(x), \qquad \phi_2(x) = |x|^{r_1+1}\sigma_2(x) + (\log|x|)\phi_1(x),$$

where σ_1, σ_2 *have power series expansions which are convergent for* $|x| < r_0$, *and* $\sigma_1(0) \neq 0$.

If $r_1 - r_2$ *is a positive integer there are two linearly independent solutions* ϕ_1, ϕ_2 *for* $0 < |x| < r_0$ *of the form*

$$\phi_1(x) = |x|^{r_1}\sigma_1(x),$$

$$\phi_2(x) = |x|^{r_2}\sigma_2(x) + c(\log|x|)\phi_1(x),$$

where σ_1, σ_2 *have power series expansions which are convergent for*

$$|x| < r_0, \quad \sigma_1(0) \neq 0, \quad \sigma_2(0) \neq 0,$$

and c *is a constant. It may happen that* c $= 0$.

The proof of the linear independence of ϕ_1, ϕ_2 will be left as an exercise for the student.

In our reasoning before the statement of Theorem 4 we have shown how the coefficients in the power series σ_1, σ_2 may be computed in each of the exceptional cases. In trying to solve a particular equation, an alternate procedure is to determine the appropriate form of the solutions (by analysing the roots of the indicial equation), and then to substitute these back into the equation to determine the required constants. We illustrate this method with an important equation in the next two sections.

EXERCISES

1. Consider the following three equations near $x = 0$:

    (i) $2x^2y'' + (5x + x^2)y' + (x^2 - 2)y = 0$

    (ii) $4x^2y'' - 4xe^xy' + 3(\cos x)y = 0$

    (iii) $(1 - x^2)x^2y'' + 3(x + x^2)y' + y = 0$

(a) Compute the roots r_1, r_2 of the indicial equation for each relative to $x = 0$.

(b) Describe (do not compute) the nature of two linearly independent solutions of each equation near $x = 0$. Using the notation of Theorem 4, determine the first non-zero coefficient in $\sigma_2(x)$ if $r_1 = r_2$, and determine whether $c = 0$ in case $r_1 - r_2$ is a positive integer.

2. Consider the equation

$$x^2y'' + xy' + (x^2 - \alpha^2)y = 0,$$

where α is a non-negative constant.

(a) Compute the indicial polynomial and its two roots.

(b) Discuss the nature of the solutions near the origin. Consider all cases carefully. Do not compute the solutions.

3. Obtain two linearly independent solutions of the following equations which are valid near $x = 0$:

    (a) $x^2y'' + 3xy' + (1 + x)y = 0$

    (b) $x^2y'' + 2x^2y' - 2y = 0$

    (c) $x^2y'' + 5xy' + (3 - x^3)y = 0$

    (d) $x^2y'' - 2x(x + 1)y' + 2(x + 1)y = 0$

    (e) $x^2y'' + xy' + (x^2 - 1)y = 0$

    (f) $x^2y'' - 2x^2y' + (4x - 2)y = 0$

4. Show that the solutions ϕ_1, ϕ_2 in Theorem 4 are linearly independent for $0 < x < r_0$.

5. Show that $\Psi(x, r_2)$, where Ψ is given by (6.5), is a constant times $\phi_1(x)$, where ϕ_1 is given by

$$\phi_1(x) = x^{r_1} \sum_{k=0}^{\infty} C_k(r_1)x^k, \qquad (C_0(r_1) = 1).$$

6. Consider the equation

$$xy' + a(x)y = 0,$$

where

$$a(x) = \sum_{k=0}^{\infty} \alpha_k x^k,$$

and the series converges for $|x| < r_0$, $r_0 > 0$.

(a) Show formally that there is a solution ϕ of the form

$$\phi(x) = x^r \sum_{k=0}^{\infty} c_k x^k, \qquad (c_0 = 1),$$

where $r + \alpha_0 = 0$, and $x > 0$.

(b) Prove that the series obtained converges for $|x| < r_0$. (*Hint*: Use the method of Sec. 5.)

7. Consider the equation

$$x^2 y'' + a(x)xy' + b(x)y = 0, \tag{*}$$

where a, b have power series expansions which are convergent for $|x| < r_0$, $r_0 > 0$. Let r_1, r_2 be the roots of the indicial polynomial, $\mathrm{Re}\ r_1 \geqq \mathrm{Re}\ r_2$. Let ϕ_1 be a solution for $x > 0$ corresponding to r_1:

$$\phi_1(x) = x^{r_1}\sigma_1(x), \qquad (\sigma_1(0) = 1),$$

where σ_1 has a power series expansion valid for $|x| < r_0$.

(a) Let ϕ be any other solution of (*), and suppose $\phi = u\phi_1$. Show that $v = u'$ satisfies the equation

$$xv' + \left[2r_1 + a(x) + \frac{2x\sigma_1'(x)}{\sigma_1(x)}\right]v = 0. \tag{**}$$

(b) Since σ_1'/σ_1 has a power series expansion on some interval $|x| < \bar{r}_0$, $\bar{r}_0 > 0$, show that the v satisfying (**) has the form

$$v(x) = x^{-2r_1-\alpha_0}\sum_{k=0}^{\infty} d_k x^k,$$

where the power series converges for $|x| < \rho_0$, where ρ_0 is the smaller of the two numbers, r_0, $\bar{r}_0$. (*Hint*: Ex. 6.)

(c) Using the results of (a), (b) show that a second solution ϕ_2 of (*) exists of the form

$$\phi_2(x) = c\ (\log x)\phi_1(x) + x^{r_2}\sigma_2(x), \qquad (x > 0),$$

where c is a constant, and σ_2 has a power series expansion which converges for $|x| < \rho_0$.

7. The Bessel equation

If α is a constant, Re $\alpha \geqq 0$, the *Bessel equation of order* α is the equation

$$x^2 y'' + xy' + (x^2 - \alpha^2)y = 0.$$

This has the form

$$x^2 y'' + xa(x)y' + b(x)y = 0,$$

with

$$a(x) = 1, \qquad b(x) = x^2 - \alpha^2.$$

Since a, b are analytic at $x = 0$, the Bessel equation has the origin as a regular singular point. The indicial polynomial q is given by

$$q(r) = r(r - 1) + r - \alpha^2 = r^2 - \alpha^2,$$

whose two roots r_1, r_2 are

$$r_1 = \alpha, \qquad r_2 = -\alpha.$$

We shall construct solutions for $x > 0$.

Let us consider the case $\alpha = 0$ first. Since the roots are both equal to zero in this case it follows from Theorem 4 that there are two solutions ϕ_1, ϕ_2 of the form

$$\phi_1(x) = \sigma_1(x), \qquad \phi_2(x) = x\sigma_2(x) + (\log x)\phi_1(x),$$

where σ_1, σ_2 have power series expansions which converge for all finite x. Let us compute σ_1, σ_2. Let for the moment

$$L(y) = x^2 y'' + x y' + x^2 y,$$

and suppose

$$\sigma_1(x) = \sum_{k=0}^{\infty} c_k x^k, \qquad (c_0 \neq 0).$$

We find

$$\sigma_1'(x) = \sum_{k=1}^{\infty} k c_k x^{k-1},$$

$$\sigma_1''(x) = \sum_{k=2}^{\infty} k(k - 1) c_k x^{k-2},$$

and hence

$$x^2 \sigma_1''(x) = \sum_{k=2}^{\infty} k(k - 1) c_k x^k,$$

$$x\sigma_1'(x) = \sum_{k=1}^{\infty} k c_k x^k = c_1 x + \sum_{k=2}^{\infty} k c_k x^k,$$

$$x^2 \sigma_1(x) = \sum_{k=0}^{\infty} c_k x^{k+2} = \sum_{k=2}^{\infty} c_{k-2} x^k.$$

Thus

$$L(\sigma_1)(x) = c_1 x + \sum_{k=2}^{\infty} \{[k(k-1) + k]c_k + c_{k-2}\}x^k = 0.$$

We see that

$$c_1 = 0,$$

$$[k(k-1) + k]c_k + c_{k-2} = 0, \qquad (k = 2, 3, \cdots).$$

The second set of equations is the same as

$$c_k = -\frac{c_{k-2}}{k^2}, \qquad (k = 2, 3, \cdots).$$

The choice $c_0 = 1$ implies

$$c_2 = -\frac{1}{2^2}, \qquad c_4 = -\frac{c_2}{4^2} = \frac{1}{2^2 \cdot 4^2}, \cdots,$$

and in general

$$c_{2m} = \frac{(-1)^m}{2^2 \cdot 4^2 \cdots (2m)^2} = \frac{(-1)^m}{2^{2m}(m!)^2}, \qquad (m = 1, 2, \cdots).$$

Since $c_1 = 0$ we have

$$c_3 = c_5 = \cdots = 0.$$

Thus σ_1 contains only even powers of x, and we obtain

$$\sigma_1(x) = \sum_{m=0}^{\infty} \frac{(-1)^m x^{2m}}{2^{2m}(m!)^2},$$

where as usual $0! = 1$, and $2^0 = 1$. The function defined by this series is called the *Bessel function of zero order of the first kind* and is denoted by J_0. Thus

$$J_0(x) = \sum_{m=0}^{\infty} \frac{(-1)^m}{(m!)^2}\left(\frac{x}{2}\right)^{2m}$$

It is easily checked by the ratio test that this series indeed converges for all finite x.

We now determine a second solution ϕ_2 for the Bessel equation of order zero. Letting $\phi_1 = J_0$ this solution has the form

$$\phi_2(x) = \sum_{k=0}^{\infty} c_k x^k + (\log x)\phi_1(x), \qquad (c_0 = 0).$$

We obtain

$$\phi_2'(x) = \sum_{k=1}^{\infty} kc_k x^{k-1} + \frac{\phi_1(x)}{x} + (\log x)\phi_1'(x),$$

$$\phi_2''(x) = \sum_{k=2}^{\infty} k(k-1)c_k x^{k-2} - \frac{\phi_1(x)}{x^2} + \frac{2}{x}\phi_1'(x) + (\log x)\phi_1''(x).$$

Thus

$$L(\phi_2)(x) = x^2\phi_2''(x) + x\phi_2'(x) + x^2\phi_2(x)$$

$$= c_1 x + 2^2 c_2 x^2 + \sum_{k=3}^{\infty} (k^2 c_k + c_{k-2})x^k$$

$$+ 2x\phi_1'(x) + (\log x)L(\phi_1)(x),$$

and since $L(\phi_1)(x) = 0$ we have

$$c_1 x + 2^2 c_2 x^2 + \sum_{k=3}^{\infty} (k^2 c_k + c_{k-2})x^k = -2\sum_{m=1}^{\infty} \frac{(-1)^m 2m x^{2m}}{2^{2m}(m!)^2}.$$

Hence

$$c_1 = 0, \quad 2^2 c_2 = 1, \quad 3^2 c_3 + c_1 = 0, \quad \cdots,$$

and we see that since the series on the right has only even powers of x,

$$c_1 = c_3 = c_5 = \cdots = 0.$$

The recursion relation for the other coefficients is

$$(2m)^2 c_{2m} + c_{2m-2} = \frac{(-1)^{m+1}m}{2^{2m-2}(m!)^2}, \quad (m = 2, 3, \cdots).$$

We have

$$c_2 = \frac{1}{2^2}, \quad c_4 = \frac{1}{4^2}\left(-\frac{1}{2^2} - \frac{1}{2\cdot 2^2}\right) = -\frac{1}{2^2 4^2}\left(1 + \frac{1}{2}\right),$$

$$c_6 = \frac{1}{6^2}\left[\frac{1}{2^2\cdot 4^2}\left(1 + \frac{1}{2}\right) + \frac{1}{2^2\cdot 4^2}\left(\frac{1}{3}\right)\right] = \frac{1}{2^2 4^2 6^2}\left(1 + \frac{1}{2} + \frac{1}{3}\right), \quad \cdots,$$

and it can be shown by induction that

$$c_{2m} = \frac{(-1)^{m-1}}{2^{2m}(m!)^2}\left(1 + \frac{1}{2} + \cdots + \frac{1}{m}\right), \quad (m = 1, 2, \cdots).$$

The solution thus determined is called a *Bessel function of zero order of the second kind*, and is denoted by K_0. Hence

$$K_0(x) = -\sum_{m=1}^{\infty} \frac{(-1)^m}{(m!)^2}\left(1 + \frac{1}{2} + \cdots + \frac{1}{m}\right)\left(\frac{x}{2}\right)^{2m} + (\log x)J_0(x).$$

Using the ratio test it is easy to check that the series on the right is convergent for all finite x.

EXERCISES

1. Prove that the series defining J_0 and K_0 converge for $|x| < \infty$.

2. Suppose ϕ is any solution of

$$x^2 y'' + xy' + x^2 y = 0$$

for $x > 0$, and let $\psi(x) = x^{1/2}\phi(x)$. Show that ψ satisfies the equation

$$x^2 y'' + (x^2 + \tfrac{1}{4})y = 0$$

for $x > 0$.

3. Show that J_0 has an infinity of positive zeros. (*Hint*: If $\psi_0(x) = x^{1/2}J_0(x)$, then ψ_0 satisfies

$$y'' + \left[1 + \frac{1}{4x^2}\right]y = 0, \qquad (x > 0).$$

The function χ given by $\chi(x) = \sin x$ satisfies $y'' + y = 0$. Apply Ex. 4 of Sec. 4, Chap. 3, to show that there is a zero of J_0 between any two positive zeros of χ.)

4. (a) If $\lambda > 0$ and $\phi_\lambda(x) = x^{1/2}J_0(\lambda x)$, show that

$$\phi_\lambda'' + \frac{1}{4x^2}\phi_\lambda = -\lambda^2\phi_\lambda. \qquad (*)$$

(*Hint*: $\lambda^{1/2}\phi_\lambda(x) = \psi_0(\lambda x)$, where ψ_0 is defined in Ex. 3.)

(b) If λ, μ are positive constants, show that

$$(\lambda^2 - \mu^2)\int_0^1 \phi_\lambda(x)\phi_\mu(x)\, dx = \phi_\lambda(1)\phi_\mu'(1) - \phi_\mu(1)\phi_\lambda'(1).$$

(*Hint*: Multiply (*) by ϕ_μ, and multiply

$$\phi_\mu'' + \frac{1}{4x^2}\phi_\mu = -\mu^2\phi_\mu$$

by ϕ_λ, and subtract to obtain

$$(\phi_\lambda\phi_\mu' - \phi_\mu\phi_\lambda')' = (\lambda^2 - \mu^2)\phi_\lambda\phi_\mu. \qquad (**)$$

Integrate from 0 to 1.)

(c) If $\lambda \neq \mu$ and $J_0(\lambda) = 0$, $J_0(\mu) = 0$, show that

$$\int_0^1 \phi_\lambda(x)\phi_\mu(x)\, dx = \int_0^1 x J_0(\lambda x)J_0(\mu x)\, dx = 0.$$

5. Using the notation of Ex. 4 show that if $J_0(\lambda) = 0$ then

$$\int_0^1 \phi_\lambda^2(x)\, dx = \int_0^1 x J_0^2(\lambda x)\, dx = \frac{1}{2}[J_0'(\lambda)]^2.$$

(*Hint*: Relation (**) in Ex. 4, (b), is valid for any positive λ and μ. Differentiate this with respect to λ, and then set $\mu = \lambda$ to obtain

$$\left[\frac{\partial \phi_\lambda}{\partial \lambda}' \phi_\lambda' - \phi_\lambda \left(\frac{\partial \phi_\lambda}{\partial \lambda} \right)' \right]' = 2\lambda \phi_\lambda^2.$$

Integrate from 0 to 1.)

6. If $\lambda > 0$ is such that $J_0(\lambda) = 0$, prove that $J_0'(\lambda) \neq 0$. (*Hint*: If $J_0(\lambda) = J_0'(\lambda) = 0$ the uniqueness theorem would imply $J_0(x) = 0$ for $x > 0$. Alternately, use Ex. 5.) (*Remark*: The result of this exercise can be used to show that the positive zeros of J_0 are denumerable, that is, they may be put into a one-to-one correspondence with the positive integers.)

7. Show that J_0' satisfies the Bessel equation of order one

$$x^2 y'' + xy' + (x^2 - 1)y = 0.$$

8. Since $J_0(0) = 1$, and J_0 is continuous, $J_0(x) \neq 0$ in some interval $0 < x < a$, for some $a > 0$. Let $0 < x_0 < a$.

(a) Show that there is a second solution ϕ_2 of the Bessel equation of order zero which has the form

$$\phi_2(x) = J_0(x) \int_{x_0}^x \left[\frac{1}{t J_0^2(t)} \right] dt, \qquad (0 < x < a).$$

(b) Show that J_0 and ϕ_2 are linearly independent on $0 < x < a$.

8. The Bessel equation (continued)

Now we compute solutions for the Bessel equation of order α, where $\alpha \neq 0$, and Re $\alpha \geqq 0$:

$$L(y) = x^2 y'' + xy' + (x^2 - \alpha^2)y = 0.$$

As before we restrict attention to the case $x > 0$. The roots of the indicial equation are

$$r_1 = \alpha, \qquad r_2 = -\alpha.$$

First we determine a solution corresponding to the root $r_1 = \alpha$. From Theorem 3 such a solution ϕ_1 has the form

$$\phi_1(x) = x^\alpha \sum_{k=0}^\infty c_k x^k, \qquad (c_0 \neq 0).$$

We find, after a little calculation, that

$$L(\phi_1)(x) = 0 \cdot c_0 x^\alpha + [(\alpha + 1)^2 - \alpha^2] c_1 x^{\alpha+1}$$

$$+ x^\alpha \sum_{k=2}^{\infty} \{[(\alpha + k)^2 - \alpha^2] c_k + c_{k-2}\} x^k = 0.$$

Thus we have

$$c_1 = 0,$$

$$[(\alpha + k)^2 - \alpha^2] c_k + c_{k-2} = 0, \qquad (k = 2, 3, \cdots).$$

Since

$$(\alpha + k)^2 - \alpha^2 = k(2\alpha + k) \neq 0 \quad \text{for} \quad k = 2, 3, \cdots,$$

and $c_1 = 0$, it follows that

$$c_1 = c_3 = c_5 = \cdots = 0.$$

We find

$$c_2 = -\frac{c_0}{2(2\alpha + 2)} = -\frac{c_0}{2^2(\alpha + 1)},$$

$$c_4 = -\frac{c_2}{4(2\alpha + 4)} = \frac{c_0}{2^4 2!(\alpha + 1)(\alpha + 2)},$$

$$c_6 = -\frac{c_4}{6(2\alpha + 6)} = -\frac{c_0}{2^6 3!(\alpha + 1)(\alpha + 2)(\alpha + 3)},$$

and, in general,

$$c_{2m} = \frac{(-1)^m c_0}{2^{2m} m!(\alpha + 1)(\alpha + 2) \cdots (\alpha + m)}.$$

Our solution thus becomes

$$\phi_1(x) = c_0 x^\alpha + c_0 x^\alpha \sum_{m=1}^{\infty} \frac{(-1)^m x^{2m}}{2^{2m} m!(\alpha + 1) \cdots (\alpha + m)}. \tag{8.1}$$

For $\alpha = 0$, $c_0 = 1$, this reduces to $J_0(x)$.

It is usual to choose

$$c_0 = \frac{1}{2^\alpha \Gamma(\alpha + 1)}, \tag{8.2}$$

where Γ is the *gamma function* defined by

$$\Gamma(z) = \int_0^\infty e^{-x} x^{z-1} \, dx, \qquad (\text{Re } z > 0).$$

It is readily seen that

$$\Gamma(z + 1) = z\Gamma(z). \tag{8.3}$$

Indeed, integrating by parts, we have

$$\Gamma(z + 1) = \lim_{T \to \infty} \int_0^T e^{-x} x^z \, dx$$

$$= \lim_{T \to \infty} \left[-x^z e^{-x} \Big|_0^T + z \int_0^T e^{-x} x^{z-1} \, dx \right]$$

$$= z \lim_{T \to \infty} \int_0^T e^{-x} x^{z-1} \, dx = z\Gamma(z),$$

since $T^z e^{-T} \to 0$ as $T \to \infty$. Also, since

$$\Gamma(1) = \int_0^\infty e^{-x} \, dx = 1,$$

if z is a positive integer n,

$$\Gamma(n + 1) = n!.$$

Thus the gamma function is an extension of the factorial function to numbers which are not integers.

The relation (8.3) can be used to define $\Gamma(z)$ for z such that $\mathrm{Re}\, z < 0$, provided z is not a negative integer. To see this suppose N is the positive integer such that

$$-N < \mathrm{Re}\, z \leqq -N + 1.$$

Then $\mathrm{Re}\, (z + N) > 0$, and we can define $\Gamma(z)$ in terms of $\Gamma(z + N)$ by

$$\Gamma(z) = \frac{\Gamma(z + N)}{z(z + 1) \cdots (z + N - 1)}, \qquad (\mathrm{Re}\, z < 0),$$

provided $z \neq -N + 1$. The gamma function is not defined at $0, -1, -2, \cdots$.

Returning to (8.1), if we use the c_0 given by (8.2) we obtain a solution of the Bessel equation of order α which is denoted by J_α, and is called the *Bessel function of order α of the first kind*:

$$J_\alpha(x) = \left(\frac{x}{2}\right)^\alpha \sum_{m=0}^\infty \frac{(-1)^m}{m!\,\Gamma(m + \alpha + 1)} \left(\frac{x}{2}\right)^{2m}, \qquad (\mathrm{Re}\, \alpha \geqq 0). \quad (8.4)$$

Notice that this formula for J_α reduces to J_0 when $\alpha = 0$, since $\Gamma(m + 1) = m!$.

There are now two cases according as $r_1 - r_2 = 2\alpha$ is a positive integer or not. If 2α is not a positive integer, by Theorem 4 there is another solution ϕ_2 of the form

$$\phi_2(x) = x^{-\alpha} \sum_{k=0}^\infty c_k x^k.$$

We find that our calculations for the root $r_1 = \alpha$ carry over provided only that we replace α by $-\alpha$ everywhere. Thus

$$J_{-\alpha}(x) = \left(\frac{x}{2}\right)^{-\alpha} \sum_{m=0}^{\infty} \frac{(-1)^m}{m!\,\Gamma(m - \alpha + 1)} \left(\frac{x}{2}\right)^{2m}$$

gives a second solution in case 2α is not a positive integer.

Since $\Gamma(m - \alpha + 1)$ exists for $m = 0, 1, 2, \cdots$, provided α is not a positive integer, we see that $J_{-\alpha}$ exists in this case, even if $r_1 - r_2 = 2\alpha$ is a positive integer. This is the rather special case we mentioned in the proof of Theorem 4. Thus, if α is not zero or a positive integer, J_α and $J_{-\alpha}$ form a basis for the solutions of the Bessel equation of order α for $x > 0$.

The only remaining case is that for which α is a positive integer, say $\alpha = n$. According to Theorem 4 there is a solution ϕ_2 of the form

$$\phi_2(x) = x^{-n} \sum_{k=0}^{\infty} c_k x^k + c\,(\log x)\,J_n(x).$$

We find that

$$L(\phi_2)(x) = x^2 \phi_2''(x) + x\phi_2'(x) + (x^2 - n^2)\phi_2(x)$$

$$= 0 \cdot c_0 x^{-n} + [(1 - n)^2 - n^2]c_1 x^{1-n}$$

$$+ x^{-n} \sum_{k=2}^{\infty} \{[(k - n)^2 - n^2]c_k + c_{k-2}\}x^k$$

$$+ 2cx J_n'(x) + c\,(\log x)\,L(J_n)(x) = 0,$$

and since $L(J_n)(x) = 0$ we have, on multiplying by x^n,

$$(1 - 2n)c_1 x + \sum_{k=2}^{\infty} [k(k - 2n)c_k + c_{k-2}]x^k$$

$$= -2c \sum_{m=0}^{\infty} (2m + n)d_{2m}x^{2m+2n}. \qquad (8.5)$$

Here we have put

$$J_n(x) = \sum_{m=0}^{\infty} d_{2m}x^{2m+n},$$

and hence

$$d_{2m} = \frac{(-1)^m}{2^{2m+n}m!\,(m + n)!}. \qquad (8.6)$$

The series on the right side of (8.5) begins with x^{2n}, and since n is a positive integer we have $c_1 = 0$. Further, if $n > 1$,

$$k(k - 2n)c_k + c_{k-2} = 0, \qquad (k = 2, 3, \cdots, 2n - 1),$$

and this implies

$$c_1 = c_3 = c_5 = \cdots = c_{2n-1} = 0,$$

whereas

$$c_2 = \frac{c_0}{2^2(n-1)}, \qquad c_4 = \frac{c_0}{2^4 2!(n-1)(n-2)},$$

and in general

$$c_{2j} = \frac{c_0}{2^{2j}j!(n-1)\cdots(n-j)}, \qquad (j = 1, 2, \cdots, n-1). \quad (8.7)$$

Comparing the coefficients of x^{2n} in (8.5) we obtain

$$c_{2n-2} = -2cnd_0 = -\frac{c}{2^{n-1}(n-1)!}.$$

On the other hand from (8.7) it follows that

$$c_{2n-2} = \frac{c_0}{2^{2n-2}(n-1)!(n-1)!},$$

and therefore

$$c = -\frac{c_0}{2^{n-1}(n-1)!}. \qquad (8.8)$$

Since the series on the right side of (8.5) contains only even powers of x the same must be true of the series on the left side of (8.5), and this implies

$$c_{2n+1} = c_{2n+3} = \cdots = 0.$$

The coefficient c_{2n} is undetermined, but the remaining coefficients

$$c_{2n+2}, \qquad c_{2n+4}, \qquad \cdots$$

are obtained from the equations

$$2m(2n+2m)c_{2n+2m} + c_{2n+2m-2} = -2c(n+2m)d_{2m}, \qquad (m = 1, 2, \cdots).$$

For $m = 1$ we have

$$c_{2n+2} = -\frac{cd_2}{2}\left(1 + \frac{1}{n+1}\right) - \frac{c_{2n}}{4(n+1)}.$$

We now choose c_{2n} so that

$$\frac{c_{2n}}{4(n+1)} = \frac{cd_2}{2}\left(1 + \frac{1}{2} + \cdots + \frac{1}{n}\right).$$

Since $4(n + 1)d_2 = -d_0$,

$$c_{2n} = -\frac{cd_0}{2}\left(1 + \frac{1}{2} + \cdots + \frac{1}{n}\right).$$

With this choice of c_{2n} we have

$$c_{2n+2} = -\frac{cd_2}{2}\left(1 + 1 + \frac{1}{2} + \cdots + \frac{1}{n+1}\right).$$

For $m = 2$ we obtain

$$c_{2n+4} = -\frac{cd_4}{2}\left(\frac{1}{2} + \frac{1}{n+2}\right) - \frac{c_{2n+2}}{2^2 \cdot 2 \cdot (n+2)}.$$

Since $2^2 \cdot 2 \cdot (n + 2)d_4 = -d_2$,

$$\frac{c_{2n+2}}{2^2 \cdot 2 \cdot (n+2)} = \frac{cd_4}{2}\left(1 + 1 + \frac{1}{2} + \cdots + \frac{1}{n+1}\right),$$

and therefore

$$c_{2n+4} = -\frac{cd_4}{2}\left(1 + \frac{1}{2} + 1 + \frac{1}{2} + \cdots + \frac{1}{n+2}\right).$$

It can be shown by induction that

$$c_{2n+2m} = -\frac{cd_{2m}}{2}\left[\left(1 + \frac{1}{2} + \cdots + \frac{1}{m}\right) + \left(1 + \frac{1}{2} + \cdots + \frac{1}{n+m}\right)\right],$$

$$(m = 1, 2, \cdots).$$

Finally, we obtain for our solution ϕ_2 the function given by

$$\phi_2(x) = c_0 x^{-n} + c_0 x^{-n} \sum_{j=1}^{n-1} \frac{x^{2j}}{2^{2j}j!(n-1)\cdots(n-j)}$$

$$- \frac{cd_0}{2}\left(1 + \frac{1}{2} + \cdots + \frac{1}{n}\right)x^n$$

$$- \frac{c}{2}\sum_{m=1}^{\infty} d_{2m}\left[\left(1 + \frac{1}{2} + \cdots + \frac{1}{m}\right) + \left(1 + \frac{1}{2} + \cdots + \frac{1}{n+m}\right)\right]x^{n+2m}$$

$$+ c\,(\log x)\,J_n(x);$$

where c_0 and c are constants related by (8.8), and d_{2m} is given by (8.6). When $c = 1$ the resulting function ϕ_2 is often denoted by K_n. In this case

$$c_0 = -2^{n-1}(n-1)!,$$

and therefore we may write

$$K_n(x) = -\frac{1}{2}\left(\frac{x}{2}\right)^{-n} \sum_{j=0}^{n-1} \frac{(n-j-1)!}{j!}\left(\frac{x}{2}\right)^{2j}$$

$$-\frac{1}{2}\frac{1}{n!}\left(1 + \frac{1}{2} + \cdots + \frac{1}{n}\right)\left(\frac{x}{2}\right)^n$$

$$-\frac{1}{2}\left(\frac{x}{2}\right)^n \sum_{m=1}^{\infty} \frac{(-1)^m}{m!(m+n)!}\left[\left(1 + \frac{1}{2} + \cdots + \frac{1}{m}\right)\right.$$

$$\left. + \left(1 + \frac{1}{2} + \cdots + \frac{1}{m+n}\right)\right]\left(\frac{x}{2}\right)^{2m} + (\log x) J_n(x).$$

This formula reduces to the one for $K_0(x)$ when $n = 0$, provided we interpret the first two sums on the right as zero in this case. The function K_n is called a *Bessel function of order n of the second kind.*

EXERCISES

1. (a) Prove that the series defining J_α and $J_{-\alpha}$ converge for $|x| < \infty$.
(b) Prove that the infinite series involved in the definition of K_n converges for $|x| < \infty$.

2. Let ϕ be any solution for $x > 0$ of the Bessel equation of order α

$$x^2 y'' + xy' + (x^2 - \alpha^2)y = 0,$$

and put $\psi(x) = x^{1/2}\phi(x)$. Show that ψ satisfies the equation

$$y'' + \left[1 + \frac{\frac{1}{4} - \alpha^2}{x^2}\right]y = 0$$

for $x > 0$.

3. (a) Show that

$$x^{1/2}J_{1/2}(x) = \frac{\sqrt{2}}{\Gamma(\frac{1}{2})}\sin x.$$

(b) Show that

$$x^{1/2}J_{-1/2}(x) = \frac{\sqrt{2}}{\Gamma(\frac{1}{2})}\cos x.$$

(*Hint:* From Ex. 2, $\psi(x) = x^{1/2}J_{1/2}(x)$ satisfies $\psi'' + \psi = 0$ for $x > 0$, and hence $\psi(x) = c_1 \cos x + c_2 \sin x$, where c_1, c_2 are constants. Show that $c_1 = 0$ and $c_2 = \sqrt{2}/\Gamma(\frac{1}{2})$.) (*Note:* It can be shown that $\Gamma(\frac{1}{2}) = \sqrt{\pi}$.)

4. (a) Show that $J_0'(x) = -J_1(x)$.
(b) Show that $K_0'(x) = -K_1(x)$.

5. Prove that between any two positive zeros of J_0 there is a zero of J_1. (*Hint*: Use Ex. 4(a), and Rolle's theorem.)

6. Show that if $\alpha > 0$ then J_α has an infinity of positive zeros. (*Hint*: If $\psi(x) = x^{1/2}J_\alpha(x)$ then ψ satisfies

$$y'' + \beta(x)y = 0, \tag{*}$$

where

$$\beta(x) = 1 + \frac{\frac{1}{4} - \alpha^2}{x^2};$$

see Ex. 2. For all large enough x, say $x > x_0$, $\beta(x) > \frac{1}{4}$. Compare (*) with the equation

$$y'' + \tfrac{1}{4}y = 0$$

satisfied by $\chi(x) = \sin(x/2)$. Apply the result of Ex. 4 of Sec. 4, Chap. 3.)

7. For α fixed, $\alpha > 0$, and $\lambda > 0$, let $\phi_\lambda(x) = x^{1/2}J_\alpha(\lambda x)$. Show that

$$\phi_\lambda'' + \left[\frac{\frac{1}{4} - \alpha^2}{x^2}\right]\phi_\lambda = -\lambda^2\phi_\lambda.$$

(*Hint*: $\lambda^{1/2}\phi_\lambda(x) = \psi(\lambda x)$, where ψ is defined in Ex. 6.)

8. If λ, μ are positive, show that

$$(\lambda^2 - \mu^2)\int_0^1 \phi_\lambda(x)\phi_\mu(x)\, dx = \phi_\lambda(1)\phi_\mu'(1) - \phi_\mu(1)\phi_\lambda'(1).$$

(*Hint*: Use Ex. 7 to show that

$$\phi_\lambda\phi_\mu'' - \phi_\mu\phi_\lambda'' = (\phi_\lambda\phi_\mu' - \phi_\mu\phi_\lambda')' = (\lambda^2 - \mu^2)\phi_\lambda\phi_\mu, \tag{*}$$

and then integrate from 0 to 1.)

9. If $\alpha > 0$, and λ, μ are positive zeros of J_α, show that

$$\int_0^1 \phi_\lambda(x)\phi_\mu(x)\, dx = \int_0^1 xJ_\alpha(\lambda x)J_\alpha(\mu x)\, dx = 0,$$

if $\lambda \neq \mu$. (*Hint*: Ex. 8.)

10. If $\alpha > 0, \lambda > 0$, and $J_\alpha(\lambda) = 0$ show that

$$\int_0^1 \phi_\lambda^2(x)\, dx = \int_0^1 xJ_\alpha^2(\lambda x)\, dx = \tfrac{1}{2}[J_\alpha'(\lambda)]^2.$$

(*Hint*: Differentiate (*), in Ex. 8, with respect to λ and then put $\mu = \lambda$ to obtain

$$\left[\frac{\partial\phi_\lambda}{\partial\lambda}\phi_\lambda' - \phi_\lambda\left(\frac{\partial\phi_\lambda}{\partial\lambda}\right)'\right]' = 2\lambda\phi_\lambda^2.$$

Integrate from 0 to 1.)

11. Define $1/\Gamma(k)$, when k is a non-positive integer, to be zero. Show that if n is a positive integer the formula for $J_{-n}(x)$ gives

$$J_{-n}(x) = (-1)^n J_n(x).$$

12. (a) Use the formula for $J_\alpha(x)$ to show that

$$(x^\alpha J_\alpha)'(x) = x^\alpha J_{\alpha-1}(x).$$

(b) Prove that

$$(x^{-\alpha} J_\alpha)'(x) = -x^{-\alpha} J_{\alpha+1}(x).$$

13. Show that

$$J_{\alpha-1}(x) - J_{\alpha+1}(x) = 2J_\alpha'(x),$$

and

$$J_{\alpha-1}(x) + J_{\alpha+1}(x) = 2\alpha x^{-1} J_\alpha(x).$$

(*Hint*: Use the results of Ex. 12.)

14. (a) Show that between any two positive zeros of J_α there is a zero of $J_{\alpha+1}$. (*Hint*: Use Ex. 12(b), and Rolle's theorem.)

(b) Show that between any two positive zeros of $J_{\alpha+1}$ there is a zero of J_α. (*Hint*: Use Ex. 12 (a), and Rolle's theorem.)

9. Regular singular points at infinity

Often it is of interest to investigate solutions of an equation

$$L(y) = y'' + a_1(x)y' + a_2(x)y = 0 \tag{9.1}$$

for large values of $|x|$. A simple way of doing this is to make the substitution $x = 1/t$, and study the solutions of the resulting equation near $t = 0$. Then, for example, the results on analytic equations and equations with a regular singular point at $t = 0$ can be applied.

If ϕ is a solution of (9.1) for $|x| > r_0$, for some $r_0 > 0$, let

$$\tilde{\phi}(t) = \phi\left(\frac{1}{t}\right), \qquad \tilde{a}_1(t) = a_1\left(\frac{1}{t}\right), \qquad \tilde{a}_2(t) = a_2\left(\frac{1}{t}\right).$$

These functions will exist for $|t| < 1/r_0$, and

$$\frac{d\phi}{dx}\left(\frac{1}{t}\right) = -t^2 \frac{d\tilde{\phi}}{dt}(t),$$

$$\frac{d^2\phi}{dx^2}\left(\frac{1}{t}\right) = t^4 \frac{d^2\tilde{\phi}}{dt^2}(t) + 2t^3 \frac{d\tilde{\phi}}{dt}(t).$$

Since from (9.1)

$$\frac{d^2\phi}{dx}\left(\frac{1}{t}\right) + a_1\left(\frac{1}{t}\right)\frac{d\phi}{dx}\left(\frac{1}{t}\right) + a_2\left(\frac{1}{t}\right)\phi\left(\frac{1}{t}\right) = 0,$$

we have

$$t^4\bar{\phi}''(t) + [2t^3 - t^2\tilde{a}_1(t)]\bar{\phi}'(t) + \tilde{a}_2(t)\bar{\phi}(t) = 0,$$

where now the prime denotes differentiation with respect to t. Thus $\bar{\phi}$ satisfies the equation

$$\tilde{L}(y) = t^4 y'' + [2t^3 - t^2\tilde{a}_1(t)]y' + \tilde{a}_2(t)y = 0. \qquad (9.2)$$

Conversely, if $\bar{\phi}$ satisfies $\tilde{L}(y) = 0$ the function ϕ will satisfy $L(y) = 0$. The equation (9.2) is called the *induced equation* associated with $L(y) = 0$ and the substitution $x = 1/t$.

We say that *infinity is a regular singular point* for (9.1) if the induced equation (9.2) has the origin $t = 0$ as a regular singular point. Writing (9.2) as

$$t^2 y'' + \left(2 - \frac{\tilde{a}_1(t)}{t}\right)ty' + \frac{\tilde{a}_2(t)}{t^2}\, y = 0$$

we see that $\tilde{L}(y) = 0$ has $t = 0$ as a regular singular point if and only if $\tilde{a}_1/t$ and $\tilde{a}_2/t^2$ are analytic at $t = 0$. This means that

$$\tilde{a}_1(t) = t\sum_{k=0}^{\infty}\alpha_k t^k, \qquad \tilde{a}_2(t) = t^2\sum_{k=0}^{\infty}\beta_k t^k,$$

where the series converge for $|t| < 1/r_0$, $r_0 > 0$. Translated into a condition involving a_1, a_2 this means that

$$a_1(x) = \frac{1}{x}\sum_{k=0}^{\infty}\frac{\alpha_k}{x^k}, \qquad a_2(x) = \frac{1}{x^2}\sum_{k=0}^{\infty}\frac{\beta_k}{x^k},$$

where these series converge for $|x| > r_0$. Thus infinity is a regular singular point for (9.1) if and only if (9.1) can be written in the form

$$x^2 y'' + a(x)xy' + b(x)y = 0,$$

where a, b have convergent power series expansions in powers of $1/x$ for $|x| > r_0$ for some $r_0 > 0$.

The simplest example of an equation with a regular singular point at infinity is

$$x^2 y'' + a\, x\, y' + b\, y = 0,$$

where a, b are constants; namely, the Euler equation. Thus this equation has the origin and infinity as regular singular points, and it is clear that there are no other possible singular points.

An example of an equation with three regular singular points (and no others) is the *hypergeometric equation*

$$(x - x^2)y'' + [\gamma - (\alpha + \beta + 1)x]y' - \alpha\beta y = 0,$$

where α, β, γ are constants. It is readily checked that 0, 1, and infinity are regular singular points.

EXERCISES

1. Show that infinity is not a regular singular point for the equation

$$y'' + ay' + by = 0,$$

where a, b are constants, not both zero.

2. Show that infinity is not a regular singular point for the Bessel equation

$$x^2y'' + xy' + (x^2 - \alpha^2)y = 0.$$

3. (a) Show that infinity is a regular singular point for the Legendre equation

$$(1 - x^2)y'' - 2xy' + \alpha(\alpha + 1)y = 0.$$

(b) Compute the induced equation associated with the Legendre equation and the substitution $x = 1/t$.

(c) Compute the indicial polynomial, and its roots, of the induced equation.

4. Find two linearly independent solutions of the equation

$$(1 - x^2)y'' - 2xy' + 2y = 0$$

of the form

$$x^{-r} \sum_{k=0}^{\infty} c_k x^{-k}$$

valid for $|x| > 1$. (*Hint*: Use Ex. 3 with $\alpha = 1$.)

5. (a) Suppose ϕ is a solution of the Legendre equation of order p

$$(1 - x^2)y'' - 2xy' + p(p + 1)y = 0,$$

and let $\psi(t) = \phi(2t - 1)$. Show that ψ satisfies the equation

$$(t - t^2)y'' + (1 - 2t)y' + p(p + 1)y = 0. \qquad (*)$$

(b) Verify that the equation (*) is a hypergeometric equation with

$$\alpha = p + 1, \qquad \beta = -p, \qquad \gamma = 1.$$

6. (a) Compute the indicial polynomial relative to the origin for the hypergeometric equation.

(b) Obtain a solution ϕ of the hypergeometric equation of the form

$$\phi(x) = x^r \sum_{k=0}^{\infty} c_k x^k,$$

if γ is not zero or a negative integer.

7. Consider the equation

$$x^2 y'' + 2xy' - n(n+1)y = 0,$$

where n is a non-negative integer.

(a) Show that infinity is a regular singular point.

(b) Compute a solution ϕ of the form

$$\phi(x) = x^{-r} \sum_{k=0}^{\infty} c_k x^{-k}.$$

CHAPTER 5

Existence and Uniqueness
of Solutions to First
Order Equations

1. Introduction

In this chapter we consider the general first order equation

$$y' = f(x, y), \tag{1.1}$$

where f is some continuous function. Only in rather special cases is it possible
to find explicit analytic expressions for the solutions of (1.1). We have already considered one such special case; namely, the linear equation

$$y' = g(x)y + h(x), \tag{1.2}$$

where g, h are continuous on some interval I. Any solution ϕ of (1.2) can
be written in the form

$$\phi(x) = e^{Q(x)} \int_{x_0}^{x} e^{-Q(t)} h(t) \, dt + c e^{Q(x)}, \tag{1.3}$$

where

$$Q(x) = \int_{x_0}^{x} g(t) \, dt,$$

x_0 is in I, and c is a constant (see Chap. 1). In Secs. 2 and 3 we indicate
procedures which can be used to solve other important special cases of
(1.1).*

* An excellent compendium of special equations and their solutions appears in the
book by E. Kamke, "*Differentialgleichungen—Lösungsmethoden und Lösungen*, vol. I,
reprinted by J. W. Edwards, Ann Arbor, Mich. (1945).

Our main purpose is to prove that a wide class of equations of the form (1.1) have solutions, and that solutions to initial value problems are unique. If f is not a linear equation there are certain limitations which must be expected concerning any general existence theorem. To illustrate this consider the equation

$$y' = y^2.$$

Here $f(x, y) = y^2$, and we see f has derivatives of all orders with respect to x and y at every point in the (x, y)-plane. A solution ϕ of this equation satisfying the initial condition

$$\phi(1) = -1$$

is given by

$$\phi(x) = -\frac{1}{x},$$

as can be readily checked. However this solution ceases to exist at $x = 0$, even though f is a nice function there. This example shows that any general existence theorem for (1.1) can only assert the existence of a solution on some interval *near-by* the initial point.

The above phenomenon does not occur in the case of the linear equation (1.2), for it is clear from (1.3) that any solution ϕ exists on *all* of the interval I. This points up one of the fundamental difficulties we encounter when we consider nonlinear equations. The equation often gives no clue as to how far a solution will exist.

We prove that initial value problems for equation (1.1) have unique solutions which can be obtained by an approximation process, provided f satisfies an additional condition, the Lipschitz condition. We first concentrate our attention on the case when f is real-valued, and later show how the results carry over to the situation when f is complex-valued.

2. Equations with variables separated

A first order equation

$$y' = f(x, y)$$

is said to have the *variables separated* if f can be written in the form

$$f(x, y) = \frac{g(x)}{h(y)},$$

where g, h are functions of a single argument. In this case we may write our equation as

$$h(y) \frac{dy}{dx} = g(x), \tag{2.1}$$

or

$$h(y) dy = g(x) dx,$$

and we readily see the origin of the term "variables separated".

For simplicity let us discuss the equation (2.1) in the case g and h are continuous real-valued functions defined for real x and y, respectively. If ϕ is a real-valued solution of (2.1) on some interval I containing a point x_0, then

$$h(\phi(x))\phi'(x) = g(x)$$

for all x in I, and therefore

$$\int_{x_0}^{x} h(\phi(t))\phi'(t) \, dt = \int_{x_0}^{x} g(t) \, dt \tag{2.2}$$

for all x in I. Letting $u = \phi(t)$ in the integral on the left in (2.2), we see that (2.2) may be written as

$$\int_{\phi(x_0)}^{\phi(x)} h(u) \, du = \int_{x_0}^{x} g(t) \, dt.$$

Conversely, suppose x and y are related by the formula

$$\int_{y_0}^{y} h(u) \, du = \int_{x_0}^{x} g(t) \, dt, \tag{2.3}$$

and that this defines implicitly a differentiable function ϕ for x in I.* Then this function satisfies

$$\int_{y_0}^{\phi(x)} h(u) \, du = \int_{x_0}^{x} g(t) \, dt$$

for all x in I, and differentiating we obtain

$$h(\phi(x))\phi'(x) = g(x),$$

which shows that ϕ is a solution of (2.1) on I.

In practice the usual way of dealing with (2.1) is to write it as

$$h(y) dy = g(x) dx$$

---

* We say that a relation $F(x, y) = 0$ defines a function ϕ implicitly for x in some interval I, if for each x in I there is a y such that $F(x, y) = 0$; this y being denoted by $\phi(x)$.

(thus separating the variables), and then integrate to obtain

$$\int h(y) \, dy = \int g(x) \, dx + c,$$

where c is a constant, and the integrals are anti-derivatives. Thus

$$H(y) = \int h(y) \, dy, \qquad G(x) = \int g(x) \, dx,$$

represent any two functions H, G such that

$$H' = h, \qquad G' = g.$$

Then any differentiable function ϕ which is defined implicitly by the relation

$$H(y) = G(x) + c \tag{2.4}$$

will be a solution of (2.1). Therefore it is usual to identify any solution thus obtained with the relation (2.4). We summarize in the following theorem.

Theorem 1. *Let* g, h *be continuous real-valued functions for* a $\leqq$ x $\leqq$ b, c $\leqq$ y $\leqq$ d *respectively, and consider the equation*

$$h(y)y' = g(x). \tag{2.1}$$

If G, H *are any functions such that* G' = g, H' = h, *and* c *is any constant such that the relation*

$$H(y) = G(x) + c$$

defines a real-valued differentiable function ϕ for x *in some interval* I *contained in* a $\leqq$ x $\leqq$ b, *then ϕ will be a solution of* (2.1) *on* I. *Conversely, if ϕ is a solution of* (2.1) *on* I, *it satisfies the relation*

$$H(y) = G(x) + c$$

on I, *for some constant* c.*

The simplest example is that case in which $h(y) = 1$. Then $y' = g(x)$, and every solution ϕ has the form

$$\phi(x) = G(x) + c, \tag{2.5}$$

where G is any function on $a \leqq x \leqq b$ such that $G' =, g$, and c is a constant. Moreover, if c is any constant, (2.5) defines a solution of $y' = g(x)$. Thus we have found *all* solutions of $y' = g(x)$ on $a \leqq x \leqq b$.

---

* The function ϕ will be a solution of $y' = g(x)/h(y)$ on I, provided $h(\phi(x)) \neq 0$ for all x in I.

Another simple case occurs when $g(x) = 1$, for then we have

$$y' = \frac{1}{h(y)}, \tag{2.6}$$

or

$$h(y)\,dy = dx.$$

Thus, if $H' = h$, any differentiable function defined implicitly by the relation

$$H(y) = x + c, \tag{2.7}$$

where c is a constant, will be a solution of (2.6). As an example, let us consider the equation

$$y' = y^2. \tag{2.8}$$

Here $h(y) = 1/y^2$, which we note is *not* continuous at $y = 0$. We have

$$\frac{dy}{y^2} = dx,$$

and thus the relation (2.7) becomes

$$-\frac{1}{y} = x + c, \quad \text{or} \quad y = \frac{-1}{x + c}.$$

Thus, if c is any constant, the function ϕ given by

$$\phi(x) = \frac{-1}{x + c} \tag{2.9}$$

is a solution of (2.8), provided $x \neq -c$.

It is important to remark that the separation of variables method of finding solutions may not yield *all* solutions of an equation. For example, it is clear from (2.8) that the function ψ which is identically zero for all x is a solution of (2.8). However, for no constant c will the ϕ of (2.9) yield this solution. Careful attention to the possibilities of dividing by zero will often alert the student to missing solutions.

Let us consider one more example:

$$y' = 3y^{2/3}. \tag{2.10}$$

This leads to

$$\frac{dy}{y^{2/3}} = 3\,dx$$

if $y \neq 0$, and hence to

$$y^{1/3} = x + c, \quad \text{or} \quad y = (x + c)^3,$$

where c is a constant. Thus the function ϕ given by

$$\phi(x) = (x + c)^3 \tag{2.11}$$

will be a solution of (2.10) for any constant c. Again we note that the identically zero function is a solution of (2.10) which can not be obtained from (2.11).

The example (2.10) illustrates one more difficulty we encounter when we deal with nonlinear equations; namely, there may be several solutions satisfying a given initial condition. Thus the two functions ϕ and ψ given by

$$\phi(x) = x^3, \qquad \psi(x) = 0, \qquad (- \infty < x < \infty),$$

are solutions of (2.10) which pass through the origin. Actually the situation is much worse than appears, for there are *infinitely* many functions which are solutions of (2.10) passing through the origin. To see this let k be any positive number, and define ϕ_k by

$$\phi_k(x) = 0, \qquad (- \infty < x \leq k),$$

$$\phi_k(x) = (x - k)^3, \qquad (k < x < \infty).$$

Then it is not difficult to see that ϕ_k is a solution of (2.10) for all real x, and clearly $\phi_k(0) = 0$. It might be instructive for the student to make a sketch of these solutions.

EXERCISES

1. Find all real-valued solutions of the following equations:

(a) $y' = x^2 y$ (b) $yy' = x$

(c) $y' = \dfrac{x + x^2}{y - y^2}$ (d) $y' = \dfrac{e^{x-y}}{1 + e^x}$

(e) $y' = x^2 y^2 - 4x^2$

2. (a) Show that the solution ϕ of

$$y' = y^2$$

which passes through the point (x_0, y_0) is given by

$$\phi(x) = \frac{y_0}{1 - y_0(x - x_0)}.$$

(*Note*: The identically zero solution can be obtained from this formula by letting $y_0 = 0$.)

(b) For which x is ϕ a well-defined function?

(c) For which x is ϕ a solution of the problem

$$y' = y^2, \qquad y(x_0) = y_0?$$

3. (a) Find the solution of $y' = 2y^{1/2}$ passing through the point (x_0, y_0), where $y_0 > 0$.

(b) Find all solutions of this equation passing through $(x_0, 0)$.

4. A function f defined for real x, y is said to be *homogeneous of degree k* if

$$f(tx, ty) = t^k f(x, y)$$

for all t, x, y. In case f is homogeneous of degree zero we have

$$f(tx, ty) = f(x, y),$$

and then we say the equation $y' = f(x, y)$ is *homogeneous*. (Unfortunately this terminology, which is rather standard, conflicts with the use of the word homogeneous in connection with linear equations.) Such equations can be reduced to ones with variables separated. To see this, let $y = ux$ in $y' = f(x, y)$. Then we obtain

$$xu' + u = f(x, ux) = f(1, u),$$

and hence

$$u' = \frac{f(1, u) - u}{x},$$

which is an equation for u with variables separated.

Find all real-valued solutions of the following equations:

(a) $y' = \dfrac{x + y}{x - y}$ (b) $y' = \dfrac{y^2}{xy + x^2}$

(c) $y' = \dfrac{x^2 + xy + y^2}{x^2}$ (d) $y' = \dfrac{y + xe^{-2y/x}}{x}$

5. The equation

$$y' = \frac{a_1 x + b_1 y + c_1}{a_2 x + b_2 y + c_2}, \tag{*}$$

where $a_1, b_1, c_1, a_2, b_2, c_2$ are constants (c_1, c_2 not both 0) can be reduced to a homogeneous equation. Assume we do not have the simple equation $y' = c_1/c_2$, and let $x = \xi + h$, $y = \eta + k$, where h, k are constants. Then (*) becomes

$$\frac{d\eta}{d\xi} = \frac{a_1 \xi + b_1 \eta + (a_1 h + b_1 k + c_1)}{a_2 \xi + b_2 \eta + (a_2 h + b_2 k + c_2)}.$$

If h, k satisfy

$$a_1 h + b_1 k + c_1 = 0, \qquad a_2 h + b_2 k + c_2 = 0, \tag{**}$$

the equation becomes homogeneous. If the equations (**) have no solution, then $a_1 b_2 - a_2 b_1 = 0$, and in this case either the substitution

$$u = a_1 x + b_1 y + c_1 \quad \text{or} \quad u = a_2 x + b_2 y + c_2,$$

leads to a separation of variables.

Solve the following equations:

(a) $y' = \dfrac{x - y + 2}{x + y - 1}$

(b) $y' = \dfrac{2x + 3y + 1}{x - 2y - 1}$

(c) $y' = \dfrac{x + y + 1}{2x + 2y - 1}$

6. (a) Show that the method of Ex. 5 can be used to reduce an equation of the form

$$y' = f\left(\frac{a_1 x + b_1 y + c_1}{a_2 x + b_2 y + c_2}\right)$$

to a homogeneous equation.

(b) Solve the equation

$$y' = \frac{1}{2}\left(\frac{x + y - 1}{x + 2}\right)^2.$$

7. Suppose there is a family F of curves in a region S in the plane with the property that through each point (x, y) of S there passes one, and only one, curve C of F, and that the slope of the tangent of C at (x, y) is given by $f(x, y)$, where f is continuous. If a curve in F can be written as $(x, \phi(x))$, where x runs over some interval I, then ϕ is a solution of $y' = f(x, y)$. If ψ is any solution of the equation $y' = -1/f(x, y)$, then the curve $C^{\perp}$ given by the points $(x, \psi(x))$ will have a tangent at each of its points (x, y) which is perpendicular to the curve in F passing through (x, y). The set G of all curves $C^{\perp}$ is called the set of *orthogonal trajectories* to the family F.

The following relations determine a family of curves, one curve for each value of the constant c. Find the orthogonal trajectories of these families.

(a) $x^2 + y^2 = c$, $(c > 0)$

(b) $y = cx$

(c) $y = cx^2$

(d) $\dfrac{x^2}{2} + \dfrac{y^2}{3} = c$, $(c > 0)$

(e) $x^2 - y^2 = c$

(f) $y = ce^{x^2}$

3. Exact equations

Suppose the first order equation $y' = f(x, y)$ is written in the form

$$y' = \frac{-M(x, y)}{N(x, y)},$$

or equivalently

$$M(x, y) + N(x, y)y' = 0, \tag{3.1}$$

where M, N are real-valued functions defined for real x, y on some rectangle R. The equation (3.1) is said to be *exact* in R if there exists a function F having continuous first partial derivatives there such that

$$\frac{\partial F}{\partial x} = M, \qquad \frac{\partial F}{\partial y} = N, \qquad (3.2)$$

in R.

If (3.1) is exact in R, and F is a function satisfying (3.2), then (3.1) becomes

$$\frac{\partial F}{\partial x}(x, y) + \frac{\partial F}{\partial y}(x, y)y' = 0.$$

If ϕ is any solution on some interval I, then

$$\frac{\partial F}{\partial x}(x, \phi(x)) + \frac{\partial F}{\partial y}(x, \phi(x))\phi'(x) = 0 \qquad (3.3)$$

for all x in I. If $\Phi(x) = F(x, \phi(x))$, then equation (3.3) just says that $\Phi'(x) = 0$, and hence

$$F(x, \phi(x)) = c,$$

where c is some constant. Thus the solution ϕ must be a function which is given implicitly by the relation

$$F(x, y) = c. \qquad (3.4)$$

Looking at this argument in reverse we see that if ϕ is a differentiable function on some interval I defined implicitly by the relation (3.4) then

$$F(x, \phi(x)) = c$$

for all x in I, and a differentiation yields (3.3). Thus ϕ is a solution of (3.1).

Theorem 2. *Suppose the equation*

$$M(x, y) + N(x, y)y' = 0 \qquad (3.1)$$

is exact in a rectangle R, *and* F *is a real-valued function such that*

$$\frac{\partial F}{\partial x} = M, \qquad \frac{\partial F}{\partial y} = N \qquad (3.2)$$

in R. *Every differentiable function* ϕ *defined implicitly by a relation*

$$F(x, y) = c, \qquad (c = constant),$$

is a solution of (3.1), *and every solution of* (3.1) *whose graph lies in* R *arises this way.*

The problem of solving an exact equation is now reduced to the problem of determining a function F satisfying (3.2). If (3.1) is exact and we write it as

$$M(x, y)\ dx + N(x, y)\ dy = \frac{\partial F}{\partial x}(x, y)\ dx + \frac{\partial F}{\partial y}(x, y)\ dy = 0,$$

we recognize that the left side of this equation is the differential dF of F. This is the explanation of the term "exact"; the left side is an exact differential of a function F.

Sometimes an F can be determined by inspection. For example, if the equation

$$y' = -\frac{x}{y} \tag{3.5}$$

is written in the form

$$x\ dx + y\ dy = 0,$$

it is clear that the left side is the differential of $(x^2 + y^2)/2$. Thus any differentiable function which is defined by the relation

$$x^2 + y^2 = c, \qquad (c = \text{constant}),$$

is a solution of (3.5). Note that the equation (3.5) does not make sense when $y = 0$.

The above example is also a special case of an equation with variables separated. Indeed *any* such equation is a special case of an exact equation, for if we write the equation as

$$g(x)dx = h(y)dy,$$

it is clear that an F is given by

$$F(x, y) = G(x) - H(y),$$

where $G' = g$, $H' = h$.

How do we recognize when an equation is exact? To see how, suppose

$$M(x, y)dx + N(x, y)dy = 0$$

is exact, and F is a function which has continuous second derivatives such that

$$\frac{\partial F}{\partial x} = M, \qquad \frac{\partial F}{\partial y} = N.$$

Then

$$\frac{\partial^2 F}{\partial y \partial x} = \frac{\partial M}{\partial y}, \qquad \frac{\partial^2 F}{\partial x \partial y} = \frac{\partial N}{\partial x},$$

and, since for such a function

$$\frac{\partial^2 F}{\partial y \partial x} = \frac{\partial^2 F}{\partial x \partial y},$$

we must have

$$\frac{\partial M}{\partial y} = \frac{\partial N}{\partial x}.$$

This is the condition we are looking for, since it is true that if this equality is valid, the equation is exact.

Theorem 3. *Let* M, N *be two real-valued functions which have continuous first partial derivatives on some rectangle*

$$R: \qquad | x - x_0 | \leqq a, \qquad | y - y_0 | \leqq b.$$

Then the equation

$$M(x, y) + N(x, y)y' = 0$$

is exact in R *if, and only if,*

$$\frac{\partial M}{\partial y} = \frac{\partial N}{\partial x} \qquad (3.6)$$

in R.

Proof. We have already seen that if the equation is exact, then (3.6) is satisfied.

Now suppose (3.6) is satisfied in R. We need to find a function F satisfying

$$\frac{\partial F}{\partial x} = M, \qquad \frac{\partial F}{\partial y} = N.$$

To see how to do this, we note that if we had such a function then

$$F(x, y) - F(x_0, y_0) = F(x, y) - F(x_0, y) + F(x_0, y) - F(x_0, y_0)$$

$$= \int_{x_0}^{x} \frac{\partial F}{\partial x}(s, y)\, ds + \int_{y_0}^{y} \frac{\partial F}{\partial y}(x_0, t)\, dt$$

$$= \int_{x_0}^{x} M(s, y)\, ds + \int_{y_0}^{y} N(x_0, t)\, dt.$$

Similarly we would have

$$F(x, y) - F(x_0, y_0) = F(x, y) - F(x, y_0) + F(x, y_0) - F(x_0, y_0)$$

$$= \int_{y_0}^{y} \frac{\partial F}{\partial y}(x, t)\, dt + \int_{x_0}^{x} \frac{\partial F}{\partial x}(s, y_0)\, ds$$

$$= \int_{y_0}^{y} N(x, t)\, dt + \int_{x_0}^{x} M(s, y_0)\, ds. \qquad (3.7)$$

We now *define* F by the formula

$$F(x, y) = \int_{x_0}^{x} M(s, y) \, ds + \int_{y_0}^{y} N(x_0, t) \, dt. \tag{3.8}$$

This definition implies that $F(x_0, y_0) = 0$, and that

$$\frac{\partial F}{\partial x}(x, y) = M(x, y)$$

for all (x, y) in R. From (3.7) we would guess that F is also given by

$$F(x, y) = \int_{y_0}^{y} N(x, t) \, dt + \int_{x_0}^{x} M(s, y_0) \, ds. \tag{3.9}$$

This is in fact true, and is a consequence of the assumption (3.6). Once this has been shown, it is clear from (3.9) that

$$\frac{\partial F}{\partial y}(x, y) = N(x, y)$$

for all (x, y) in R, and we have found our F.

In order to show that (3.9) is valid, where F is the function given by (3.8), let us consider the difference

$$F(x, y) - \left[\int_{y_0}^{y} N(x, t) \, dt + \int_{x_0}^{x} M(s, y_0) \, ds \right]$$

$$= \int_{x_0}^{x} [M(s, y) - M(s, y_0)] \, ds - \int_{y_0}^{y} [N(x, t) - N(x_0, t)] \, dt$$

$$= \int_{x_0}^{x} \left[\int_{y_0}^{y} \frac{\partial M}{\partial y}(s, t) \, dt \right] ds - \int_{y_0}^{y} \left[\int_{x_0}^{x} \frac{\partial N}{\partial x}(s, t) \, ds \right] dt$$

$$= \int_{x_0}^{x} \int_{y_0}^{y} \left[\frac{\partial M}{\partial y}(s, t) - \frac{\partial N}{\partial x}(s, t) \right] ds \, dt,$$

which is zero by virtue of (3.6). This completes our proof of Theorem 3.

As an example let us consider the equation

$$y' = \frac{3x^2 - 2xy}{x^2 - 2y}, \tag{3.10}$$

which we write as

$$(3x^2 - 2xy) \, dx + (2y - x^2) \, dy = 0.$$

Here

$$M(x, y) = 3x^2 - 2xy, \qquad N(x, y) = 2y - x^2,$$

and a computation shows that

$$\frac{\partial M}{\partial y}(x, y) = \frac{\partial N}{\partial x}(x, y) = -2x,$$

which shows that our equation is exact for all x, y. To find an F we could use either of the two formulas (3.8) or (3.9), but the following way is often simpler. We know there is an F such that

$$\frac{\partial F}{\partial x} = M, \qquad \frac{\partial F}{\partial y} = N.$$

Thus F satisfies

$$\frac{\partial F}{\partial x}(x, y) = 3x^2 - 2xy,$$

which implies that for each fixed y,

$$F(x, y) = x^3 - x^2y + f(y), \tag{3.11}$$

where f is independent of x. Now $\partial F/\partial y = N$ tells us that

$$-x^2 + f'(y) = 2y - x^2,$$

or that

$$f'(y) = 2y.$$

Thus a choice for f is given by $f(y) = y^2$, and placing this back into (3.11) we obtain finally

$$F(x, y) = x^3 - x^2y + y^2.$$

Any differentiable function ϕ which is defined implicitly by a relation

$$x^3 - x^2y + y^2 = c, \tag{3.12}$$

where c is a constant, will be a solution of (3.10), and all solutions of (3.10) arise in this way. Often the solutions are identified with the relations (3.12).

It is proved in advanced calculus texts that (3.12) will define a unique differentiable function ϕ near, and passing through, a given point (x_0, y_0) provided that

$$F(x_0, y_0) = c,$$

and that

$$\frac{\partial F}{\partial y}(x_0, y_0) \neq 0.$$

Notice that the only points (x_0, y_0) satisfying (3.12) for which

$$\frac{\partial F}{\partial y}(x_0, y_0) = 0$$

are those satisfying

$$-x_0^2 + 2y_0 = 0,$$

and these are precisely the points where the given equation (3.10) is not defined. Thus, if (x_0, y_0) is a point for which $(3x^2 - 2xy)/(x^2 - 2y)$ is defined, there will be a unique solution of (3.10) whose graph passes through (x_0, y_0).

EXERCISES

1. The equations below are written in the form $M(x, y)\, dx + N(x, y)\, dy = 0$, where M, N exist on the whole plane. Determine which equations are exact there, and solve these.

(a) $2xy\, dx + (x^2 + 3y^2)\, dy = 0$

(b) $(x^2 + xy)\, dx + xy\, dy = 0$

(c) $e^x\, dx + (e^y(y + 1))\, dy = 0$

(d) $\cos x \cos^2 y\, dx - \sin x \sin 2y\, dy = 0$

(e) $x^2 y^3\, dx - x^3 y^2\, dy = 0$

(f) $(x + y)\, dx + (x - y)\, dy = 0$

(g) $(2y e^{2x} + 2x \cos y)\, dx + (e^{2x} - x^2 \sin y)\, dy = 0$

(h) $(3x^2 \log |x| + x^2 + y)\, dx + x\, dy = 0$

2. Even though an equation $M(x, y)\, dx + N(x, y)\, dy = 0$ may not be exact, sometimes it is not too difficult to find a function u, nowhere zero, such that

$$u(x, y)M(x, y)\, dx + u(x, y)N(x, y)\, dy = 0$$

is exact. Such a function is called an *integrating factor*. For example,

$$y\, dx - x\, dy = 0$$

is not exact, but multiplying the equation by $u(x, y) = 1/y^2$, makes it exact for $y \neq 0$. Solutions are then given by $y = cx$.

Find an integrating factor for each of the following equations, and solve them.

(a) $(2y^3 + 2)\, dx + 3xy^2\, dy = 0$

(b) $\cos x \cos y\, dx - 2 \sin x \sin y\, dy = 0$

(c) $(5x^3 y^2 + 2y)\, dx + (3x^4 y + 2x)\, dy = 0$

(d) $(e^y + xe^y)\, dx + xe^y\, dy = 0$

(*Note:* If you have trouble discovering integrating factors, do Exs. 3–5 first.)

3. Consider the equation

$$M(x, y)\, dx + N(x, y)\, dy = 0,$$

where M, N have continuous first partial derivatives on some rectangle R.

Prove that a function u on R, having continuous first partial derivatives, is an integrating factor if and only if,

$$u \left(\frac{\partial M}{\partial y} - \frac{\partial N}{\partial x} \right) = N \frac{\partial u}{\partial x} - M \frac{\partial u}{\partial y}$$

on R. (*Hint*: Theorem 3.)

4. (a) Under the same conditions as in Ex. 3, show that if the equation

$$M(x, y) \, dx + N(x, y) \, dy = 0$$

has an integrating factor u, which is a function of x alone, then

$$p = \frac{1}{N} \left(\frac{\partial M}{\partial y} - \frac{\partial N}{\partial x} \right)$$

is a continuous function of x alone.

(b) If p is continuous and independent of y, show that an integrating factor is given by

$$u(x) = e^{P(x)},$$

where P is any function satisfying $P' = p$.

5. (a) Under the same conditions as in Ex. 3, show that if

$$M(x, y) \, dx + N(x, y) \, dy = 0$$

has an integrating factor u, which is a function of y alone, then

$$q = \frac{1}{M} \left(\frac{\partial N}{\partial x} - \frac{\partial M}{\partial y} \right)$$

is a continuous function of y alone.

(b) If q is continuous, and independent of x, show that an integrating factor is given by

$$u(y) = e^{Q(y)},$$

where Q is any function such that $Q' = q$.

6. Consider the linear equation of the first order

$$y' + a(x)y = b(x),$$

where a, b are continuous on some interval I.

(a) Show that there is an integrating factor which is a function of x alone. (*Hint*: Ex. 4.)

(b) Solve this equation, using an integrating factor. (Compare this procedure with that followed in Chap. 1, Sec. 7.)

4. The method of successive approximations

We now face up to the general problem of finding solutions of the equation

$$y' = f(x, y), \tag{4.1}$$

where f is any continuous *real-valued* function defined on some rectangle

$$R: \quad |x - x_0| \leqq a, \quad |y - y_0| \leqq b, \quad (a, b > 0),$$

in the real (x, y)-plane. Our object is to show that on some interval I containing x_0 there is a solution ϕ of (4.1) satisfying

$$\phi(x_0) = y_0. \tag{4.2}$$

By this we mean there is a *real-valued* differentiable function ϕ satisfying (4.2) such that the points $(x, \phi(x))$ are in R for x in I, and

$$\phi'(x) = f(x, \phi(x))$$

for all x in I. Such a function ϕ is called a solution to the *initial value problem*

$$y' = f(x, y), \qquad y(x_0) = y_0, \tag{4.3}$$

on I.

Our first step will be to show that the initial value problem is equivalent to an integral equation, namely

$$y = y_0 + \int_{x_0}^{x} f(t, y) \, dt \tag{4.4}$$

on I. By a solution of this equation on I is meant a real-valued continuous function ϕ on I such that $(x, \phi(x))$ is in R for all x in I, and

$$\phi(x) = y_0 + \int_{x_0}^{x} f(t, \phi(t)) \, dt \tag{4.5}$$

for all x on I.

Theorem 4. *A function ϕ is a solution of the initial value problem* (4.3) *on an interval* I *if and only if it is a solution of the integral equation* (4.4) *on* I.

Proof. Suppose ϕ is a solution of the initial value problem on I. Then

$$\phi'(t) = f(t, \phi(t)) \tag{4.6}$$

on I. Since ϕ is continuous on I, and f is continuous on R, the function F defined by

$$F(t) = f(t, \phi(t))$$

is continuous on I. Integrating (4.6) from x_0 to x we obtain

$$\phi(x) = \phi(x_0) + \int_{x_0}^{x} f(t, \phi(t))\, dt,$$

and since $\phi(x_0) = y_0$ we see that ϕ is a solution of (4.4).

Conversely, suppose ϕ satisfies (4.5) on I. Differentiating we find, using the fundamental theorem of integral calculus, that

$$\phi'(x) = f(x, \phi(x))$$

for all x on I. Moreover from (4.5) it is clear that $\phi(x_0) = y_0$, and thus ϕ is a solution of the initial value problem (4.3).

We now turn our attention to solving (4.4). As a first approximation to a solution we consider the function ϕ_0 defined by

$$\phi_0(x) = y_0.$$

This function satisfies the initial condition $\phi_0(x_0) = y_0$, but does not in general satisfy (4.4). However, if we compute

$$\phi_1(x) = y_0 + \int_{x_0}^{x} f(t, \phi_0(t))\, dt$$

$$= y_0 + \int_{x_0}^{x} f(t, y_0)\, dt,$$

we might expect that ϕ_1 is a closer approximation to a solution than ϕ_0. In fact, if we continue the process and define successively

$$\phi_0(x) = y_0,$$

$$\phi_{k+1}(x) = y_0 + \int_{x_0}^{x} f(t, \phi_k(t))\, dt, \qquad (k = 0, 1, 2, \cdots),$$

(4.7)

we might expect, on taking the limit as $k \to \infty$, that we would obtain

$$\phi_k(x) \to \phi(x),$$

where ϕ would satisfy

$$\phi(x) = y_0 + \int_{x_0}^{x} f(t, \phi(t))\, dt.$$

Thus ϕ would be our desired solution.

We call the functions $\phi_0, \phi_1, \cdots$ defined by (4.7) *successive approximations* to a solution of the integral equation (4.4), or the initial value problem (4.3). One way to picture the successive approximations is to think of a machine S (for *solving*) which converts functions ϕ into new functions $S(\phi)$ defined by

$$S(\phi)(x) = y_0 + \int_{x_0}^{x} f(t, \phi(t)) \, dt.$$

A solution of the initial value problem (4.3) would then be a function ϕ which moves through the machine untouched, that is, a function satisfying $S(\phi) = \phi$. Starting with $\phi_0(x) = y_0$, we see that S converts ϕ_0 into ϕ_1, and then ϕ_1 into ϕ_2. In general $S(\phi_k) = \phi_{k+1}$, and ultimately we end up with a ϕ such that $S(\phi) = \phi$; see Fig. 5.

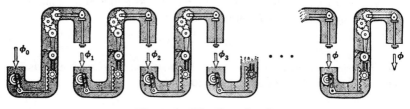

Figure 5. The "S-machine"

Of course we need to show that the ϕ_k merit the name, that is, we need to show that all the ϕ_k exist on some interval I containing x_0, and that they converge there to a solution of (4.4), or of (4.3). Before doing this let us consider an example:

$$y' = xy, \qquad y(0) = 1. \tag{4.8}$$

The integral equation corresponding to this problem is

$$y = 1 + \int_0^x ty \, dt,$$

and the successive approximations are given by

$$\phi_0(x) = 1,$$

$$\phi_{k+1}(x) = 1 + \int_0^x t\phi_k(t) \, dt, \qquad (k = 0, 1, 2, \cdots).$$

Thus

$$\phi_1(x) = 1 + \int_0^x t \, dt = 1 + \frac{x^2}{2},$$

$$\phi_2(x) = 1 + \int_0^x t\left(1 + \frac{t^2}{2}\right) dt = 1 + \frac{x^2}{2} + \frac{x^4}{2\cdot4},$$

and it may be established by induction that

$$\phi_k(x) = 1 + \left(\frac{x^2}{2}\right) + \frac{1}{2!}\left(\frac{x^2}{2}\right)^2 + \cdots + \frac{1}{k!}\left(\frac{x^2}{2}\right)^k.$$

We recognize $\phi_k(x)$ as a partial sum for the series expansion of the function

$$\phi(x) = e^{x^2/2}.$$

We know that this series converges for all real x and this just means that

$$\phi_k(x) \to \phi(x), \qquad (k \to \infty),$$

for all real x. The function ϕ is the solution of the problem (4.8).

Let us now show that there is an interval I containing x_0 where all the functions ϕ_k, $k = 0, 1, 2, \cdots$, defined by (4.7) exist. Since f is continuous on R, it is bounded there, that is, there exists a constant $M > 0$ such that

$$|f(x, y)| \leq M$$

for all (x, y) in R^*. Let α be the smaller of the two numbers a, b/M. Then we prove that the ϕ_k are all defined on $|x - x_0| \leq \alpha$.

Theorem 5. *The successive approximations ϕ_k, defined by* (4.7), *exist as continuous functions on*

$$I: \qquad |x - x_0| \leq \alpha = minimum \{a, b/M\},$$

and $(x, \phi_k(x))$ *is in* R *for* x *in* I. *Indeed, the* ϕ_k *satisfy*

$$|\phi_k(x) - y_0| \leq M|x - x_0| \tag{4.9}$$

for all x *in* I.

Note: Since for x in I, $|x - x_0| \leq b/M$, the inequality (4.9) implies that

$$|\phi_k(x) - y_0| \leq b$$

for x in I, which shows that the points $(x, \phi_k(x))$ are in R for x in I. The precise geometric interpretation of the inequality (4.9) is that the graph of each ϕ_k lies in the region T in R bounded by the two lines

$$y - y_0 = M(x - x_0), \qquad y - y_0 = -M(x - x_0),$$

and the lines

$$x - x_0 = \alpha, \qquad x - x_0 = -\alpha;$$

see Figs. 6 and 7.

---

* This result is usually proved in advanced calculus texts. The student may assume that f satisfies the additional condition $|f(x, y)| \leq M$ if he wishes.

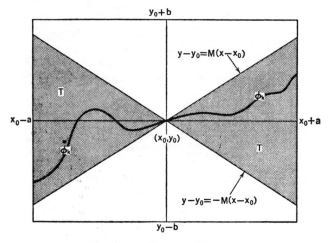

Figure 6. The region $T(\alpha = a)$

Proof of Theorem 5. Clearly ϕ_0 exists on I as a continuous function, and satisfies (4.9) with $k = 0$. Now

$$\phi_1(x) = y_0 + \int_{x_0}^{x} f(t, y_0) \, dt,$$

and hence

$$|\phi_1(x) - y_0| = \left| \int_{x_0}^{x} f(t, y_0) \, dt \right| \leqq \left| \int_{x_0}^{x} |f(t, y_0)| \, dt \right| \leqq M \, |x - x_0|,$$

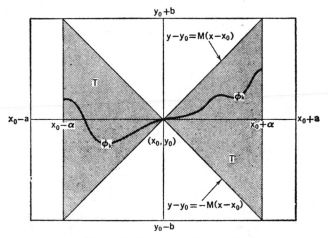

Figure 7. The region $T(\alpha = b/M)$

which shows that ϕ_1 satisfies the inequality (4.9). Since f is continuous on R the function F_0 defined by

$$F_0(t) = f(t, y_0)$$

is continuous on I. Thus ϕ_1, which is given by

$$\phi_1(x) = y_0 + \int_{x_0}^{x} F_0(t) \, dt,$$

is continuous on I.

Now assume the theorem has been proved for the functions ϕ_0, ϕ_1, $\cdots$, ϕ_k. We prove it is valid for ϕ_{k+1}. Indeed the proof is just a repetition of the above. We know that $(t, \phi_k(t))$ is in R for t in I. Thus the function F_k given by

$$F_k(t) = f(t, \phi_k(t))$$

exists for t in I. It is continuous on I since f is continuous on R, and ϕ_k is continuous on I. Therefore ϕ_{k+1}, which is given by

$$\phi_{k+1}(x) = y_0 + \int_{x_0}^{x} F_k(t) \, dt,$$

exists as a continuous function on I. Moreover

$$| \phi_{k+1}(x) - y_0 | \leq \left| \int_{x_0}^{x} | F_k(t) | \, dt \right| \leq M | x - x_0 |,$$

which shows that ϕ_{k+1} satisfies (4.9). The theorem is thus proved by induction.

Our next step is to show that the successive approximations converge on I to a solution of our initial value problem. In order to do this we must impose a further restriction on f. We discuss such a restriction in the next section.

EXERCISES

1. Consider the initial value problem

$$y' = 3y + 1, \qquad y(0) = 2.$$

(a) Show that all the successive approximations ϕ_0, ϕ_1, $\cdots$ exist for all real x.

(b) Compute the first four approximations ϕ_0, ϕ_1, ϕ_2, ϕ_3 to the solution.

(c) Compute the solution by using the method of Chap. 1.

(d) Compare the results of (b) and (c).

2. For each of the following problems compute the first four successive approximations ϕ_0, ϕ_1, ϕ_2, ϕ_3:

(a) $y' = x^2 + y^2$, $y(0) = 0$ (b) $y' = 1 + xy$, $y(0) = 1$

(c) $y' = y^2$, $y(0) = 0$ (d) $y' = y^2$, $y(0) = 1$

3. (a) Show that all the successive approximations for the problem

$$y' = y^2, \qquad y(0) = 1,$$

exist for all real x.

(b) Find a solution of the initial value problem in (a). On what interval does it exist?

(c) Assuming there is just one solution of the problem in (a), indicate why the successive approximations found in (a) can not converge to a solution for all real x.

4. Consider the problem

$$y' = x^2 + y^2, \qquad y(0) = 0,$$

on

$$R: \qquad |x| \leq 1, \qquad |y| \leq 1.$$

(a) Compute an upper bound M for the function $f(x, y) = x^2 + y^2$ on R.

(b) On what interval containing $x = 0$ will all the successive approximations exist, and be such that their graphs are in R?

5. Let f be a real-valued continuous function defined on the rectangle

$$R: \qquad |x - x_0| \leq a, \quad |y - y_0| \leq b, \qquad (a, b > 0).$$

Let ϕ be a real-valued function defined on an interval I containing x_0.

(a) Define carefully what it would mean to say that ϕ is a solution on I of the initial value problem

$$y'' = f(x, y), \qquad y(x_0) = y_0, \qquad y'(x_0) = y_1. \tag{*}$$

(b) Define carefully what it would mean to say that ϕ is a solution on I of the integral equation

$$y = y_0 + (x - x_0)y_1 + \int_{x_0}^{x} (x - t)f(t, y) \, dt. \tag{**}$$

(c) Show that ϕ is a solution of the initial value problem (*) on I if and only if ϕ is a solution of the integral equation (**) on I. (*Hint:* In proving the statement one way it is useful to use the rule that

$$\frac{d}{dx} \int_{a}^{x} F(t, x) \, dt = F(x, x) + \int_{a}^{x} \frac{\partial F}{\partial x}(t, x) \, dt,$$

if F and $\partial F/\partial x$ are continuous. In proving the statement the other way, let $F(x) = f(x, \phi(x))$, and solve

$$y'' = F(x), \qquad y(x_0) = y_0, \qquad y'(x_0) = y_1,$$

by the variation of constants method.)

6. Let f be the same as in Ex. 5.

(a) Define a sequence of successive approximations $\phi_0, \phi_1, \phi_2, \cdots$ for (*), or (**), in Ex. 5. (*Hint*: Let $\phi_0(x) = y_0$.)

(b) Prove Theorem 5 for the sequence $\{\phi_k\}$ of (a), where now $\alpha =$ minimum $\{a, b/M_1\}$, with $M_1 = |y_1| + (Ma/2)$.

7. (a) Find a sequence of successive approximations for the problem

$$y'' = x - y, \qquad y(0) = 1, \qquad y'(0) = 0,$$

and show that the sequence tends to a limit for all real x. (*Hint*: Ex. 6.)

(b) Compare the limit obtained in (a) with the solution of this problem obtained by the methods of Chapter 2.

8. Let f be a real-valued continuous function defined on the strip

$$S: \qquad |x| \le a, \qquad |y| < \infty, \qquad (a > 0),$$

and let I denote the interval $|x| \le a$. Suppose ϕ is a real-valued function on I.

(a) Define what it would mean to say that ϕ is a solution on I of the initial value problem

$$y'' + \lambda^2 y = f(x, y), \qquad y(0) = 0, \qquad y'(0) = 1, \qquad (\lambda > 0). \tag{*}$$

(b) Show that ϕ is a solution of (*) on I if and only if ϕ is a solution of the integral equation

$$y = \frac{\sin \lambda x}{\lambda} + \int_0^x \frac{\sin \lambda(x - t)}{\lambda} f(t, y)\, dt \tag{**}$$

on I. (*Hint*: See the Hint in Ex. 5, (c).)

(c) Define a sequence of successive approximations $\phi_0, \phi_1, \phi_2, \cdots$ for the initial value problem (*), or the integral equation (**), and show that each ϕ_k is defined as a continuous function on I. (*Hint*: Let $\phi_0(x) = 0$. It is a result in advanced calculus that if a function g is continuous in (t, x), then

$$\int_a^x g(t, x)\, dt$$

is continuous in x.)

5. The Lipschitz condition

Let f be a function defined for (x, y) in a set S. We say f satisfies a *Lipschitz condition* on S if there exists a constant $K > 0$ such that

$$| f(x, y_1) - f(x, y_2) | \leq K | y_1 - y_2 |$$

for all (x, y_1), (x, y_2) in S. The constant K is called a *Lipschitz constant*.

If f is continuous and satisfies a Lipschitz condition on the rectangle R, then the successive approximations converge to a solution of the initial value problem on $| x - x_0 | \leq \alpha$. Before we prove this, let us remark that a Lipschitz condition is a rather mild restriction on f.

Theorem 6. *Suppose* S *is either a rectangle*

$$| x - x_0 | \leq a, \qquad | y - y_0 | \leq b, \qquad (a, b > 0),$$

or a strip

$$| x - x_0 | \leq a, \qquad | y | < \infty, \qquad (a > 0),$$

and that f *is a real-valued function defined on* S *such that* $\partial f / \partial y$ *exists, is continuous on* S, *and*

$$\left| \frac{\partial f}{\partial y}(x, y) \right| \leq K, \qquad ((x, y) \ in \ S),$$

for some K > 0. *Then* f *satisfies a Lipschitz condition on* S *with Lipschitz constant* K.

Proof. We have

$$f(x, y_1) - f(x, y_2) = \int_{y_2}^{y_1} \frac{\partial f}{\partial y}(x, t) \, dt,$$

and hence

$$| f(x, y_1) - f(x, y_2) | \leq \left| \int_{y_2}^{y_1} \left| \frac{\partial f}{\partial y}(x, t) \right| dt \right| \leq K | y_1 - y_2 |,$$

for all (x, y_1), (x, y_2) in S.

An example of a function satisfying a Lipschitz condition is

$$f(x, y) = xy^2$$

on

$$R: \qquad | x | \leq 1, \qquad | y | \leq 1.$$

Here

$$\left| \frac{\partial f}{\partial y}(x, y) \right| = | 2xy | \leq 2$$

for (x, y) on R. This function does not satisfy a Lipschitz condition on the strip

$$S: \quad |x| \leq 1, \quad |y| < \infty,$$

since

$$\left| \frac{f(x, y_1) - f(x, 0)}{y_1 - 0} \right| = |x||y_1|,$$

which tends to infinity as $|y_1| \to \infty$, if $|x| \neq 0$.

An example of a continuous function not satisfying a Lipschitz condition on a rectangle is

$$f(x, y) = y^{2/3}$$

on

$$R: \quad |x| \leq 1, \quad |y| \leq 1.$$

Indeed, if $y_1 > 0$,

$$\frac{|f(x, y_1) - f(x, 0)|}{|y_1 - 0|} = \frac{y_1^{2/3}}{y_1} = \frac{1}{y_1^{1/3}},$$

which is unbounded as $y_1 \to 0$.

EXERCISES

1. By computing appropriate Lipschitz constants, show that the following functions satisfy Lipschitz conditions on the sets S indicated:

(a) $f(x, y) = 4x^2 + y^2$, on $S: \quad |x| \leq 1, |y| \leq 1$

(b) $f(x, y) = x^2 \cos^2 y + y \sin^2 x$, on $S: \quad |x| \leq 1, |y| < \infty$

(c) $f(x, y) = x^3 e^{-xy^2}$, on $S: \quad 0 \leq x \leq a, |y| < \infty, (a > 0)$

(d) $f(x, y) = a(x)y^2 + b(x)y + c(x)$, on $S: \quad |x| \leq 1, |y| \leq 2$, ($a, b, c$ are continuous functions on $|x| \leq 1$)

(e) $f(x, y) = a(x)y + b(x)$, on $S: \quad |x| \leq 1, |y| < \infty$, ($a, b$ are continuous functions on $|x| \leq 1$)

2. (a) Show that the function f given by

$$f(x, y) = y^{1/2}$$

does not satisfy a Lipschitz condition on

$$R: \quad |x| \leq 1, \quad 0 \leq y \leq 1.$$

(b) Show that this f satisfies a Lipschitz condition on any rectangle R of the form

$$R: \quad |x| \leq a, \quad b \leq y \leq c, \quad (a, b, c > 0).$$

3. (a) Show that the function f given by

$$f(x, y) = x^2 |y|$$

satisfies a Lipschitz condition on
$$R: \quad |x| \leq 1, \quad |y| \leq 1.$$

(b) Show that $\partial f/\partial y$ does not exist at $(x, 0)$ if $x \neq 0$.

4. Show that the assumption that $\partial f/\partial y$ be continuous on S is superfluous in Theorem 6. (*Hint*: For each fixed x the mean value theorem implies that

$$f(x, y_1) - f(x, y_2) = \frac{\partial f}{\partial y}(x, \eta)(y_1 - y_2),$$

where η (which may depend on x, y_1, y_2) is between y_1 and y_2.)

6. Convergence of the successive approximations

We now prove the main existence theorem.

Theorem 7. (*Existence Theorem*). *Let* f *be a continuous real-valued function on the rectangle*

$$R: \quad |x - x_0| \leq a, \quad |y - y_0| \leq b, \quad (a, b > 0),$$

and let

$$|f(x, y)| \leq M$$

for all (x, y) *in* R. *Further suppose that* f *satisfies a Lipschitz condition with constant* K *in* R. *Then the successive approximations*

$$\phi_0(x) = y_0, \quad \phi_{k+1}(x) = y_0 + \int_{x_0}^{x} f(t, \phi_k(t))\, dt, \quad (k = 0, 1, 2, \cdots),$$

converge on the interval

$$I: \quad |x - x_\lrcorner| \leq \alpha = \min\{a, b/M\}$$

to a solution ϕ *of the initial value problem*

$$y' = f(x, y), \quad y(x_0) = y_0$$

on I.

Note: If f is just continuous on R it is possible to show that there is a solution of the initial value problem on I. Since more sophisticated methods from advanced calculus are required for the proof of this, we shall forego such a proof. However, in order to show that the successive approximations converge to a solution, something more than the continuity of f must be assumed; see Ex. 3.

Proof of Theorem 7. (a) *Convergence of* $\{\phi_k(x)\}$. The key to the proof is the observation that ϕ_k may be written as

$$\phi_k = \phi_0 + (\phi_1 - \phi_0) + (\phi_2 - \phi_1) + \cdots + (\phi_k - \phi_{k-1}),$$

and hence $\phi_k(x)$ is a partial sum for the series

$$\phi_0(x) + \sum_{p=1}^{\infty} [\phi_p(x) - \phi_{p-1}(x)]. \tag{6.1}$$

Therefore to show that the sequence $\{\phi_k(x)\}$ converges is equivalent to showing that the series (6.1) converges. To prove the latter we must estimate the terms $\phi_p(x) - \phi_{p-1}(x)$ of this series.

By Theorem 5 the functions ϕ_p all exist as continuous functions on I, and $(x, \phi_p(x))$ is in R for x in I. Moreover, as shown in Theorem 5,

$$|\phi_1(x) - \phi_0(x)| \leq M |x - x_0| \tag{6.2}$$

for x in I. Writing down the relations defining ϕ_2 and ϕ_1, and subtracting, we obtain

$$\phi_2(x) - \phi_1(x) = \int_{x_0}^{x} [f(t, \phi_1(t)) - f(t, \phi_0(t))] \, dt.$$

Therefore

$$|\phi_2(x) - \phi_1(x)| \leq \left| \int_{x_0}^{x} |f(t, \phi_1(t)) - f(t, \phi_0(t))| \, dt \right|,$$

and since f satisfies the Lipschitz condition

$$|f(x, y_1) - f(x, y_2)| \leq K |y_1 - y_2|,$$

we have

$$|\phi_2(x) - \phi_1(x)| \leq K \left| \int_{x_0}^{x} |\phi_1(t) - \phi_0(t)| \, dt \right|.$$

Using (6.2) we obtain

$$|\phi_2(x) - \phi_1(x)| \leq KM \left| \int_{x_0}^{x} |t - x_0| \, dt \right|.$$

Thus, if $x \geq x_0$,

$$|\phi_2(x) - \phi_1(x)| \leq KM \int_{x_0}^{x} (t - x_0) \, dt = KM \frac{(x - x_0)^2}{2}.$$

The same result is valid in case $x \leq x_0$.

We shall prove by induction that

$$|\phi_p(x) - \phi_{p-1}(x)| \leq \frac{MK^{p-1} |x - x_0|^p}{p!} \tag{6.3}$$

for x in I. We have seen that this is true for $p = 1$ and $p = 2$. Let us assume $x \geq x_0$; the proof is similar for $x \leq x_0$. Assume (6.3) for $p = m$. Using the definition of ϕ_{m+1} and ϕ_m we obtain

$$\phi_{m+1}(x) \ - \ \phi_m(x) \ = \ \int_{x_0}^{x} \left[f(t, \phi_m(t)) \ - f(t, \phi_{m-1}(t)) \right] dt,$$

and thus

$$\left| \phi_{m+1}(x) \ - \ \phi_m(x) \right| \ \leqq \ \int_{x_0}^{x} \left| f(t, \phi_m(t)) \ - f(t, \phi_{m-1}(t)) \right| dt.$$

Using the Lipschitz condition we get

$$\left| \phi_{m+1}(x) \ - \ \phi_m(x) \right| \ \leqq \ K \int_{x_0}^{x} \left| \phi_m(t) \ - \phi_{m-1}(t) \right| dt.$$

Since we have assumed (6.3) for $p = m$ this yields

$$\left| \phi_{m+1}(x) \ - \ \phi_m(x) \right| \ \leqq \ \frac{MK^m}{m!} \int_{x_0}^{x} \left| t \ - x_0 \right|^m dt = \frac{MK^m \left| x \ - x_0 \right|^{m+1}}{(m+1)!}.$$

This is just (6.3) for $p = m + 1$, and hence (6.3) is valid for all $p = 1, 2,$ $\cdots$, by induction.

It follows from (6.3) that the infinite series

$$\phi_0(x) \ + \ \sum_{p=1}^{\infty} \left[\phi_p(x) \ - \ \phi_{p-1}(x) \right] \tag{6.1}$$

is absolutely convergent on I, that is, the series

$$\left| \phi_0(x) \right| \ + \ \sum_{p=1}^{\infty} \left| \phi_p(x) \ - \ \phi_{p-1}(x) \right| \tag{6.4}$$

is convergent on I. Indeed, from (6.3) we see that

$$\left| \phi_p(x) \ - \ \phi_{p-1}(x) \right| \ \leqq \ \frac{M}{K} \frac{K^p \left| x \ - x_0 \right|^p}{p!},$$

which shows that the p-th term of the series in (6.4) is less than or equal to M/K times the p-th term of the power series for $e^{K|x-x_0|}$. Since the power series for $e^{K|x-x_0|}$ is convergent, the series (6.4) is convergent for x in I. This implies that the series (6.1) is convergent on I. Therefore the k-th partial sum of (6.1), which is just $\phi_k(x)$, tends to a limit $\phi(x)$ as $k \to \infty$, for each x in I.

(b) *Properties of the limit ϕ.* This limit function ϕ is a solution to our problem on I. First, let us show that ϕ is continuous on I. This may be seen in the following way. If x_1, x_2 are in I

$$\left| \phi_{k+1}(x_1) \ - \ \phi_{k+1}(x_2) \right| \ = \ \left| \int_{x_2}^{x_1} f(t, \phi_k(t)) dt \right| \ \leqq \ M \left| x_1 \ - x_2 \right|,$$

which implies, by letting $k \to \infty$,

$$| \phi(x_1) - \phi(x_2) | \leqq M | x_1 - x_2 |. \tag{6.5}$$

This shows that as $x_2 \to x_1$, $\phi(x_2) \to \phi(x_1)$, that is, ϕ is continuous on I. Also, letting $x_1 = x$, $x_2 = x_0$ in (6.5) we obtain

$$| \phi(x) - y_0 | \leqq M | x - x_0 |, \qquad (x \text{ in } I),$$

which implies that the points $(x, \phi(x))$ are in R for all x in I.

(c) *Estimate for* $| \phi(x) - \phi_k(x) |$. We now estimate $| \phi(x) - \phi_k(x) |$. We have

$$\phi(x) = \phi_0(x) + \sum_{.p=1}^{\infty} [\phi_p(x) - \phi_{p-1}(x)],$$

and

$$\phi_k(x) = \phi_0(x) + \sum_{p=1}^{k} [\phi_p(x) - \phi_{p-1}(x)].$$

Therefore, using (6.3), we find that

$$| \phi(x) - \phi_k(x) | = \left| \sum_{p=k+1}^{\infty} [\phi_p(x) - \phi_{p-1}(x)] \right|$$

$$\leqq \sum_{p=k+1}^{\infty} | \phi_p(x) - \phi_{p-1}(x) |$$

$$\leqq \frac{M}{K} \sum_{p=k+1}^{\infty} \frac{(K\alpha)^p}{p!}$$

$$\leqq \frac{M}{K} \frac{(K\alpha)^{k+1}}{(k+1)!} \sum_{p=0}^{\infty} \frac{(K\alpha)^p}{p!}$$

$$= \frac{M}{K} \frac{(K\alpha)^{k+1}}{(k+1)!} e^{K\alpha}. \tag{6.6}$$

Letting $\epsilon_k = (K\alpha)^{k+1}/(k+1)!$, we see that $\epsilon_k \to 0$ as $k \to \infty$, since ϵ_k is a general term for the series for $e^{K\alpha}$. In terms of ϵ_k (6.6) may be written as

$$| \phi(x) - \phi_k(x) | \leqq \frac{M}{K} e^{K\alpha} \epsilon_k, \qquad (\epsilon_k \to 0, \quad k \to \infty). \tag{6.7}$$

(d) *The limit ϕ is a solution.* To complete the proof we must show that

$$\phi(x) = y_0 + \int_{x_0}^{x} f(t, \phi(t)) \, dt \tag{6.8}$$

for all x in I. The right side of (6.8) makes sense for ϕ is continuous on I, f

is continuous on R, and thus the function F given by

$$F(t) = f(t, \phi(t))$$

is continuous on I. Now

$$\phi_{k+1}(x) = y_0 + \int_{x_0}^{x} f(t, \phi_k(t)) \, dt,$$

and $\phi_{k+1}(x) \to \phi(x)$, as $k \to \infty$. Thus to prove (6.8) we must show that for each x in I

$$\int_{x_0}^{x} f(t, \phi_k(t)) dt \to \int_{x_0}^{x} f(t, \phi(t)) \, dt, \qquad (k \to \infty). \tag{6.9}$$

We have

$$\left| \int_{x_0}^{x} f(t, \phi(t)) \, dt - \int_{x_0}^{x} f(t, \phi_k(t)) \, dt \right|$$

$$\leq \left| \int_{x_0}^{x} |f(t, \phi(t)) - f(t, \phi_k(t))| \, dt \right|$$

$$\leq K \left| \int_{x_0}^{x} |\phi(t) - \phi_k(t)| \, dt \right|, \tag{6.10}$$

using the fact that f satisfies a Lipschitz condition. The estimate (6.7) can now be used in (6.10) to obtain

$$\left| \int_{x_0}^{x} f(t, \phi(t)) \, dt - \int_{x_0}^{x} f(t, \phi_k(t)) \, dt \right| \leq M e^{K\alpha} \, \epsilon_k \, | x - x_0 |$$

which tends to zero as $k \to \infty$, for each x in I. This proves (6.9), and hence that ϕ satisfies (6.8). Thus our proof of Theorem 7 is now complete.

The estimate (6.6) of how well the k-th approximation ϕ_k approximates the solution ϕ is worthy of special attention.

Theorem 8. *The* k-*th successive approximation* ϕ_k *to the solution* ϕ *of the initial value problem of Theorem 7 satisfies*

$$| \phi(x) - \phi_k(x) | \leq \frac{M}{K} \frac{(K\alpha)^{k+1}}{(k + 1)!} e^{K\alpha}$$

for all x *in* I.

EXERCISES

1. Consider the problem

$$y' = 1 - 2xy, \qquad y(0) = 0.$$

(a) Since the differential equation is linear, an expression can be found for the solution. Find it.

(b) Consider the above problem on

$$R: \qquad |x| \leq \tfrac{1}{2}, \qquad |y| \leq 1.$$

If $f(x, y) = 1 - 2xy$, show that

$$|f(x, y)| \leq 2, \qquad ((x, y) \text{ in } R),$$

and that all the successive approximations to the solution exist on $|x| \leq \tfrac{1}{2}$, and their graphs remain in R.

(c) Show that f satisfies a Lipschitz condition on R with Lipschitz constant $K = 1$, and therefore by Theorem 7 the successive approximations converge to a solution ϕ of the initial value problem on $|x| \leq \tfrac{1}{2}$.

(d) Show that the approximation ϕ_3 satisfies

$$|\phi(x) - \phi_3(x)| < .01$$

for $|x| \leq \tfrac{1}{2}$.

(e) Compute ϕ_3.

2. Consider the problem

$$y' = 1 + y^2, \qquad y(0) = 0.$$

(a) Using separation of variables, find the solution ϕ of this problem. (It is not difficult to convince oneself that the separation of variables technique gives the only solution of the problem.) On what interval does ϕ exist?

(b) Show that all the successive approximations $\phi_0, \phi_1, \phi_2, \cdots$ exist for all real x.

(c) Show that $\phi_k(x) \to \phi(x)$ for each x satisfying $|x| \leq \tfrac{1}{2}$. (*Hint*: Consider $f(x, y) = 1 + y^2$ on

$$R: \qquad |x| \leq \tfrac{1}{2}, \qquad |y| \leq 1.$$

Show that $\alpha = \tfrac{1}{2}$.)

3. On the square

$$R: \qquad |x| \leq 1, \qquad |y| \leq 1,$$

let f be defined by

$$
\begin{aligned}
f(x, y) &= 0, & \text{if } x = 0, & \qquad |y| \leq 1, \\
&= 2x, & \text{if } 0 < |x| \leq 1, & \qquad -1 \leq y < 0, \\
&= 2x - \frac{4y}{x}, & \text{if } 0 < |x| \leq 1, & \qquad 0 \leq y \leq x^2, \\
&= -2x, & \text{if } 0 < |x| \leq 1, & \qquad x^2 \leq y \leq 1.
\end{aligned}
$$

(a) Show that this f is continuous on R, and $|f(x, y)| \leq 2$ on R. (It might help to make a sketch.)

(b) Show that this f does not satisfy a Lipschitz condition on R.

(c) Show that the successive approximations $\phi_0, \phi_1, \phi_2, \cdots$ for the problem

$$y' = f(x, y), \qquad y(0) = 0,$$

satisfy

$$\phi_0(x) = 0, \quad \phi_{2m-1}(x) = x^2, \quad \phi_{2m}(x) = -x^2, \quad (m = 1, 2, \cdots).$$

(d) Prove that neither of the convergent subsequences in (c) converge to a solution of the initial value problem. (*Note*: This problem has a solution, but the above shows that it can not be obtained by using successive approximations.)

4. Consider f, R as in Theorem 7. Let ϕ_0 be *any* continuous function on $|x - x_0| \leq a$ such that the points $(x, \phi_0(x))$ are in R for $|x - x_0| \leq a$. Let

$$\phi_1(x) = y_0 + \int_{x_0}^{x} f(t, \phi_0(t)) \, dt,$$

and

$$\phi_{k+1}(x) = y_0 + \int_{x_0}^{x} f(t, \phi_k(t)) \, dt, \qquad (k = 1, 2, \cdots).$$

(a) Show that all the functions $\phi_1, \phi_2, \cdots$ exist and are continuous for $|x - x_0| \leq \alpha$, and satisfy

$$|\phi_k(x) - y_0| \leq M |x - x_0|, \qquad (k = 1, 2, \cdots).$$

(b) Show that $\phi_k(x) \to \phi(x)$ on $|x - x_0| \leq \alpha$, where ϕ is a solution to the initial value problem

$$y' = f(x, y), \qquad y(x_0) = y_0.$$

(*Hint*: Show that the proof is a repetition of most of the proof of Theorem 7.)
(*Note*: This shows that we may start our successive approximation procedure with any function ϕ_0 with the above properties, instead of with the particular one $\phi_0(x) = y_0$.)

(c) Show that an estimate like that in Theorem 8 is valid, namely

$$|\phi(x) - \phi_k(x)| \leq \frac{2M}{K} \frac{(K\alpha)^k}{k!} e^{K\alpha}.$$

5. Let f satisfy the conditions of Theorem 7. Show that the successive approximations

$$\phi_0(x) = y_0,$$

$$\psi_{k+1}(x) = y_0 + (x - x_0)y_1 + \int_{x_0}^{x} (x - t)f(t, \phi_k(t)) \, dt, \qquad (k = 0, 1, 2, \cdots),$$

converge on the interval

$$I: \qquad |x - x_0| \leq \alpha = \text{minimum } \{a, b/M_1\},$$

where $M_1 = |y_1| + (Ma/2)$, to a solution of the initial value problem

$$y'' = f(x, y), \qquad y(x_0) = y_0, \qquad y'(x_0) = y_1.$$

(*Note*: From Exs. 5, 6 of Sec. 4 it follows that each ϕ_k exists, is continuous on I, and that $(x, \phi_k(x))$ is in R for all x in I.)

7. Non-local existence of solutions

Theorem 7 is called a *local* existence theorem since it guarantees a solution only for x near the initial point x_0. There are many cases when a solution to the initial value problem exists on the entire interval $|x - x_0| \leqq a$, and in such cases we say that a solution exists *non-locally*.

As seen in Sec. 1, an example of non-local existence is furnished by the linear equation

$$y' + g(x)y = h(x). \qquad (7.1)$$

The solutions exist on every interval where g and h are continuous. Suppose g and h are continuous on $|x - x_0| \leqq a$, and that K is a positive constant such that

$$|g(x)| \leqq K, \qquad (|x - x_0| \leqq a).$$

Then if we write (7.1) as

$$y' = f(x, y) = -g(x)y + h(x),$$

we see that

$$|f(x, y_1) - f(x, y_2)| = |-g(x)(y_1 - y_2)| \leqq K|y_1 - y_2|,$$

for all (x, y_1), (x, y_2) in the strip

$$S: \qquad |x - x_0| \leqq a, \qquad |y| < \infty.$$

By looking carefully at the proof of Theorem 7 we can show that if f satisfies a Lipschitz condition in a strip S, instead of in a rectangle R, then solutions will exist on the entire interval.

Theorem 9. *Let* f *be a real-valued continuous function on the strip*

$$S: \qquad |x - x_0| \leqq a, \qquad |y| < \infty, \qquad (a > 0),$$

and suppose that f *satisfies on* S *a Lipschitz condition with constant* K > 0. *The successive approximations* $\{\phi_k\}$ *for the problem*

$$y' = f(x, y), \qquad y(x_0) = y_0, \qquad (7.2)$$

exist on the entire interval $|x - x_0| \leqq a$, *and converge there to a solution* ϕ *of* (7.2).

Proof. The successive approximations are given by

$$\phi_0(x) = y_0,$$

$$\phi_{k+1}(x) = y_0 + \int_{x_0}^{x} f(t, \phi_k(t)) \, dt, \qquad (k = 0, 1, 2, \cdots).$$

An induction argument establishes the existence of each ϕ_k for

$$|x - x_0| \leqq a;$$

see the proof of Theorem 5.

Since f is continuous on S, the function F_0 given by

$$F_0(x) = f(x, y_0)$$

is continuous for $|x - x_0| \leqq a$, and hence bounded there. Let M be any positive constant such that

$$|f(x, y_0)| \leqq M, \qquad (|x - x_0| \leqq a). \tag{7.3}$$

The proof of the convergence of $\{\phi_k(x)\}$ now follows that of part (a) of the proof of Theorem 7, once we note that

$$|\phi_1(x) - \phi_0(x)| = \left| \int_{x_0}^{x} f(t, y_0) \, dt \right|$$

$$\leqq \left| \int_{x_0}^{x} |f(t, y_0)| \, dt \right| \leqq M |x - x_0|,$$

due to (7.3).

The limit function ϕ need no longer satisfy the inequality (6.5) for the M given in (7.3). However, we note that (6.3) is valid, and this implies that

$$|\phi_k(x) - y_0| = \left| \sum_{p=1}^{k} [\phi_p(x) - \phi_{p-1}(x)] \right| \leqq \sum_{p=1}^{k} |\phi_p(x) - \phi_{p-1}(x)|$$

$$\leqq \frac{M}{K} \sum_{p=1}^{k} \frac{K^p |x - x_0|^p}{p!} \leqq \frac{M}{K} \sum_{p=1}^{\infty} \frac{K^p |x - x_0|^p}{p!}$$

$$\leqq \frac{M}{K} (e^{Ka} - 1),$$

for $|x - x_0| \leqq a$. If we let

$$b = \frac{M}{K} (e^{Ka} - 1),$$

we see that the approximations satisfy

$$|\phi_k(x) - y_0| \leqq b, \qquad (|x - x_0| \leqq a),$$

and taking the limit as $k \to \infty$ we obtain

$$| \phi(x) - y_0 | \leqq b, \qquad (| x - x_0 | \leqq a).$$

Now since f is continuous on

$$R: \qquad | x - x_0 | \leqq a, \qquad | y - y_0 | \leqq b,$$

it is bounded there, that is, there is a positive constant N such that

$$| f(x, y) | \leqq N$$

for (x, y) in R. The continuity of ϕ may now be exhibited just as in part (b) of the proof of Theorem 7. Indeed, for x_1, x_2 in our interval $| x - x_0 | \leqq a$,

$$| \phi_{k+1}(x_1) - \phi_{k+1}(x_2) | = \left| \int_{x_2}^{x_1} f(t, \phi_k(t) \, dt \right| \leq N | x_1 - x_2 |,$$

which implies, on letting $k \to \infty$,

$$| \phi(x_1) - \phi(x_2) | \leqq N | x_1 - x_2 |.$$

The remainder of the proof is a repetition of parts (c) and (d) of the proof of Theorem 7, with α replaced by a everywhere.

Corollary. *Suppose* f *is a real-valued continuous function on the plane*

$$| x | < \infty, \qquad | y | < \infty,$$

which satisfies a Lipschitz condition on each strip

$$S_a: \qquad | x | \leqq a, \qquad | y | < \infty,$$

where a *is any positive number.* * *Then every initial value problem*

$$y' = f(x, y), \qquad y(x_0) = y_0,$$

has a solution which exists for all real x.

Proof. If x is any real number there is an $a > 0$ such that x is contained inside an interval $| x - x_0 | \leqq a$. For this a the function f satisfies the conditions of Theorem 9 on the strip

$$| x - x_0 | \leqq a, \qquad | y | < \infty,$$

since this strip is contained in the strip

$$| x | \leqq | x_0 | + a, \qquad | y | < \infty.$$

Thus $\{\phi_k(x)\}$ tends to $\phi(x)$, where ϕ is a solution to the initial-value problem.

---

* The Lipschitz constant K_a for f in S_a may depend on a.

An example of a nonlinear equation satisfying the conditions of this corollary is

$$y' = \frac{y^3 e^x}{1 + y^2} + x^2 \cos y. \tag{7.4}$$

If we let $f(x, y)$ denote the right side of (7.4) we see that f is continuous on the plane. Since

$$\frac{\partial f}{\partial y}(x, y) = \frac{(y^4 + 3y^2)}{(1 + y^2)^2} e^x - x^2 \sin y,$$

we have

$$\left| \frac{\partial f}{\partial y}(x, y) \right| \leq 3e^a + a^2$$

for all (x, y) in the strip

$$S_a: \qquad |x| \leq a, \qquad |y| < \infty.$$

Hence, by Theorem 6, f satisfies a Lipschitz condition on S_a with Lipschitz constant $K_a = 3e^a + a^2$. Therefore equation (7.4), together with any initial condition $y(x_0) = y_0$, is a problem which has a solution existing for all real x.

Note that the function f given by

$$f(x, y) = y^2$$

does not satisfy a Lipschitz condition on any strip S_a, although it satisfies one on any rectangle R. As we have seen in Sec. 1 the problem

$$y' = y^2, \qquad y(1) = -1,$$

has a solution ϕ which exists only for $x > 0$.

EXERCISES

1. Consider the equation

$$y' = (3x^2 + 1) \cos^2 y + (x^3 - 2x) \sin 2y$$

on the strip $S_a: |x| \leq a$ $(a > 0)$. If $f(x, y)$ denotes the right side of this equation, show that f satisfies a Lipschitz condition on the strip S_a, and hence every initial value problem

$$y' = f(x, y), \qquad y(x_0) = y_0,$$

has a solution which exists for all real x.

2. Let

$$f(x, y) = \frac{\cos y}{1 - x^2}, \qquad (|x| < 1).$$

(a) Show that f satisfies a Lipschitz condition on every strip S_a: $|x| \leqq a$, where $0 < a < 1$.

(b) Show that every initial value problem

$$y' = f(x, y), \qquad y(0) = y_0, \qquad (|y_0| < \infty),$$

has a solution which exists for $|x| < 1$.

3. Consider the equation

$$y' = f(x)p(\cos y) + g(x)q(\sin y),$$

where f, g are continuous for all real x, and p, q are polynomials. Show that every initial value problem for this equation has a solution which exists for all real x.

4. Let f be a real-valued continuous function on the strip

$$S: \qquad |x - x_0| \leqq a, \qquad |y| < \infty, \qquad (a > 0),$$

and suppose that f satisfies on S a Lipschitz condition with constant $K > 0$. Show that the successive approximations

$$\phi_0(x) = y_0,$$

$$\phi_{k+1}(x) = y_0 + (x - x_0)y_1 + \int_{x_0}^{x} (x - t)f(t, \phi_k(t)) \, dt, \qquad (k = 0, 1, 2, \cdots),$$

exist as continuous functions on the whole interval $I : |x - x_0| \leqq a$, and converge on I to a solution ϕ of the initial value problem

$$y'' = f(x, y), \qquad y(x_0) = y_0, \qquad y'(x_0) = y_1.$$

5. Prove the Corollary to Theorem 9 for the initial value problem

$$y'' = f(x, y), \qquad y(x_0) = y_0, \qquad y'(x_0) = y_1.$$

6. Let f be a real-valued continuous function on the strip

$$S: \qquad |x| \leqq a, \qquad |y| < \infty, \qquad (a > 0),$$

and suppose f satisfies a Lipschitz condition on S with constant $K > 0$. Show that the successive approximations

$$\phi_0(x) = 0,$$

$$\phi_{k+1}(x) = \frac{\sin \lambda x}{\lambda} + \int_0^x \frac{\sin \lambda(x - t)}{\lambda} f(t, \phi_k(t)) \, dt, \qquad (\lambda > 0),$$

$$(k = 0, 1, 2, \cdots),$$

exist as continuous functions on I: $|x| \leqq a$, and converge there to a solution ϕ of the initial value problem

$$y'' + \lambda^2 y = f(x, y), \qquad y(0) = 0, \qquad y'(0) = 1.$$

(*Hint*: See Ex. 8, Sec. 4.) (*Note*: The existence of a solution to the initial value problem can also be demonstrated by applying Ex. 4 to the problem

$$y'' = f(x, y) - \lambda^2 y, \qquad y(0) = 0, \qquad y'(0) = 1.)$$

7. Prove the Corollary to Theorem 9 for the initial value problem

$$y'' + \lambda^2 y = f(x, y), \qquad y(0) = 0, \qquad y'(0) = 1.$$

8. Let q be a real-valued continuous function on I: $|x| \leq a$, where $a > 0$. Consider the initial value problem

$$y'' + \lambda^2 y = q(x)y, \qquad (\lambda \geq 0), \qquad y(0) = 0, \qquad y'(0) = 1. \qquad (*)$$

(a) Show that there is a solution ϕ of (*) on I, and give an integral equation which ϕ also satisfies.

(b) If q is continuous for all real x, show that there is a solution of (*) for all real x. (*Hint*: See Exs. 4, 5, 6, 7.)

8. Approximations to, and uniqueness of, solutions

Under the same assumptions as in Theorem 7 we can show that the solution obtained there is the only one satisfying the initial value problem on I. The method of proof can be adapted to yield other important information concerning approximations to solutions. Suppose we have two initial value problems

$$y' = f(x, y), \qquad y(x_0) = y_1, \qquad (8.1)$$

and

$$y' = g(x, y), \qquad y(x_0) = y_2, \qquad (8.2)$$

where f, g are both continuous real-valued functions on

$$R: \qquad |x - x_0| \leq a, \qquad |y - y_0| \leq b, \qquad (a, b > 0),$$

and (x_0, y_1), (x_0, y_2) are points in R. We shall show that if g is close to f, and y_2 close to y_1, then any solution ψ of (8.2) on an interval I containing x_0 is close to a solution ϕ of (8.1) on I. Suppose there exist non-negative constants ϵ, δ such that

$$|f(x, y) - g(x, y)| \leq \epsilon, \qquad ((x, y) \text{ in } R), \qquad (8.3)$$

and

$$|y_1 - y_2| \leq \delta. \qquad (8.4)$$

Then we have the following result.

Theorem 10. *Let* f, g *be continuous on* R, *and suppose* f *satisfies a Lipschitz condition there with Lipschitz constant* K. *Let* ϕ, ψ *be solutions of*

(8.1), (8.2) *respectively on an interval* I *containing* x_0, *with graphs contained in* R. *If the inequalities* (8.3), (8.4) *are valid, then*

$$| \phi(x) - \psi(x) | \leqq \delta e^{K|x-x_0|} + \frac{\epsilon}{K}(e^{K|x-x_0|} - 1) \tag{8.5}$$

for all x *in* I.

Before proving Theorem 10 let us note some consequences of this inequality (8.5). If we take $g = f$ and $y_0 = y_1 = y_2$ we see that we may choose $\epsilon = 0$, $\delta = 0$, and we have

Corollary 1. (*Uniqueness theorem*) *Let* f *be continuous and satisfy a Lipschitz condition on* R. *If* ϕ *and* ψ *are two solutions of*

$$y' = f(x, y), \qquad y(x_0) = y_0,$$

on an interval I *containing* x_0, *then* $\phi(x) = \psi(x)$ *for all* x *in* I.

We remark that some restriction on f, in addition to continuity, is required in order to guarantee uniqueness. The initial value problem

$$y' = 3y^{2/3}, \qquad y(0) = 0,$$

considered in Sec. 2, illustrates this. Here $f(x, y) = 3y^{2/3}$, and thus f is continuous for all (x, y). The two functions ϕ, ψ given by

$$\phi(x) = x^3, \qquad \psi(x) = 0, \qquad (-\infty < x < \infty),$$

are both solutions of this problem. Of course, as we have seen in Sec. 5, this f does not satisfy a Lipschitz condition on any rectangle containing the origin.

Intuitively, if we have a sequence of functions $g_k \to f$ on R, and a sequence $y_k \to y_0$, we would expect that the solutions ψ_k of

$$y' = g_k(x, y), \qquad y(x_0) = y_k, \tag{8.6}$$

would tend to the solution ϕ of

$$y' = f(x, y), \qquad y(x_0) = y_0. \tag{8.7}$$

This is a direct consequence of (8.5). Suppose the g_k are continuous on R and there are constants ϵ_k such that

$$| f(x, y) - g_k(x, y) | \leqq \epsilon_k, \qquad (\text{all } (x, y) \text{ in } R), \tag{8.8}$$

and constants δ_k such that

$$| y_k - y_0 | \leqq \delta_k,$$

where ϵ_k and δ_k tend to 0 as $k \to \infty$. Applying (8.5) we obtain

Corollary 2. *Let* f *be continuous and satisfy a Lipschitz condition on* R. *Let the* g_k *(k = 1, 2, $\cdots$) be continuous on* R *and satisfy* (8.8) *for some constants*

$$\epsilon_k \to 0 \quad (k \to \infty), \quad and \; let \quad y_k \to y_0 \quad (k \to \infty).$$

If ψ_k *is a solution of* (8.6) *on an interval* I *containing* x_0, *and* ϕ *is the solution of* (8.7) *on* I, *then* $\psi_k(x) \to \phi(x)$ *on* I.

Proof of Theorem 10. From (8.1), (8.2) we see that

$$\phi(x) = y_1 + \int_{x_0}^{x} f(t, \phi(t)) \; dt,$$

$$\psi(x) = y_2 + \int_{x_0}^{x} g(t, \psi(t)) \; dt,$$

and hence

$$\phi(x) - \psi(x) = y_1 - y_2 + \int_{x_0}^{x} [f(t, \phi(t)) - g(t, \psi(t))] \, dt$$

$$= y_1 - y_2 + \int_{x_0}^{x} [f(t, \phi(t)) - f(t, \psi(t))] \, dt$$

$$+ \int_{x_0}^{x} [f(t, \psi(t)) - g(t, \psi(t))] \, dt.$$

Using (8.3), (8.4), and the fact that f satisfies a Lipschitz condition with constant K, we obtain for $x \geqq x_0$

$$| \phi(x) - \psi(x) | \leqq \delta + K \int_{x_0}^{x} | \phi(t) - \psi(t) | \, dt + \epsilon(x - x_0). \tag{8.9}$$

If

$$E(x) = \int_{x_0}^{x} | \phi(t) - \psi(t) | \, dt$$

we see that (8.9) may be written as

$$E'(x) - KE(x) \leqq \delta + \epsilon(x - x_0). \tag{8.10}$$

This is a first order differential inequality which we may "solve" in the same way we solve first order linear differential equations. Multiplying

(8.10) by $e^{-K(x-x_0)}$ we get, after changing x to t,

$$[e^{-K(t-x_0)}E]'(t) \leqq \delta e^{-K(t-x_0)} + \epsilon(t - x_0)e^{-K(t-x_0)}.$$

An integration from x_0 to x yields*

$$e^{-K(x-x_0)}E(x) \leqq \frac{\delta}{K}[1 - e^{-K(x-x_0)}]$$

$$+ \frac{\epsilon}{K^2}[-K(x - x_0) - 1]e^{-K(x-x_0)} + \frac{\epsilon}{K^2}.$$

Multiplying both sides of this inequality by $e^{K(x-x_0)}$ we find

$$E(x) \leqq \frac{\delta}{K}[e^{K(x-x_0)} - 1] - \frac{\epsilon}{K^2}[K(x - x_0) + 1] + \frac{\epsilon}{K^2}e^{K(x-x_0)},$$

and using this in (8.9) we obtain finally

$$|\phi(x) - \psi(x)| \leqq \delta e^{K(x-x_0)} + \frac{\epsilon}{K}[e^{K(x-x_0)} - 1].$$

This is just (8.5) for $x \geqq x_0$. A similar proof holds in case $x \leqq x_0$.

EXERCISES

1. Consider the initial value problem

$$y' = xy + y^{10}, \qquad y(0) = \tfrac{1}{10}. \qquad (*)$$

(a) Show that a solution ψ of this problem exists for $|x| \leqq \frac{1}{2}$. (*Hint*: Consider this problem on

$$R: \qquad |x| \leqq \tfrac{1}{2}, \qquad |y - \tfrac{1}{10}| \leqq \tfrac{1}{10}.$$

If $g(x, y) = xy + y^{10}$, show that

$$|g(x, y)| < \tfrac{1}{5}$$

for (x, y) in R, and hence that the α of Theorem 7 may be taken to be $\frac{1}{2}$.)

(b) For small $|y|$ the problem (*) can be approximated by the problem

$$y' = xy, \qquad y(0) = \tfrac{1}{10}.$$

Compute a solution ϕ of this problem, and show that its graph is in R for $|x| \leqq \frac{1}{2}$.

---

* Recall that if c is a constant ($c \neq 0$)

$$\int te^{ct}\,dt = \frac{1}{c^2}(ct - 1)e^{ct}.$$

We have also used the fact that $E(x_0) == 0$.

(c) Show that

$$| \phi(x) - \psi(x) | \leqq \frac{2}{5^{10}} (e^{|x|/2} - 1)$$

for $| x | \leqq \frac{1}{2}$. (*Hint*: Apply Theorem 10 with $f(x, y) = xy$ on R.)

(d) Prove also that

$$| \phi(x) - \psi(x) | \leqq \frac{1}{5^{10}} (e^{|x|} - 1).$$

2. Consider the problem

$$y' = y + \lambda x^2 \sin y, \qquad y(0) = 1,$$

where λ is some real parameter, $| \lambda | \leqq 1$.

(a) Show that the solution ψ of this problem exists for $| x | \leqq 1$.

(b) Prove that

$$| \psi(x) - e^x | \leqq | \lambda | (e^{|x|} - 1)$$

for $| x | \leqq 1$.

3. Let f be a continuous function for (x, y, λ) in

$$R : \qquad | x - x_0 | \leqq a, \qquad | y - y_0 | \leqq b, \qquad | \lambda - \lambda_0 | \leqq c,$$

where $a, b, c > 0$, and suppose there is a constant $K > 0$ such that

$$| f(x, y_1, \lambda) - f(x, y_2, \lambda) | \leqq K | y_1 - y_2 |$$

for all (x, y_1, λ), (x, y_2, λ) in R. Further suppose that $\partial f / \partial \lambda$ exists and there is a constant $L > 0$ such that

$$\left| \frac{\partial f}{\partial \lambda} (x, y, \lambda) \right| \leqq L$$

for all (x, y, λ) in R. If ϕ_λ represents the solution of

$$y' = f(x, y, \lambda), \qquad y(x_0) = y_0,$$

show that

$$| \phi_\lambda(x) - \phi_\mu(x) | \leqq \frac{L | \lambda - \mu |}{K} (e^{K|x-x_0|} - 1)$$

for all x for which ϕ_λ, ϕ_μ exist.

4. (a) Apply Ex. 3 to the initial value problem

$$y' + \lambda^2 y = q(x)y, \qquad y(x_0) = y_0, \qquad \text{(*)}$$

where λ is real, and q is continuous for $| x - x_0 | \leqq a$.

(b) Solve (*) using the method of Chap. 1.

5. Let f, g be as in Theorem 10, and consider the two initial value problems

$$y'' = f(x, y), \qquad y(x_0) = y_0, \qquad y'(x_0) = y_1, \tag{*}$$

$$y'' = g(x, y), \qquad y(x_0) = z_0, \qquad y'(x_0) = z_1, \tag{**}$$

Suppose ϕ, ψ are solutions of (*) and (**), respectively, on an interval I containing x_0. State, and prove, analogues of Theorem 10, and Corollaries 1 and 2. (*Hint*: From Ex. 5, Sec. 4,

$$\phi(x) = y_0 + (x - x_0)y_1 + \int_{x_0}^{x} (x - t)f(t, \phi(t))\, dt,$$

$$\psi(x) = z_0 + (x - x_0)z_1 + \int_{x_0}^{x} (x - t)g(t, \psi(t))\, dt.$$

If $|y_0 - z_0| \leqq \delta_0$, $|y_1 - z_1| \leqq \delta_1$, show that the estimate (8.5) is valid with $\delta = \delta_0$, K replaced by Ka, ϵ replaced by $\delta_1 + (\epsilon a/2)$.)

6. Let f be a real-valued continuous function on the strip

$$S: \qquad |x| \leqq a, \qquad |y| < \infty, \qquad (a > 0),$$

and suppose f satisfies a Lipschitz condition on S. Show that the solution of the initial value problem

$$y'' + \lambda^2 y = f(x, y), \qquad y(0) = 0, \qquad y'(0) = 1, \qquad (\lambda > 0),$$

is unique. (*Hint*: Apply Ex. 5.) (*Note*: From Ex. 6, Sec. 7, it follows that a solution exists on $|x| \leqq a$.)

7. Let ϕ and ψ be solutions of the two problems

$$y'' + \lambda^2 y = f(x, y), \qquad y(x_0) = y_0, \qquad y'(x_0) = y_1,$$

$$y'' + \lambda^2 y = g(x, y), \qquad y(x_0) = z_0, \qquad y'(x_0) = z_1,$$

respectively, with $\lambda > 0$. State and prove analogues of Theorem 10, and Corollaries 1 and 2, for this situation. (*Hint*: Apply Ex. 5.)

9. Equations with complex-valued functions

We now consider equations of the form

$$y' = f(x, y),$$

where f is complex-valued. In this case we must admit complex-valued solutions, and therefore f must be defined for complex y. Thus suppose f is a complex-valued continuous function in

$$R: \qquad |x - x_0| \leqq a, \qquad |y - y_0| \leqq b, \qquad (a, b > 0).$$

Here x, x_0 are real, and y, y_0 are complex. The set of points y satisfying

$|\, y - y_0 \,| \leqq b$ is now a circular disk with center y_0 and radius b, and therefore R is no longer a rectangle. A solution of the initial value problem

$$y' = f(x, y), \qquad y(x_0) = y_0,$$

on an interval I containing x_0 is now a complex-valued differentiable function ϕ on I for which $(x, \phi(x))$ is in R for x in I, and such that

$$\phi'(x) = f(x, \phi(x)), \qquad (x \text{ in } I)$$

$$\phi(x_0) = y_0.$$

With these interpretations for R, f, and ϕ, all the results of Secs. 4–8, and their proofs, remain valid in case f is complex-valued.

The proof of Theorem 6 requires an integration

$$\int_{y_2}^{y_1} \frac{\partial f}{\partial y}(x, t)\, dt,$$

where t, y_1, y_2 are now *complex*. This can be given a meaning, but it is easy to modify the proof of Theorem 6 so as to avoid this issue. For fixed x, y_1, y_2, let

$$F(s) = f(x, y_2 + s(y_1 - y_2)), \qquad (0 \leqq s \leqq 1).$$

Then if $\partial f/\partial y$ exists the function F will be differentiable, and

$$F'(s) = (y_1 - y_2)\, \frac{\partial f}{\partial y}\, (x, y_2 + s(y_1 - y_2)).$$

If $|\, \partial f/\partial y \,| \leqq K$, as in Theorem 6, then

$$|\, F'(s) \,| \leqq K \,|\, y_1 - y_2 \,|, \qquad (0 \leqq s \leqq 1).$$

Thus

$$f(x, y_1) - f(x, y_2) = F(1) - F(0) = \int_0^1 F'(s)\, ds,$$

and hence

$$|\, f(x, y_1) - f(x, y_2) \,| \leqq K \,|\, y_1 - y_2 \,|.$$

We shall henceforward assume that the results of Secs. 4–8 are valid for complex-valued f defined for real x and complex y. The student is urged to check that the proofs do carry over to this case.

CHAPTER 6

Existence and Uniqueness of

Solutions to Systems and

n-th Order Equations

1. Introduction

In this chapter we shall see how most of the general results of Chap. 5 remain valid for a wide class of systems of equations and n-th order equations. The type of system we have in mind has the form

$$
\begin{aligned}
y_1' &= f_1(x, y_1, \cdots, y_n), \\
y_2' &= f_2(x, y_1, \cdots, y_n), \\
&\quad\vdots \\
y_n' &= f_n(x, y_1, \cdots, y_n).
\end{aligned}
\tag{1.1}
$$

This is a system of n ordinary differential equations of the first order where the derivatives $y_1', \cdots, y_n'$ appear explicitly. It is the analogue of the single equation

$$
y' = f(x, y)
$$

which was studied in Chap. 5. In (1.1) $f_1, \cdots, f_n$ are given complex-valued functions defined in some set R in the $(x, y_1, \cdots, y_n)$ space, where x is real and $y_1, \cdots, y_n$ are complex. The equations (1.1) are just shorthand for the problem of finding n differentiable functions $\phi_1, \cdots, \phi_n$ on some interval I such that

(a) $(x, \phi_1(x), \cdots, \phi_n(x))$ is in R, for x in I,

(b) $\phi_1'(x) = f_1(x, \phi_1(x), \cdots, \phi_n(x))$,

.

.

$$\phi_n'(x) = f_n(x, \phi_1(x), \cdots, \phi_n(x)), \quad \text{for all } x \text{ in } I.$$

If n such functions exist we say $(\phi_1, \cdots, \phi_n)$ is a *solution* of (1.1) on I. Thus a solution is a set of n functions.

One of the most famous systems of the type (1.1) results from Newton's second law of motion for a particle of mass m. Using rectangular coordinates (x, y, z) this law is usually written as

$$mx'' = X, \qquad my'' = Y, \qquad mz'' = Z. \tag{1.2}$$

Here differentiation is with respect to the time t, and x'', y'', z'' represent the acceleration of the particle in the x, y, z directions respectively, whereas X, Y, Z represent the forces acting on the particle in these directions. In general X, Y, Z are functions of t, x, y, z, x', y', z'. To see how (1.2) can be viewed as a system of the type (1.1), let us make the following substitutions in (1.2):

$$t \to x, \quad x \to y_1, \quad y \to y_2, \quad z \to y_3,$$

$$x' \to y_4, \quad y' \to y_5, \quad z' \to y_6.$$

Then (1.2) is equivalent to the system of six equations

$$y_1' = y_4,$$

$$y_2' = y_5,$$

$$y_3' = y_6,$$

$$y_4' = \frac{1}{m} X(x, y_1, \cdots, y_6),$$

$$y_5' = \frac{1}{m} Y(x, y_1, \cdots, y_6),$$

$$y_6' = \frac{1}{m} Z(x, y_1, \cdots, y_6),$$

which is of the type (1.1)

An equation of the n-th order

$$y^{(n)} = f(x, y, y', \cdots, y^{(n-1)}) \tag{1.3}$$

may also be treated as a system of the type (1.1). To see this let in (1.3)

$$y_1 = y, \qquad y_2 = y', \qquad \cdots, \qquad y_n = y^{(n-1)}.$$

Then (1.3) is equivalent to the system

$$y_1' = y_2,$$
$$y_2' = y_3,$$
$$\vdots$$
$$y_{n-1}' = y_n,$$
$$y_n' = f(x, y_1, y_2, \cdots, y_n),$$

which is of the type (1.1).

In Sec. 2 we discuss an interesting example of a system of equations which has historical interest. This is the system which gives a model for the motion of the planets about the sun. Sec. 3 is devoted to some special equations which are either solvable, or can be easily reduced to first order equations. The remainder of the chapter is devoted to showing how the arguments used in Chap. 5 can be adapted to prove existence and uniqueness of solutions to initial value problems for systems of the type (1.1), and for n-th order equations of the type (1.3). It is just a matter of introducing a convenient notation in order to see that this is possible.

2. An example — central forces and planetary motion

In this section we give an example of a system of equations which arise in the study of dynamics. Suppose a particle of mass m moves in a plane, and is subjected to a force which is directed along the line joining the particle to the origin, and which has a magnitude depending only on the distance between the particle and the origin. We then say we have a *central force*. The functions x, y (of the time t) which describe the path the particle takes satisfy, according to Newton's second law,

$$mx'' = \frac{x}{r} F(r),$$
$$\tag{2.1}$$
$$my'' = \frac{y}{r} F(r),$$

where $r = \sqrt{x^2 + y^2}$, and $| F(r) |$ represents the magnitude of the force on the particle when it is at the distance r from the origin.

The system (2.1) is equivalent to a system of four first order equations in x, y, x', y'. However, since F is a function of r alone, it is advantageous to introduce polar coordinates

$$x = r \cos \theta, \qquad y = r \sin \theta$$

It is shown in calculus texts that the components of acceleration in the radial and angular directions are given by

$$r'' - r(\theta')^2, \qquad 2r'\theta' + r\theta''$$

respectively. Since the components of the force in these directions are $F(r)$ and 0, equations (2.1) are replaced by

$$m[r'' - r(\theta')^2] = F(r),$$
$$m[2r'\theta' + r\theta''] = 0. \tag{2.2}$$

Upon multiplying the second equation in (2.2) by r/m, this equation becomes

$$(r^2\theta')' = 0,$$

and hence

$$r^2\theta' = h, \tag{2.3}$$

where h is a constant. (For some reason or other this constant is almost always denoted by h !) The equation (2.3) has an interesting geometrical meaning. The area $A(t)$ traversed by the line segment from the origin to $(r(s), \theta(s))$ as s goes from t_0 to t is given by

$$A(t) = \int_{t_0}^{t} \tfrac{1}{2}r^2(s)\theta'(s) \, ds,$$

since the element of area in polar coordinates is

$$dA = \tfrac{1}{2}r^2 \, d\theta;$$

see Fig. 8. Since $r^2\theta' = h$ we see that

$$A(t) = \tfrac{1}{2}h(t - t_0). \tag{2.4}$$

Thus, if $h \neq 0$, the line segment from the origin to the particle sweeps out equal areas in equal times.

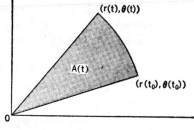

$(r(t), \theta(t))$

$A(t)$

$(r(t_0), \theta(t_0))$

Figure 8

Now, supposing that $h > 0$, let us analyse the first equation in (2.2). We introduce a function v defined for θ of the form $\theta(t)$ by*

$$v(\theta(t)) = \frac{1}{r(t)}. \tag{2.5}$$

Then

$$r'(t) = -\frac{1}{v^2(\theta(t))}\left[\frac{dv}{d\theta}(\theta(t))\right]\theta'(t) = -h\frac{dv}{d\theta}(\theta(t)),$$

and

$$r''(t) = -h\frac{d^2v}{d\theta^2}(\theta(t))\theta'(t) = -h^2v^2(\theta(t))\frac{d^2v}{d\theta^2}(\theta(t)),$$

where we have used (2.3). Thus the first equation in (2.2) becomes the following equation for v:

$$\frac{d^2v}{d\theta^2} + v = -\frac{F(1/v)}{mh^2v^2}. \tag{2.6}$$

Now let us assume that $F(r)$ is inversely proportional to r^2, and that the force is directed toward the origin (the inverse square law of Newton). Thus let k be a positive constant such that

$$F(r) = -\frac{km}{r^2}, \quad \text{or} \quad F(1/v) = -kmv^2.$$

Then (2.6) becomes

$$\frac{d^2v}{d\theta^2} + v = \frac{k}{h^2}. \tag{2.7}$$

All solutions of this linear equation may be written in the form

$$v(\theta) = \frac{k}{h^2} + B\cos(\theta - \omega),$$

where B, ω are constants. Returning to the definition of v in (2.5) we see that r is related to θ in the following way:

$$r = \frac{(h^2/k)}{1 + e\cos(\theta - \omega)}, \tag{2.8}$$

where $e = Bh^2/k$. For $h^2/k > 0$ and $e \geqq 0$ the equation (2.8) is the equation of a conic with the focus at the origin and with eccentricity e. The conic is an ellipse, parabola, or hyperbola according as $0 \leqq e < 1$, $e = 1$, or $e > 1$ respectively.

---

* The equation (2.3) implies that θ is an increasing function, if $r \neq 0$, and this in turn implies that v exists.

Let us analyse further the case when the conic is an ellipse having major and minor semi-axes a and b; see Fig. 9. Then $2a$ must be the sum of the largest and smallest values that r can assume, namely,

$$2a = \frac{h^2}{k}\left(\frac{1}{1-e} + \frac{1}{1+e}\right) = \frac{2h^2}{k(1-e^2)}.$$

The eccentricity is related to a, b via

$$b^2 = a^2(1-e^2),$$

and hence

$$b^2 = \frac{h^2 a}{k}. \qquad (2.9)$$

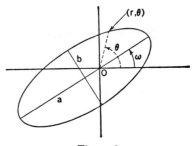

Now the area of the ellipse is πab, and this is related to the time T required for the particle to traverse the ellipse once by

$$\tfrac{1}{2}hT = \pi ab;$$

Figure 9

see (2.4). Thus, using (2.9) we obtain

$$T^2 = \frac{4\pi^2}{k}a^3. \qquad (2.10)$$

Kepler, on the basis of observations of Tycho Brahe on the motions of the planets about the sun, deduced his famous three laws of planetary motion:

(1) *the line segment from the sun to a planet sweeps out equal areas in equal times,*

(2) *the planets move along ellipses with the sun as a focus,*

(3) *the squares of the periods are proportional to the cubes of the major axes of the ellipses.*

If we idealize the motion of a planet about the sun as a plane motion, with the sun fixed at the origin and exerting an attractive central force on the planet (thought of as a particle of mass m), then we see that Newton's second law implies that the motion of the planet is governed by the system of equations (2.2). Kepler's first law is a consequence of the central force assumption. His second and third laws then result from the assumption that the central force is proportional to $1/r^2$.

Newton discovered that Kepler's first two laws imply the inverse square law. Indeed, it was this that led Newton to the formulation of his famous law of universal gravitation. The first law

$$r^2\theta' = h$$

implies that there is no force acting perpendicular to the line segment from the origin to the particle, i.e., the second equation of (2.2) is valid. Hence the particle is acted on by a force which acts in the radial direction only. If $F(r, \theta)$ is the radial component of this force at (r, θ), we have the equation

$$m[r'' - r(\theta')^2] = F(r, \theta) \qquad (2.11)$$

as the analogue of the first equation in (2.2). Introducing v as in (2.5), we see that (2.11) implies the following equation for v:

$$\frac{d^2v}{d\theta^2} + v = -\frac{F(1/v, \theta)}{mh^2v^2}. \qquad (2.12)$$

Now Kepler's second law implies that r is related to θ via an equation of the form (2.8) with $0 \leqq e < 1$, and then v will satisfy the equation (2.7). A comparison of equations (2.7) and (2.12) then shows that

$$F(1/v, \theta) = -kmv^2,$$

or that

$$F(r, \theta) = -\frac{km}{r^2}.$$

Thus F depends only on r according to Newton's inverse square law.

EXERCISES

1. A particle of mass m moves in a plane, and is attracted to the origin with a force proportional to its distance r from the origin. Thus if

$$F(r) = -k^2mr, \qquad (k > 0),$$

in (2.1) the equations (2.1) become

$$x'' = -k^2x, \qquad y'' = -k^2y.$$

(a) Show that the path of the particle is an ellipse, if it satisfies the initial conditions $x(0) = a$, $x'(0) = 0$, $y(0) = 0$, $y'(0) = b$, $(a, b > 0)$.
(b) Compute the period of the motion.

2. A particle of mass m moves in a vertical plane near the surface of the earth, and is acted on by the force of gravity alone. The equations for the motion assume the form

$$mx'' = 0, \qquad my'' = -mg,$$

where g is a constant.
(a) Find the solution of these equations satisfying

$$x(0) = 0, \quad y(0) = 0, \quad x'(0) = v_0 \cos \alpha, \quad y'(0) = v_0 \sin \alpha,$$

where $v_0 > 0$ and α are constants, $0 < \alpha < \pi/2$.
(b) Show that the particle path is a parabola.

(c) Compute the vertex of this parabola, and the time required to reach this vertex.

(d) Compute $x(T)$ for that $T > 0$ for which $y(T) = 0$. (This is called the *horizontal range*.) For what α is this range a maximum?

3. Suppose a particle moves on a circle through the origin, and is acted on by a central force $F(r)$. Show that $F(r)$ is proportional to r^{-5}.

4. (a) Determine the equations of motion of the particle in Ex. 2, given that the resistance of the air is proportional to the velocity of the particle. For simplicity express the constant of proportionality as ϵm.

(b) Find the solutions of these equations satisfying

$$x(0) = 0, \quad y(0) = 0, \quad x'(0) = v_0 \cos \alpha, \quad y'(0) = v_0 \sin \alpha,$$

where $v_0 > 0$ and α are constants.

(c) Show that for each fixed t the solutions of (b) approach the solutions of Ex. 2(a) as $\epsilon \to 0$.

5. What initial conditions are sufficient to completely determine the solutions of the equations (2.2)? Give a reason for your answer.

3. Some special equations

There are a number of problems which lead to rather special types of second order equations, or systems of such equations. We consider two of these types in this section.

(a) *The equation $y'' = f(x, y')$.* This second order equation has an f which is independent of y, and is hence really a first order equation in y'. Indeed this equation is equivalent to the system of two equations of the first order

$$y' = z, \quad z' = f(x, z), \tag{3.1}$$

in that ϕ will be a solution of $y'' = f(x, y')$ on an interval I if, and only if, the functions ϕ, ϕ' satisfy the system (3.1) on I. Now the system (3.1) can be solved by first solving the first order equation

$$z' = f(x, z)$$

for ϕ', and then integrating to obtain ϕ.

As a simple example consider the equation

$$xy'' - y' = 0, \quad (x > 0).$$

Letting $y' = z$ we obtain the first order linear equation

$$z' - \frac{z}{x} = 0,$$

which has as solutions

$$\phi'(x) = cx, \qquad (x > 0),$$

where c may be any constant. Thus

$$\phi(x) = \frac{cx^2}{2} + d, \qquad (x > 0),$$

where c, d are constants. Note that for $x > 0$ the equation is equivalent to

$$x^2 y'' - xy' = 0,$$

which is an Euler equation.

(b) *The equation* $y'' = f(y, y')$. Here f is independent of x, and the strategy is somewhat different than in (a). Suppose we have a solution ϕ of $y'' = f(y, y')$, and there is a differentiable function ψ, defined for all y of the form $y = \phi(x)$, such that

$$\phi'(x) = \psi(\phi(x)).$$

Then ϕ would be a solution of the first order equation

$$\frac{dy}{dx} = \psi(y). \tag{3.2}$$

Also

$$\phi''(x) = \phi'(x) \frac{d\psi}{dy}(\phi(x)) = \psi(\phi(x)) \frac{d\psi}{dy}(\phi(x)),$$

and moreover

$$\phi''(x) = f(\phi(x), \phi'(x)) = f(\phi(x), \psi(\phi(x))).$$

Thus ψ must satisfy the equation

$$\psi(y) \frac{d\psi}{dy}(y) = f(y, \psi(y))$$

for all $y = \phi(x)$, and hence must be a solution of

$$z \frac{dz}{dy} = f(y, z). \tag{3.3}$$

The argument can be reversed. If ψ is a solution of (3.3), then any solution ϕ of (3.2) will be a solution of the given equation $y'' = f(y, y')$. Thus solutions to the original equation can be found by first solving (3.3) to obtain ψ, and then solving (3.2).

As an example consider the equation

$$yy'' = (y')^2,$$

and suppose we seek a solution ϕ satisfying

$$\phi(0) = 1, \qquad \phi'(0) = 2.$$

The equation (3.3) becomes in this case

$$z\frac{dz}{dy} = \frac{z^2}{y},$$

which has for solutions functions ψ given by

$$\psi(y) = cy,$$

where c may be any constant. The equation (3.2) then becomes

$$\frac{dy}{dx} = cy$$

which has for solutions

$$\phi(x) = de^{cx},$$

where d may be any constant. The solution satisfying the given initial conditions is given by

$$\phi(x) = e^{2x}.$$

EXERCISES

1. Solve the following equations:

   (a) $y'' + y' = 1$ (b) $y'' + e^x y' = e^x$

   (c) $yy'' + 4(y')^2 = 0$ (d) $y'' + k^2 y = 0, \qquad (k > 0)$

   (e) $y'' = yy'$ (f) $xy'' - 2y' = x^3$

   (g) $y^2 y'' = y'$

2. Find the solution ϕ of

$$y'' = 1 + (y')^2$$

which satisfies $\phi(0) = 0, \phi'(0) = 0$.

3. Find a solution ϕ of

$$y'' = -\frac{1}{2y^2}$$

satisfying $\phi(0) = 1, \phi'(0) = -1$.

4. Suppose that f is a continuous function on an interval

$$|x - x_0| \leqq a.$$

Show that the solution ϕ of the initial value problem

$$y'' = f(x), \qquad y(x_0) = \alpha, \qquad y'(x_0) = \beta,$$

can be written as

$$\phi(x) = \alpha + \beta(x - x_0) + \int_{x_0}^{x} (x - t)f(t) \, dt.$$

5. (a) Let f be a continuous function for $|y - y_0| \leq b$, $(b > 0)$, and consider
the equation

$$y'' = f(y).$$

Show that the equation (3.3) has a solution ψ, in this case, given by

$$\psi^2(y) = \psi^2(y_0) + 2 \int_{y_0}^{y} f(t) \, dt.$$

(b) Consider the special case

$$y'' + \sin y = 0,$$

which is an equation associated with the oscillations of a pendulum. If ϕ is
a solution satisfying

$$\phi(0) = 0, \qquad \phi'(0) = \beta > 0,$$

show that ϕ satisfies the equation

$$y' = \beta\sqrt{1 - k^2 \sin^2(y/2)}, \tag{*}$$

where $k = 2/\beta$.

(c) Solve the equation (*) in the case $k = 1$.

(d) Can you solve this equation if $k \neq 1$?

4. Complex n-dimensional space

It is clear that one of the main differences between the one equation

$$y' = f(x, y),$$

and the system of n equations

$$y_1' = f_1(x, y_1, \cdots, y_n),$$

$$\cdot$$

$$\cdot$$

$$\cdot$$

$$y_n' = f_n(x, y_1, \cdots, y_n),$$

is that instead of one complex number y we have now to deal with n such numbers $y_1, \cdots, y_n$. Let us call such an ordered n-tuple of complex numbers $y_1, \cdots, y_n$ a *vector* and denote it by **y**, and write*

$$\mathbf{y} = (y_1, \cdots, y_n).$$

The complex number y_k is called the k-th *component* of **y**. The set of all such vectors we call the *complex n-dimensional space*, and denote it by $\mathbf{C}_n$. This certainly is a nice abbreviation, but in order to make use of this abbreviation we must define how to operate with such vectors. We define the *zero vector* **0** (also called the *origin* in $\mathbf{C}_n$) by

$$\mathbf{0} = (0, \cdots, 0),$$

and the negative of **y** by

$$-\mathbf{y} = (-y_1, \cdots, -y_n).$$

If c is any complex number $c\mathbf{y}$ is the vector

$$c\mathbf{y} = (cy_1, \cdots, cy_n).$$

Two vectors $\mathbf{y} = (y_1, \cdots, y_n)$ and $\mathbf{z} = (z_1, \cdots, z_n)$ in $\mathbf{C}_n$ are said to be *equal*, and we write $\mathbf{y} = \mathbf{z}$, provided that

$$y_1 = z_1, \quad y_2 = z_2, \quad \cdots, \quad y_n = z_n.$$

The sum $\mathbf{y} + \mathbf{z}$ is defined by

$$\mathbf{y} + \mathbf{z} = (y_1 + z_1, \cdots, y_n + z_n),$$

and the difference $\mathbf{y} - \mathbf{z}$ by

$$\mathbf{y} - \mathbf{z} = \mathbf{y} + (-\mathbf{z}).$$

Suppose, for example, that in $\mathbf{C}_2$

$$\mathbf{y} = (1, i), \quad \mathbf{z} = (-2, 1 + i).$$

Then

$$i\mathbf{y} = (i, -1), \quad \mathbf{y} + \mathbf{z} = (-1, 1 + 2i), \quad \mathbf{y} - \mathbf{z} = (3, -1).$$

The set of all vectors **y** in $\mathbf{C}_n$ of the form

$$\mathbf{y} = (y_1, \cdots, y_n),$$

where $y_1, \cdots, y_n$ are all *real* numbers, is called *real n-dimensional space*, and denoted by $\mathbf{R}_n$. If **y**, **z** are in $\mathbf{R}_n$ then so are $\mathbf{y} + \mathbf{z}$, and $c\mathbf{y}$ for any *real* number c.

---

* A convenient way of writing the bold-face **y** is to write y with a bar beneath it, like $\underset{\sim}{y}$.

The *magnitude* of a vector $\mathbf{y}$, denoted by $|\mathbf{y}|$, is defined by

$$|\mathbf{y}| = |y_1| + \cdots + |y_n|.$$

For example, in $\mathbf{C}_2$ the vector $\mathbf{y} = (3 - i, 1 + i)$ has the magnitude $|\mathbf{y}| = \sqrt{10} + \sqrt{2}$. This magnitude has the nice properties that the magnitude of a complex number has, namely

(a) $|\mathbf{y}| \geqq 0$, and $|\mathbf{y}| = 0$ if and only if $\mathbf{y} = \mathbf{0}$,

(b) $|c\mathbf{y}| = |c||\mathbf{y}|$, for any complex number c,

(c) $|\mathbf{y} + \mathbf{z}| \leqq |\mathbf{y}| + |\mathbf{z}|$.

The first two properties are obvious from the definition of $|\mathbf{y}|$, and property (c) follows from the corresponding property for complex numbers. Indeed

$$\begin{aligned}
|\mathbf{y} + \mathbf{z}| &= |y_1 + z_1| + \cdots + |y_n + z_n| \\
&\leqq |y_1| + |z_1| + \cdots + |y_n| + |z_n| \\
&= |\mathbf{y}| + |\mathbf{z}|.
\end{aligned}$$

From (b) and (c) it follows that

$$||\mathbf{y}| - |\mathbf{z}|| \leqq |\mathbf{y} + \mathbf{z}|. \tag{4.1}$$

Indeed

$$|\mathbf{y}| = |\mathbf{y} + \mathbf{z} + (-\mathbf{z})| \leqq |\mathbf{y} + \mathbf{z}| + |\mathbf{z}|,$$

and hence

$$|\mathbf{y}| - |\mathbf{z}| \leqq |\mathbf{y} + \mathbf{z}|$$

for all $\mathbf{y}, \mathbf{z}$. Similarly

$$|\mathbf{z}| - |\mathbf{y}| \leqq |\mathbf{y} + \mathbf{z}|$$

for all $\mathbf{y}, \mathbf{z}$, and these two inequalities imply (4.1).

We define the *distance* between $\mathbf{y}$ and $\mathbf{z}$ to be $|\mathbf{y} - \mathbf{z}|$. It readily follows from the properties (a), (b), (c) that the distance satisfies

(i) $|\mathbf{y} - \mathbf{z}| \geqq 0$, and $|\mathbf{y} - \mathbf{z}| = 0$ if and only if $\mathbf{y} = \mathbf{z}$,

(ii) $|\mathbf{y} - \mathbf{z}| = |\mathbf{z} - \mathbf{y}|$,

(iii) $|\mathbf{y} - \mathbf{z}| \leqq |\mathbf{y} - \mathbf{w}| + |\mathbf{w} - \mathbf{z}|$.

Indeed (i), (ii) result from (a), (b) respectively, whereas (iii) follows from (c) by replacing $\mathbf{y}, \mathbf{z}$ there by $\mathbf{y} - \mathbf{w}$ and $\mathbf{w} - \mathbf{z}$.

Using this distance we can define the concept of convergence of a sequence of vectors. We say that a sequence $\{\mathbf{y}_m\}$, $(m = 1, 2, \cdots)$, *converges* (or *tends*) to a *limit vector* **y**, and write

$$\mathbf{y}_m \to \mathbf{y}, \qquad (m \to \infty),$$

if

$$|\mathbf{y}_m - \mathbf{y}| \to 0, \qquad (m \to \infty). \tag{4.2}$$

If

$$\mathbf{y}_m = (y_{1m}, \cdots, y_{nm}), \qquad (m = 1, 2, \cdots),$$

and

$$\mathbf{y} = (y_1, \cdots, y_n),$$

then (4.2) says that

$$|y_{1m} - y_1| + |y_{2m} - y_2| + \cdots + |y_{nm} - y_n| \to 0, \quad (m \to \infty).$$

But this is true if, and only if,

$$|y_{1m} - y_1| \to 0, \qquad (m \to \infty),$$

$$\vdots$$

$$|y_{nm} - y_n| \to 0, \qquad (m \to \infty).$$

Thus we see that a sequence of vectors $\{\mathbf{y}_m\}$ tends to a limit vector **y** if, and only if, for each $k = 1, \cdots, n$ the sequence of complex numbers $\{y_{km}\}$, $(m = 1, 2, \cdots)$ tends to the complex number y_k. It is this fact which allows us to take over all results concerning limits of sequences of complex numbers. Thus if

$$\mathbf{y}_m \to \mathbf{y}, \qquad \mathbf{z}_m \to \mathbf{z}, \qquad (m \to \infty),$$

then

$$\mathbf{y}_m + \mathbf{z}_m \to \mathbf{y} + \mathbf{z}, \qquad (m \to \infty).$$

An example of a convergent sequence of vectors in $\mathbf{C}_2$ is furnished by

$$\mathbf{y}_m = \left(\frac{m+1}{m}, \frac{1}{m^2} - i \right), \qquad (m = 1, 2, \cdots).$$

Clearly

$$\mathbf{y}_m \to \mathbf{y} = (1, -i), \qquad (m \to \infty).$$

Now let us consider a function ϕ which is defined on some real interval I and has values in $\mathbf{C}_n$. Thus to each x in I there is associated just one vector $\phi(x)$ in $\mathbf{C}_n$ which we may write as

$$\phi(x) = (\phi_1(x), \cdots, \phi_n(x)).$$

Such functions are called *vector-valued functions.* For each such ϕ there are associated n complex-valued functions $\phi_1, \cdots, \phi_n$ on I, the function ϕ_k is the one which associates with each x in I the k-th component of $\phi(x)$. The functions $\phi_1, \cdots, \phi_n$ are called the *components* of ϕ, and we write

$$\phi = (\phi_1, \cdots, \phi_n).$$

An example in $\mathbf{C}_2$ is given by

$$\phi(x) = (x^2, x - ix^3), \qquad (0 \leqq x \leqq 1). \tag{4.3}$$

Here

$$\phi_1(x) = x^2, \quad (0 \leqq x \leqq 1), \quad \text{and} \quad \phi_2(x) = x - ix^3, \quad (0 \leqq x \leqq 1).$$

If ϕ is a vector-valued function defined on an interval I, we say that ϕ is *continuous,* or *differentiable,* on I if each of its components is. If ϕ is differentiable on I we define its *derivative* ϕ' by

$$\phi' = (\phi_1', \cdots, \phi_n').$$

Thus the ϕ given by (4.3) is differentiable on $0 \leqq x \leqq 1$, and

$$\phi'(x) = (2x, 1 - 3ix^2), \qquad (0 \leqq x \leqq 1).$$

We define the *integral* of a continuous vector function ϕ which is defined on an interval $c \leqq x \leqq d$ to be a vector

$$\int_c^d \phi(x) \, dx = \left(\int_c^d \phi_1(x) \, dx, \cdots, \int_c^d \phi_n(x) \, dx \right),$$

i.e., the k-th component of the integral of ϕ is the integral of the k-th component of ϕ. The important inequality satisfied by the integral is

$$\left| \int_c^d \phi(x) \, dx \right| \leqq \int_c^d |\phi(x)| \, dx.$$

The proof is easy since

$$\left| \int_c^d \phi(x) \, dx \right| = \left| \int_c^d \phi_1(x) \, dx \right| + \cdots + \left| \int_c^d \phi_n(x) \, dx \right|$$

$$\leqq \int_c^d |\phi_1(x)| \, dx + \cdots + \int_c^d |\phi_n(x)| \, dx$$

$$= \int_c^d [|\phi_1(x)| + \cdots + |\phi_n(x)|] \, dx$$

$$= \int_c^d |\phi(x)| \, dx.$$

As an example in $\mathbf{C}_2$ the function $\boldsymbol{\phi}$ in (4.3) has the integral

$$\int_0^1 \boldsymbol{\phi}(x) \; dx = \left(\int_0^1 x^2 \; dx, \int_0^1 (x - ix^3) \; dx \right)$$

$$= \left(\frac{1}{3}, \frac{1}{2} - \frac{i}{4} \right),$$

and

$$\left| \int_0^1 \boldsymbol{\phi}(x) \; dx \right| = \frac{1}{3} + \frac{\sqrt{5}}{4}.$$

EXERCISES

1. Suppose $\mathbf{y}, \mathbf{z}, \mathbf{w}$ are the following vectors in $\mathbf{C}_3$:

$$\mathbf{y} = (8 + i, 3i, -2), \quad \mathbf{z} = (i, -i, 2), \quad \mathbf{w} = (2 + i, 0, 1).$$

(a) Compute $\mathbf{y} + \mathbf{z}$.
(b) Compute $\mathbf{y} - \mathbf{z}$.
(c) Show, for some number s, that $\mathbf{w} = \mathbf{z} + s(\mathbf{y} - \mathbf{z})$.

2. If $\mathbf{y}, \mathbf{z}, \mathbf{w}$ are any vectors in $\mathbf{C}_n$ show that the following rules are valid:
(a) $\mathbf{y} + \mathbf{z} = \mathbf{z} + \mathbf{y}$ (b) $(\mathbf{y} + \mathbf{z}) + \mathbf{w} = \mathbf{y} + (\mathbf{z} + \mathbf{w})$
(c) $\mathbf{y} + \mathbf{0} = \mathbf{y}$ (d) $\mathbf{y} + (-\mathbf{y}) = \mathbf{0}$
(*Hint:* These rules are valid for $\mathbf{C}_1$.)

3. For each k, $1 \leq k \leq n$, let $\mathbf{e}_k$ be the vector with 1 as its k-th component and 0 for its other components. Thus

$$\mathbf{e}_1 = (1, 0, \cdots, 0), \quad \mathbf{e}_2 = (0, 1, 0, \cdots, 0), \quad \cdots, \quad \mathbf{e}_n = (0, \cdots, 0, 1).$$

(a) If $\mathbf{y} = (y_1, \cdots, y_n)$ show that

$$\mathbf{y} = y_1 \mathbf{e}_1 + \cdots + y_n \mathbf{e}_n.$$

(b) Show that
$$|\mathbf{e}_k| = 1, \quad (k = 1, \cdots, n).$$

4. Let $\boldsymbol{\phi}$ be the vector-valued function defined for all real x by

$$\boldsymbol{\phi}(x) = (x, x^2, ix^4).$$

Compute the following:
(a) $\boldsymbol{\phi}(1)$
(b) $\boldsymbol{\phi}'(x), \quad \boldsymbol{\phi}'(2)$
(c) $\displaystyle\int_{-1}^1 \boldsymbol{\phi}(x) \; dx$

(d) Verify that $\left| \displaystyle\int_{-1}^1 \boldsymbol{\phi}(x) \; dx \right| \leq \displaystyle\int_{-1}^1 |\boldsymbol{\phi}(x)| \; dx$

5. If ϕ is a continuously differentiable vector-valued function defined for real x in an interval $a \leq x \leq b$, and the values of ϕ are in $\mathbf{R}_n$, show that:

(a) ϕ' has values in $\mathbf{R}_n$

(b) $\displaystyle\int_a^x \phi(t)\, dt$ is in $\mathbf{R}_n$ for each x, $a \leq x \leq b$

6. For each $\mathbf{y} = (y_1, \cdots, y_n)$ in $\mathbf{C}_n$ let

$$\| \mathbf{y} \| = (y_1 \bar{y}_1 + \cdots + y_n \bar{y}_n)^{1/2},$$

the positive square root being understood. This is the *Euclidean length* of $\mathbf{y}$

(a) Show that
$$\| \mathbf{y} \| \leq | \mathbf{y} | \leq \sqrt{n}\, \| \mathbf{y} \|.$$
(*Hint*: Show that
$$\| \mathbf{y} \|^2 \leq | \mathbf{y} |^2 \leq n \, \| \mathbf{y} \|^2.$$
Use the inequality $2\, | a | \, | b | \leq | a |^2 + | b |^2$.)

(b) Show that a sequence $\{\mathbf{y}_m\}$, $(m = 1, 2, \cdots)$, of vectors in $\mathbf{C}_n$ is such that
$$| \mathbf{y}_m - \mathbf{y} | \to 0, \qquad (m \to \infty),$$
if and only if
$$\| \mathbf{y}_m - \mathbf{y} \| \to 0, \qquad (m \to \infty).$$

7. For any two vectors $\mathbf{y} = (y_1, \cdots, y_n)$ and $\mathbf{z} = (z_1, \cdots, z_n)$ in $\mathbf{C}_n$ define the *inner product* $\mathbf{y} \cdot \mathbf{z}$ to be the number given by

$$\mathbf{y} \cdot \mathbf{z} = y_1 \bar{z}_1 + \cdots + y_n \bar{z}_n.$$

(a) Show that $\mathbf{z} \cdot \mathbf{y} = \overline{(\mathbf{y} \cdot \mathbf{z})}$.

(b) Show that $(\mathbf{y}_1 + \mathbf{y}_2) \cdot \mathbf{z} = (\mathbf{y}_1 \cdot \mathbf{z}) + (\mathbf{y}_2 \cdot \mathbf{z})$.

(c) Show that if c is a complex number

$$(c\mathbf{y}) \cdot \mathbf{z} = c(\mathbf{y} \cdot \mathbf{z}) = \mathbf{y} \cdot (\bar{c}\mathbf{z}).$$

(d) Show that $\| \mathbf{y} \|^2 = \mathbf{y} \cdot \mathbf{y}$.

(e) Prove that
$$| \mathbf{y} \cdot \mathbf{z} | \leq \| \mathbf{y} \| \, \| \mathbf{z} \|.$$

This is called the *Schwarz inequality*. (*Hint*: If $\mathbf{z} = \mathbf{0}$ the result is obvious. If $\mathbf{z} \neq \mathbf{0}$ let $\mathbf{u} = \mathbf{z}/\| \mathbf{z} \|$. Then $\| \mathbf{u} \| = 1$. Use the fact that
$$\| \mathbf{y} - (\mathbf{y} \cdot \mathbf{u})\mathbf{u} \|^2 \geq 0.)$$

8. Show that the Euclidean length satisfies the same rules as the magnitude, namely:

(i) $\| \mathbf{y} \| \geq 0$, and $\| \mathbf{y} \| = 0$ if and only if $\mathbf{y} = \mathbf{0}$,

(ii) $\| c\mathbf{y} \| = | c | \, \| \mathbf{y} \|$, for any complex number c,

(iii) $\| \mathbf{y} + \mathbf{z} \| \leq \| \mathbf{y} \| + \| \mathbf{z} \|$.

(*Hint*: In terms of the inner product $\| \mathbf{y} \|^2 = \mathbf{y} \cdot \mathbf{y}$. Use the Schwarz inequality of Ex. 7 (e).)

5. Systems as vector equations

We return now to the first order system of equations

$$
\begin{aligned}
y_1' &= f_1(x, y_1, \cdots, y_n), \\
y_2' &= f_2(x, y_1, \cdots, y_n), \\
&\;\;\vdots \\
y_n' &= f_n(x, y_1, \cdots, y_n).
\end{aligned}
\tag{5.1}
$$

We assume that $f_1, \cdots, f_n$ are given complex-valued functions defined for $x, y_1, \cdots, y_n$ in some set R, where x is real and $y_1, \cdots, y_n$ are complex. Using the notations of Sec. 4 it is clear that we can consider f_1 as a function of x and the vector

$$\mathbf{y} = (y_1, \cdots, y_n) \quad \text{in} \quad \mathbf{C}_n.$$

Therefore we write

$$f_1(x, \mathbf{y}) = f_1(x, y_1, \cdots, y_n).$$

Also in (5.1) we have n functions $f_1, \cdots, f_n$ which may be considered as a vector-valued function

$$\mathbf{f} = (f_1, \cdots, f_n),$$

the value of $\mathbf{f}$ at $(x, \mathbf{y})$ being given by

$$\mathbf{f}(x, \mathbf{y}) = (f_1(x, \mathbf{y}), \cdots, f_n(x, \mathbf{y})).$$

If we let

$$\mathbf{y}' = (y_1', \cdots, y_n')$$

we see that the system (5.1) may now be written as

$$\mathbf{y}' = \mathbf{f}(x, \mathbf{y}).
\tag{5.2}$$

This vector differential equation has the same form as the equation $y' = f(x, y)$ considered in Chap. 5.

As an example let us consider the system of two equations

$$
\begin{aligned}
y_1' &= x^2 + y_1^2 + y_2, \\
y_2' &= y_1 + y_2 - y_1 y_2.
\end{aligned}
$$

Here $\mathbf{y} = (y_1, y_2)$,

$$f_1(x, \mathbf{y}) = f_1(x, y_1, y_2) = x^2 + y_1^2 + y_2,$$

$$f_2(x, \mathbf{y}) = f_2(x, y_1, y_2) = y_1 + y_2 - y_1 y_2,$$

and thus

$$\mathbf{f}(x, \mathbf{y}) = (x^2 + y_1^2 + y_2, \, y_1 + y_2 - y_1 y_2).$$

A solution of the system (5.2) may be described as a vector-valued function

$$\phi = (\phi_1, \cdots, \phi_n)$$

which is differentiable on a real interval I and such that

(a) $(x, \phi(x))$ is in R, for x in I,

(b) $\phi'(x) = f(x, \phi(x))$, for all x in I.

Thus, for example, the system

$$y_1' = y_2, \qquad y_2' = -y_1,$$

has the vector-valued function ϕ given by

$$\phi(x) = (\sin x, \cos x), \qquad (-\infty < x < \infty),$$

as a solution.

A vector-valued function f defined for (x, y) in some set S (x real, y in C_n) is said to be *continuous* on S if each of its components is continuous on S.* The definition of a Lipschitz condition is formally the same as before. We say that f satisfies a *Lipschitz condition* on S if there exists a constant $K > 0$ such that

$$|f(x, y) - f(x, z)| \leq K |y - z|$$

for all (x, y), (x, z) in S. The constant K is called a *Lipschitz constant* for f on S. For example, if

$$f(x, y) = (3x + 2y_1, y_1 - y_2) \tag{5.3}$$

for

$$S: \qquad |x| < \infty, \qquad |y| < \infty,$$

then f satisfies a Lipschitz condition on S with Lipschitz constant $K = 3$, since

$$|f(x, y) - f(x, z)| = |2(y_1 - z_1), (y_1 - z_1) - (y_2 - z_2)|$$

$$= 2|y_1 - z_1| + |(y_1 - z_1) - (y_2 - z_2)|$$

$$\leq 2|y_1 - z_1| + |y_1 - z_1| + |y_2 - z_2|$$

$$\leq 3|y - z|.$$

The analogue of Theorem 6, Chap. 5 is the following result.

---

* See Sec. 3, Chap. 0, for the definition of continuity of complex-valued functions defined on S.

Theorem 1. *Suppose* **f** *is a vector-valued function defined for* (x, y) *on a set* S *of the form*

$$|x - x_0| \leqq a, \qquad |\mathbf{y} - \mathbf{y}_0| \leqq b, \qquad (a, b > 0),$$

or of the form

$$|x - x_0| \leqq a, \qquad |\mathbf{y}| < \infty, \qquad (a > 0).$$

If $\partial \mathbf{f}/\partial y_k$ (k = 1, $\cdots$, n) *exists, is continuous on* S, *and there is a constant* K > 0 *such that*

$$\left| \frac{\partial \mathbf{f}}{\partial y_k} (x, \mathbf{y}) \right| \leqq K, \qquad (k = 1, \cdots, n), \tag{5.4}$$

for all (x, y) *in* S, *then* **f** *satisfies a Lipschitz condition on* S *with Lipschitz constant* K.

Proof. The proof is a direct copy of the proof outlined in Sec. 9 of Chap. 5. Let $(x, \mathbf{y})$, $(x, \mathbf{z})$ be fixed points in S, and define the vector-valued function **F** for real s, $0 \leqq s \leqq 1$, by

$$\mathbf{F}(s) = \mathbf{f}(x, \mathbf{z} + s(\mathbf{y} - \mathbf{z})), \qquad (0 \leqq s \leqq 1).$$

This is a well-defined function since the points $(x, \mathbf{z} + s(\mathbf{y} - \mathbf{z}))$ are in S for $0 \leqq s \leqq 1$. Clearly

$$|x - x_0| \leqq a,$$

and if

$$|\mathbf{y} - \mathbf{y}_0| \leqq b, \qquad |\mathbf{z} - \mathbf{y}_0| \leqq b,$$

then

$$\begin{aligned}
|\mathbf{z} + s(\mathbf{y} - \mathbf{z}) - \mathbf{y}_0| &= |(1 - s)(\mathbf{z} - \mathbf{y}_0) + s(\mathbf{y} - \mathbf{y}_0)| \\
&\leqq (1 - s)|\mathbf{z} - \mathbf{y}_0| + s|\mathbf{y} - \mathbf{y}_0| \\
&\leqq (1 - s)b + sb = b.
\end{aligned}$$

If $|\mathbf{y}| < \infty$, $|\mathbf{z}| < \infty$, then

$$|\mathbf{z} + s(\mathbf{y} - \mathbf{z})| \leqq (1 - s)|\mathbf{z}| + s|\mathbf{y}| \leqq |\mathbf{z}| + |\mathbf{y}| < \infty.$$

We now have

$$\mathbf{F}'(s) = (y_1 - z_1) \frac{\partial \mathbf{f}}{\partial y_1} (x, \mathbf{z} + s(\mathbf{y} - \mathbf{z})) +$$

$$\cdots + (y_n - z_n) \frac{\partial \mathbf{f}}{\partial y_n} (x, \mathbf{z} + s(\mathbf{y} - \mathbf{z})),$$

where $\mathbf{y} = (y_1, \cdots, y_n)$, $\mathbf{z} = (z_1, \cdots, z_n)$. Using (5.4) we see that

$$|\, \mathbf{F}'(s)\, | \leq K\, |\, \mathbf{y} - \mathbf{z}\, |, \qquad (0 \leq s \leq 1).$$

Thus, since

$$\mathbf{f}(x, \mathbf{y}) - \mathbf{f}(x, \mathbf{z}) = \mathbf{F}(1) - \mathbf{F}(0) = \int_0^1 \mathbf{F}'(s)\, ds,$$

we have

$$|\, \mathbf{f}(x, \mathbf{y}) - \mathbf{f}(x, \mathbf{z})\, | \leq K\, |\, \mathbf{y} - \mathbf{z}\, |,$$

which was to be proved.

In the example given in (5.3) we find that

$$\frac{\partial \mathbf{f}}{\partial y_1}(x, \mathbf{y}) = (2, 1), \qquad \frac{\partial \mathbf{f}}{\partial y_2}(x, \mathbf{y}) = (0, -1),$$

and

$$\left|\, \frac{\partial \mathbf{f}}{\partial y_1}(x, \mathbf{y})\, \right| = 3, \qquad \left|\, \frac{\partial \mathbf{f}}{\partial y_2}(x, \mathbf{y})\, \right| = 1.$$

Thus, as we have seen directly, $\mathbf{f}$ satisfies a Lipschitz condition on S with a Lipschitz constant $K = 3$.

EXERCISES

1. Let $\mathbf{f}$ be the vector-valued function defined on

$$R: \quad |\, x\, | \leq 1, \qquad |\, \mathbf{y}\, | \leq 1, \qquad (\mathbf{y} \text{ in } \mathbf{C}_2),$$

by

$$\mathbf{f}(x, \mathbf{y}) = (y_2^2 + 1,\, x + y_1^2).$$

(a) Find an upper bound M for $|\, \mathbf{f}(x, \mathbf{y})\, |$ for $(x, \mathbf{y})$ in R.

(b) Compute a Lipschitz constant K for $\mathbf{f}$ on R.

2. Consider the system of two equations

$$y_1' = ay_1 + by_2,$$
$$y_2' = cy_1 + dy_2,$$

where a, b, c, d are constants.

(a) If this system is written in the form

$$\mathbf{y}' = \mathbf{f}(x, \mathbf{y}),$$

what is $\mathbf{f}$?

(b) Show that the $\mathbf{f}$ of (a) satisfies a Lipschitz condition for all $(x, \mathbf{y})$ where x is real and $\mathbf{y}$ is in $\mathbf{C}_2$.

(c) Show that the $\mathbf{f}$ of (a) is linear in $\mathbf{y}$, that is,

$$\mathbf{f}(x, \alpha \mathbf{y} + \beta \mathbf{z}) = \alpha \mathbf{f}(x, \mathbf{y}) + \beta \mathbf{f}(x, \mathbf{z}),$$

for all real x, complex numbers α, β, and all $\mathbf{y}, \mathbf{z}$ in $\mathbf{C}_2$.

3. Find a solution ϕ of the system

$$y_1' = y_1,$$
$$y_2' = y_1 + y_2,$$

which satisfies $\phi(0) = (1, 2)$.

4. Find a solution ϕ of the system

$$y_1' = y_2,$$
$$y_2' = 6y_1 + y_2,$$

satisfying $\phi(0) = (1, -1)$.

5. Find a solution ϕ of the system

$$y_1' = y_1 + y_2,$$
$$y_2' = y_1 + y_2 + e^{3x},$$

satisfying $\phi(0) = (0, 0)$. (*Hint*: Let $z = y_1 + y_2$.)

6. Let $\mathbf{f}$ be a vector-valued function defined for $(x, \mathbf{y})$ in a set S, with x real, $\mathbf{y}$ in C_n.

(a) Show that $\mathbf{f}$ is continuous at a point $(x_0, \mathbf{y}_0)$ in S if, and only if,

$$|\mathbf{f}(x, \mathbf{y}) - \mathbf{f}(x_0, \mathbf{y}_0)| \to 0,$$

as

$$0 < |x - x_0| + |\mathbf{y} - \mathbf{y}_0| \to 0.$$

(b) Show that $\mathbf{f}$ satisfies a Lipschitz condition in S if, and only if, each component of $\mathbf{f}$ satisfies a Lipschitz condition in S.

6. Existence and uniqueness of solutions of systems

Let $\mathbf{f}$ be a continuous vector-valued function defined on

$$R: \qquad |x - x_0| \leq a, \qquad |\mathbf{y} - \mathbf{y}_0| \leq b, \qquad (a, b > 0).$$

An *initial value problem*

$$\mathbf{y}' = \mathbf{f}(x, \mathbf{y}), \qquad \mathbf{y}(x_0) = \mathbf{y}_0, \tag{6.1}$$

is the problem of finding a solution ϕ of $\mathbf{y}' = \mathbf{f}(x, \mathbf{y})$ on an interval I containing x_0 such that $\phi(x_0) = \mathbf{y}_0$. If

$$\mathbf{y}_0 = (\alpha_1, \cdots, \alpha_n),$$

the problem (6.1) written out becomes

$$y_1' = f_1(x, y_1, \cdots, y_n),$$
$$\cdot$$
$$\cdot$$
$$\cdot$$
$$y_n' = f_n(x, y_1, \cdots, y_n),$$
$$y_1(x_0) = \alpha_1, \qquad \cdots, \qquad y_n(x_0) = \alpha_n.$$

If $\mathbf{f}$ is continuous on R the problem (6.1) always has a solution on some interval containing x_0. If, in addition, $\mathbf{f}$ satisfies a Lipschitz condition on R, this fact may be demonstrated exactly as in Chap. 5 by introducing the *successive approximations* $\boldsymbol{\phi}_0$, $\boldsymbol{\phi}_1$, $\cdots$, where

$$\boldsymbol{\phi}_0(x) = \mathbf{y}_0,$$

$$\boldsymbol{\phi}_{k+1}(x) = \mathbf{y}_0 + \int_{x_0}^{x} \mathbf{f}(t, \boldsymbol{\phi}_k(t))\, dt, \qquad (k = 0, 1, 2, \cdots). \tag{6.2}$$

As an example, let us consider the problem

$$y_1' = y_2,$$

$$y_2' = -y_1,$$

$$\mathbf{y}(0) = (0, 1).$$

Here $\mathbf{f}(x, \mathbf{y}) = (y_2, -y_1)$, and

$$\boldsymbol{\phi}_0(x) = (0, 1),$$

$$\boldsymbol{\phi}_1(x) = (0, 1) + \int_0^x (1, 0)\, dt = (x, 1),$$

$$\boldsymbol{\phi}_2(x) = (0, 1) + \int_0^x (1, -t)\, dt$$

$$= (0, 1) + \left(x, -\frac{x^2}{2}\right) = \left(x, 1 - \frac{x^2}{2}\right),$$

$$\boldsymbol{\phi}_3(x) = (0, 1) + \int_0^x \left(1 - \frac{t^2}{2}, -t\right) dt = \left(x - \frac{x^3}{3!}, 1 - \frac{x^2}{2}\right)$$

It is not too difficult to show that all the $\boldsymbol{\phi}_k$ exist for all real x, and that

$$\boldsymbol{\phi}_k(x) \to \boldsymbol{\phi}(x) = (\sin x, \cos x),$$

where $\boldsymbol{\phi}$ is the solution of the problem.

We summarize the main results for systems.

Theorem 2. (*Local existence*) *Let* $\mathbf{f}$ *be a continuous vector-valued function defined on*

$$R: \qquad |x - x_0| \leqq a, \qquad |\mathbf{y} - \mathbf{y}_0| \leqq b, \qquad (a, b > 0),$$

and suppose $\mathbf{f}$ *satisfies a Lipschitz condition on* R. *If* M *is a constant such that*

$$|\mathbf{f}(x, \mathbf{y})| \leqq M$$

for all (x, y) *in* R, *the successive approximations* $\{\boldsymbol{\phi}_k\}$, (k = 0, 1, 2, $\cdots$), *given by* (6.2) *converge on the interval*

$$I: \qquad |x - x_0| \leqq \alpha = minimum \ \{a, b/M\},$$

to a solution ϕ *of the initial value problem*

$$\mathbf{y}' = \mathbf{f}(x, \mathbf{y}), \qquad \mathbf{y}(x_0) = \mathbf{y}_0,$$

on I.

The proof is the same as that of Theorems 5 and 7 of Chap. 5, with *y*, *f*, *φ* replaced everywhere by **y, f, φ**.

Theorem 3. *If* **f** *satisfies the same conditions as in Theorem 2, and* K *is a Lipschitz constant for* **f** *in* R, *then*

$$| \ \phi(x) \ - \ \phi_k(x) \ | \ \leqq \ \frac{M}{K} \ \frac{(K\alpha)^{k+1}}{(k+1)!} \ e^{K\alpha}$$

for all x *in* I.

This is the analogue of Theorem 8 of Chap. 5, and the proof is the same. The analogues of Theorem 9, Chap. 5, and its corollary are the following results.

Theorem 4. (*Non-local existence*) *Let* **f** *be a continuous vector-valued function defined on*

$$S: \qquad |x - x_0| \leqq a, \qquad |\mathbf{y}| < \infty, \qquad (a > 0),$$

and satisfy there a Lipschitz condition. Then the successive approximations $\{\phi_k\}$ *for the problem*

$$\mathbf{y}' = \mathbf{f}(x, \mathbf{y}), \qquad \mathbf{y}(x_0) = \mathbf{y}_0, \qquad (|\mathbf{y}_0| < \infty),$$

exist on $| \ x \ - \ x_0 \ | \leqq$ a, *and converge there to a solution* ϕ *of this problem.*

Corollary. *Suppose* **f** *is a continuous vector-valued function defined on*

$$|x| < \infty, \qquad |\mathbf{y}| < \infty,$$

and satisfies a Lipschitz condition on each "strip"

$$|x| \leqq a, \qquad |\mathbf{y}| < \infty,$$

where a *is any positive number. Then every initial value problem*

$$\mathbf{y}' = \mathbf{f}(x, \mathbf{y}), \qquad \mathbf{y}(x_0) = \mathbf{y}_0,$$

has a solution which exists for all real x.

The proofs carry over directly from those for Theorem 9 and its corollary in Chap. 5.

Theorem 5. (*Approximation and uniqueness*) *Let* **f**, **g** *be two continuous vector-valued functions defined on*

$$R: \quad |x - x_0| \leqq a, \quad |\mathbf{y} - \mathbf{y}_0| \leqq b, \quad (a, b > 0),$$

and suppose **f** *satisfies a Lipschitz condition on* R *with Lipschitz constant* K. *Suppose* ϕ, ψ *are solutions of the problems*

$$\mathbf{y}' = \mathbf{f}(x, \mathbf{y}), \quad \mathbf{y}(x_0) = \mathbf{y}_1,$$

$$\mathbf{y}' = \mathbf{g}(x, \mathbf{y}), \quad \mathbf{y}(x_0) = \mathbf{y}_2,$$

respectively, on some interval I *containing* x_0. *If for* ϵ, $\delta \geqq 0$

$$|\mathbf{f}(x, \mathbf{y}) - \mathbf{g}(x, \mathbf{y})| \leqq \epsilon, \quad (all \ (x, \mathbf{y}) \ in \ R),$$

and

$$|\mathbf{y}_1 - \mathbf{y}_2| \leqq \delta,$$

then

$$|\phi(x) - \psi(x)| \leqq \delta e^{K|x-x_0|} + \frac{\epsilon}{K}(e^{K|x-x_0|} - 1)$$

for all x *in* I. *In particular, the problem*

$$\mathbf{y}' = \mathbf{f}(x, \mathbf{y}), \quad \mathbf{y}(x_0) = \mathbf{y}_0,$$

has at most one solution on any interval containing x_0.

This is the analogue of Theorem 10, Chap. 5, and the Corollary 1 to this result. The proof carries over directly.

EXERCISES

1. Consider the initial value problem

$$y_1' = y_2^2 + 1,$$

$$y_2' = y_1^2,$$

$$y_1(0) = 0, \quad y_2(0) = 0.$$

(a) If this problem is denoted by

$$\mathbf{y}' = \mathbf{f}(x, \mathbf{y}), \quad \mathbf{y}(0) = \mathbf{y}_0,$$

what are **f** and $\mathbf{y}_0$?

(b) Show that the **f** of (a) satisfies the conditions of Theorem 2 on

$$R: \quad |x| \leqq 1, \quad |\mathbf{y}| \leqq 1.$$

Compute a bound M, a Lipschitz constant K, and an α.

(c) Compute the first three successive approximations ϕ_0, ϕ_1, ϕ_2.

2. Consider the system

$$y_1' = 3y_1 + xy_3,$$
$$y_2' = y_2 + x^3y_3,$$
$$y_3' = 2xy_1 - y_2 + e^xy_3.$$

Show that every initial value problem for this system has a unique solution which exists for all real x.

3. Suppose $\mathbf{y}_0$ is a vector in $\mathbf{R}_n$, and the function $\mathbf{f}$ considered in Theorem 2 has values in $\mathbf{R}_n$ for $(x, \mathbf{y})$ such that $\mathbf{y}$ is in $\mathbf{R}_n$. Show that each of the successive approximations ϕ_k has values in $\mathbf{R}_n$, and hence that the solution ϕ of

$$\mathbf{y}' = \mathbf{f}(x, \mathbf{y}), \qquad \mathbf{y}(x_0) = \mathbf{y}_0,$$

on I has values in $\mathbf{R}_n$. Thus, if $\phi = (\phi_1, \cdots, \phi_n)$, the functions $\phi_1, \cdots, \phi_n$ are real-valued. (*Note*: If the above conditions hold it is sufficient, in Theorem 2, to consider $\mathbf{f}$ defined on that part of R with $\mathbf{y}$ in $\mathbf{R}_n$, and to require $\mathbf{f}$ to be continuous, satisfy $|\mathbf{f}(x, \mathbf{y})| \leq M$, and satisfy a Lipschitz condition on that part of R. Similar statements are valid for the other theorems and corollaries in this and the preceding section.)

4. Consider the system

$$y_1' = y_1 + \epsilon y_2,$$
$$y_2' = \epsilon y_1 + y_2,$$

where ϵ is a positive constant.

(a) Show that every solution exists for all real x.

(b) Let ϕ be the solution satisfying $\phi(0) = (1, -1)$, and let ψ be the solution of

$$y_1' = y_1, \qquad y_2' = y_2,$$

satisfying $\psi(0) = (1, -1)$. Without solving the original system show that

$$|\phi(x) - \psi(x)| \to 0, \qquad (\epsilon \to 0),$$

for each real x.

(c) Find all solutions of the original system. (*Hint*: If ϕ is a solution show that $\chi(x) = e^{-x}\phi(x)$ satisfies

$$z_1' = \epsilon z_2, \qquad z_2' = \epsilon z_1.)$$

(d) Find the solutions ϕ and ψ of (b), and verify the conclusions in (b).

5. Show that all solutions with values in $\mathbf{R}_2$ of the following system exist for all real x:

$$y_1' = a(x) \cos y_1 + b(x) \sin y_2,$$
$$y_2' = c(x) \sin y_1 + d(x) \cos y_2,$$

where a, b, c, d are polynomials with real coefficients. (*Hint*: See the *Note* in Exercise 3.)

6. Prove Theorem 2.

7. Prove Theorem 4.

8. Show that the following modification of Theorem 4 is valid: Let **f** be a continuous vector-valued function defined on

$$S: \quad |x - x_0| \leq a, \quad |\mathbf{y}| < \infty, \quad (a > 0),$$

and satisfy there a Lipschitz condition. The initial value problem

$$\mathbf{y}' = \mathbf{f}(x, \mathbf{y}), \qquad \mathbf{y}(x_1) = \mathbf{y}_0,$$

where $|x_1 - x_0| \leq a$, $|\mathbf{y}_0| < \infty$, has a solution ϕ on $|x - x_0| \leq a$, which can be obtained as the limit of the successive approximations $\{\phi_k\}$. (*Hint:* Consider the two intervals $x_0 - a \leq x \leq x_1$ and $x_1 \leq x \leq x_0 + a$ separately.)

9. Show that the solution ϕ of Ex. 8 is unique.

7. Existence and uniqueness for linear systems

As an important application of the results of Sec. 5 we consider the case of a *linear system*. This is a system

$$\mathbf{y}' = \mathbf{f}(x, \mathbf{y}),$$

where the components $f_1, \cdots, f_n$ of **f** have the form

$$f_1(x, \mathbf{y}) = a_{11}(x)y_1 + a_{12}(x)y_2 + \cdots + a_{1n}(x)y_n + b_1(x),$$

$$\cdot$$
$$\cdot \tag{7.1}$$
$$\cdot$$

$$f_n(x, \mathbf{y}) = a_{n1}(x)y_1 + a_{n2}(x)y_2 + \cdots + a_{nn}(x)y_n + b_n(x).$$

Here the $a_{11}, \cdots, a_{nn}, b_1, \cdots, b_n$ are complex-valued functions defined for real x in some interval I. If all the a_{ij} are continuous on an interval $|x - x_0| \leq a$, where $a > 0$, then the corresponding vector-valued function **f** satisfies a Lipschitz condition on the "strip"

$$S: \quad |x - x_0| \leq a, \quad |\mathbf{y}| < \infty.$$

This can be seen directly, or we can invoke Theorem 1. Let K be any positive constant such that

$$\sum_{j=1}^{n} |a_{jk}(x)| \leq K, \qquad (k = 1, \cdots, n),$$

for all x satisfying $|x - x_0| \leq a$. Then from (7.1) we see that

$$\left| \frac{\partial \mathbf{f}}{\partial y_k}(x, \mathbf{y}) \right| = |(a_{1k}(x), \cdots, a_{nk}(x))| = \sum_{j=1}^{n} |a_{jk}(x)| \leq K.$$

Hence, by Theorem 1, **f** satisfies a Lipschitz condition on S with a Lipschitz constant K. Theorems 4 and 5 are thus applicable to a linear system, and we have the following result.

Theorem 6. *Consider a linear system*

$$\mathbf{y}' = \mathbf{f}(x, \mathbf{y}),$$

where the components of **f** *are given by*

$$f_j(x, \mathbf{y}) = \sum_{k=1}^{n} a_{jk}(x) y_k + b_j(x), \qquad (j = 1, \cdots, n),$$

and the functions a_{jk}, b_j *are continuous on an interval* I *containing* x_0. *If* $\mathbf{y}_0$ *is any vector in* $\mathbf{C}_n$ *there exists one, and only one, solution* $\boldsymbol{\phi}$ *of the problem*

$$\mathbf{y}' = \mathbf{f}(x, \mathbf{y}), \qquad \mathbf{y}(x_0) = \mathbf{y}_0,$$

on I.

Actually the existence Theorem 4 only applies in case I consists of all x satisfying $|x - x_0| \leq a$ for some $a > 0$. However the proof of Theorem 4 applies in case this interval is replaced there by any interval I of the form $\alpha \leq x \leq \beta$ containing x_0. On such an interval the successive approximations will converge to a solution of the initial value problem, which is unique by Theorem 5. We can then apply this modification of Theorem 4 to prove Theorem 6 in case I is *any* interval, containing x_0, on which the a_{jk}, b_j are continuous. In particular, the interval I may be infinite in length.

EXERCISES

1. Show that if the functions a_{jk} and b_j in Theorem 6 are real-valued on I, and the initial vector $\mathbf{y}_0$ is in $\mathbf{R}_n$, then the solution $\boldsymbol{\phi}$ has values in $\mathbf{R}_n$.

2. The linear system

$$y'_j = \sum_{k=1}^{n} a_{jk}(x) y_k, \qquad (j = 1, \cdots, n), \tag{*}$$

with a_{jk} continuous on some interval I, is called a *homogeneous linear system*.

(a) Show that the function $\boldsymbol{\psi}$, defined by $\boldsymbol{\psi}(x) = \mathbf{0}$ for all x in I, is a solution of the system (*). This is called the *trivial solution*.

(b) Let K be a positive constant such that

$$\sum_{j=1}^{n} |a_{jk}(x)| \leq K, \qquad (k = 1, \cdots, n),$$

for all x in I. If ϕ is any solution of (*), and x_0 is any point in I, show that

$$| \phi(x) | \leq | \phi(x_0) | \, e^{K|x-x_0|},$$

for all x in I. (*Hint:* Use Theorem 5.)

(c) Let ψ_1, ψ_2 be two solutions of the system

$$y'_j = \sum_{k=1}^{n} a_{jk}(x)y_k + b_j(x), \qquad (j = 1, \cdots, n),$$

on I. Show that $\phi = \psi_1 - \psi_2$ is a solution of (*) on I.

3. Consider the linear system

$$y'_1 = ay_1 + by_2,$$
$$y'_2 = cy_1 + dy_2,$$

where a, b, c, d are constants.

(a) Show that this system always has a solution ϕ of the form

$$\phi(x) = e^{rx}\alpha,$$

where $\alpha = (\alpha_1, \alpha_2) \neq (0, 0)$ is a constant vector, and r is a constant.

(b) Show that the r of (a) must satisfy

$$\begin{vmatrix} a - r & b \\ c & d - r \end{vmatrix} = 0.$$

(c) Compute a solution of the system

$$y'_1 = 3y_1 + 4y_2,$$
$$y'_2 = 5y_1 + 6y_2.$$

4. Consider the system

$$y'_1 = ay_1 + by_2,$$
$$y'_2 = -by_1 + ay_2,$$

where a, b are real constants.

(a) If $\phi = (\phi_1, \phi_2)$ is any solution with values in R_2 show that

$$\| \phi(x) \| = \| \phi(0) \| \, e^{ax},$$

where

$$\| \phi(x) \| = [\phi_1^2(x) + \phi_2^2(x)]^{1/2}.$$

(b) Verify that the solution satisfying $\phi(0) = (1, 0)$ is given by

$$\phi(x) = e^{ax} (\cos bx, \, - \sin bx).$$

(c) For the case $a = -1$, $b = 1$, plot the curve in the (y_1, y_2)-plane given by the solution in (b), namely the curve

$$y_1 = e^{-x} \cos x, \qquad y_2 = -e^{-x} \sin x, \qquad (-\infty < x < \infty).$$

Also plot the curves corresponding to the cases $a = 0$, $b = 1$, and $a = -1$, $b = 0$ in (b).

5. Consider the system

$$y_1' = ay_1 + by_2,$$
$$y_2' = cy_1 + dy_2, \tag{*}$$

where a, b, c, d are constants.

(a) If $\phi = (\phi_1, \phi_2)$ is any solution show that ϕ_1 and ϕ_2 satisfy the second-order equation

$$y'' - (a + d)y' + (ad - bc)y = 0. \tag{**}$$

(b) Compute the characteristic polynomial p of the equation (**). Compare this with Ex. 3 (b).

(c) Find all solutions of (**). Consider separately the cases when p has unequal, or equal, roots.

(d) Find all solutions of (*), by using (a), (b), (c).

(e) Find all solutions of

$$y_1' = 4y_1 - 3y_2,$$
$$y_2' = 2y_1 - y_2.$$

6. Suppose ϕ, ψ are two solutions of the system (*) in Ex. 5. Show that $\chi = \alpha\phi + \beta\psi$ is also a solution, for any two complex numbers α, β.

8. Equations of order n

An n-th order equation

$$y^{(n)} = f(x, y, y', \cdots, y^{(n-1)}), \tag{8.1}$$

may be viewed as a system of n equations of the first order. Indeed, if

$$y_1 = y, \quad y_2 = y', \quad \cdots, \quad y_n = y^{(n-1)},$$

we may associate with the equation (8.1) the first order system

$$y_1' = y_2,$$
$$y_2' = y_3,$$
$$\cdot$$
$$\cdot \tag{8.2}$$
$$\cdot$$
$$y_{n-1}' = y_n.$$
$$y_n' = f(x, y_1, \cdots, y_n).$$

This has the form $\mathbf{y}' = \mathbf{f}(x, \mathbf{y})$ provided we let $\mathbf{y} = (y_1, \cdots, y_n)$ and

$$f_1(x, y_1, \cdots, y_n) = y_2,$$
$$f_2(x, y_1, \cdots, y_n) = y_3,$$

$$\cdot$$
$$\cdot \qquad\qquad\qquad\qquad (8.3)$$
$$\cdot$$

$$f_{n-1}(x, y_1, \cdots, y_n) = y_n,$$
$$f_n(x, y_1, \cdots, y_n) = f(x, y_1, \cdots, y_n).$$

Moreover if ϕ is a solution of (8.1) then the vector

$$\boldsymbol{\phi} = (\phi, \phi', \cdots, \phi^{(n-1)})$$

is a solution of (8.2). Conversely if

$$\boldsymbol{\phi} = (\phi_1, \cdots, \phi_n)$$

is a solution of (8.2) the first component ϕ_1 is a solution of (8.1), since we have

$$\phi_1' = \phi_2, \quad \phi_1'' = \phi_2' = \phi_3, \quad \cdots, \quad \phi_1^{(n-1)} = \phi_n,$$
$$\phi_1^{(n)}(x) = \phi_n'(x) = f(x, \phi_1(x), \phi_1'(x), \cdots, \phi_1^{(n-1)}(x)).$$

It is thus clear that all results proved for first order systems may be applied to give results for n-th order equations of the type (8.1). In particular we have the following existence and uniqueness result.

Theorem 7. *Let* f *be a complex-valued continuous function defined on*

$$R: \qquad |x - x_0| \leqq a, \quad |\mathbf{y} - \mathbf{y}_0| \leqq b, \qquad (a, b > 0),$$

such that

$$|f(x, \mathbf{y})| \leqq N$$

for all $(\mathrm{x}, \mathbf{y})$ *in* R. *Suppose there exists a constant* L > 0 *such that*

$$|f(x, \mathbf{y}) - f(x, \mathbf{z})| \leqq L |\mathbf{y} - \mathbf{z}|$$

for all $(\mathrm{x}, \mathbf{y})$ *and* $(\mathrm{x}, \mathbf{z})$ *in* R. *Then there exists one, and only one, solution* ϕ *of*

$$y^{(n)} = f(x, y, y', \cdots, y^{(n-1)})$$

on the interval

$$I: \qquad |x - x_0| \leqq \min\{a, b/M\}, \qquad (M = N + b + |\mathbf{y}_0|),$$

which satisfies

$$\phi(x_0) = \alpha_1, \quad \phi'(x_0) = \alpha_2, \quad \cdots, \quad \phi^{(n-1)}(x_0) = \alpha_n,$$
$$(\mathbf{y}_0 = (\alpha_1, \cdots, \alpha_n)).$$

Proof. Consider the system $\mathbf{y}' = \mathbf{f}(x, \mathbf{y})$ with the components of $\mathbf{f}$ given by (8.3). Then

$$|\mathbf{f}(x, \mathbf{y})| = |y_2| + |y_3| + \cdots + |y_n| + |f(x, \mathbf{y})|$$

$$\leq |\mathbf{y}| + |f(x, \mathbf{y})| \leq |\mathbf{y}_0| + b + N,$$

since

$$|\mathbf{y}| - |\mathbf{y}_0| \leq |\mathbf{y} - \mathbf{y}_0| \leq b.$$

It is clear that $\mathbf{f}$ is continuous on R, and

$$|\mathbf{f}(x, \mathbf{y}) - \mathbf{f}(x, \mathbf{z})| = |y_2 - z_2| + \cdots + |y_n - z_n|$$

$$+ |f(x, \mathbf{y}) - f(x, \mathbf{z})|$$

$$\leq |\mathbf{y} - \mathbf{z}| + L|\mathbf{y} - \mathbf{z}|$$

$$= (1 + L)|\mathbf{y} - \mathbf{z}|.$$

Thus $\mathbf{f}$ satisfies a Lipschitz condition on R with Lipschitz constant $K = 1 + L$. We can now apply Theorems 2 and 5 to this system, and the first component of the vector solution is the solution required.

For linear equations of order n we have non-local existence.

Theorem 8. *Let* $a_1, \cdots, a_n, b$ *be continuous complex-valued functions on an interval I containing a point* x_0. *If* $\alpha_1, \cdots, \alpha_n$ *are any n constants, there exists one, and only one, solution* ϕ *of the equation*

$$y^{(n)} + a_1(x)y^{(n-1)} + \cdots + a_n(x)y = b(x)$$

on I satisfying

$$\phi(x_0) = \alpha_1, \quad \phi'(x_0) = \alpha_2, \quad \cdots, \quad \phi^{(n-1)}(x_0) = \alpha_n.$$

Proof. Let $\mathbf{y}_0 = (\alpha_1, \cdots, \alpha_n)$, and consider the linear system

$$y_1' = y_2,$$

$$y_2' = y_3,$$

$$\cdot$$
$$\cdot$$
$$\cdot$$

$$y_{n-1}' = y_n,$$

$$y_n' = -a_n(x)y_1 - a_{n-1}(x)y_2 - \cdots - a_1(x)y_n + b(x).$$

According to Theorem 6 there is a unique solution $\phi = (\phi_1, \cdots, \phi_n)$ of this system on I satisfying

$$\phi_1(x_0) = \alpha_1, \quad \phi_2(x_0) = \alpha_2, \quad \cdots, \quad \phi_n(x_0) = \alpha_n.$$

But since

$$\phi_2 = \phi_1', \quad \phi_3 = \phi_2' = \phi_1'', \quad \cdots, \quad \phi_n = \phi_1^{(n-1)},$$

the function ϕ_1 is the required solution on I.

Note that Theorem 8 includes Theorem 1 of Chapter 3.

E X E R C I S E S

1. Consider the second order equation

$$y'' + a_1 y' + a_2 y = 0, \tag{*}$$

where a_1, a_2 are constants.

(a) What system of the first order is equivalent to this equation?

(b) If the system in (a) is denoted by

$$\mathbf{y}' = \mathbf{f}(x, \mathbf{y}), \tag{**}$$

show that $\mathbf{f}$ satisfies a Lipschitz condition on the set

$$S: \quad |x| < \infty, \quad |\mathbf{y}| < \infty.$$

(c) Show that a Lipschitz constant for $\mathbf{f}$ on S can be chosen to be

$$K = 1 + |a_1| + |a_2|.$$

(d) Let ϕ be any solution of (*). Then $\boldsymbol{\phi} = (\phi, \phi')$ is a solution of (**). Show that if x_0 is any real number then

$$|\boldsymbol{\phi}(x)| \leqq |\boldsymbol{\phi}(x_0)| \, e^{K|x-x_0|}.$$

(Compare this with Theorem 3, Chap. 2.)

2. Consider the linear equation

$$y^{(n)} + a_1(x)y^{(n-1)} + \cdots + a_n(x)y = 0,$$

where $a_1, \cdots, a_n$ are continuous functions on some interval I. Suppose there are non-negative constants $b_1, \cdots, b_n$ such that

$$|a_j(x)| \leqq b_j, \qquad (j = 1, \cdots, n),$$

for all x in I. If ϕ is any solution of this equation on I, and

$$\boldsymbol{\phi} = (\phi, \phi', \cdots, \phi^{(n-1)}),$$

show that

$$|\boldsymbol{\phi}(x)| \leqq |\boldsymbol{\phi}(x_0)| \, e^{K|x-x_0|}, \qquad \text{(all } x \text{ in } I\text{)},$$

where x_0 is a fixed point in I, and

$$K = 1 + b_1 + \cdots + b_n.$$

(Compare this with Theorem 2, Chap. 3.)

3. Show that all real-valued solutions of the equation

$$y'' + \sin y = b(x),$$

where b is continuous for $-\infty < x < \infty$, exist for all real x.

4. Let ϵ, g, v_0, α be positive constants. Consider the two systems

$$y'' = -\epsilon y',$$

$$z'' = -g - \epsilon z', \tag{*}$$

and

$$y'' = 0,$$

$$z'' = -g, \tag{**}$$

each with initial conditions

$$y(0) = z(0) = 0, \quad y'(0) = v_0 \cos \alpha, \quad z'(0) = v_0 \sin \alpha.$$

(a) Determine first order systems (of four equations) which are equivalent to the problems (*) and (**). Show that each of these has a unique solution which exists for all real x.

(b) By solving the systems show that the respective solutions ϕ and ψ satisfy

$$|\phi(x) - \psi(x)| \to 0, \quad \epsilon \to 0.$$

References

1. Kaplan, Wilfred, *Ordinary Differential Equations.* Reading, Mass.: Addison-Wesley, 1958.
   This book is written at about the same level as the present one. It contains an excellent treatment of many applications.

2. Hurewicz, Witold, *Lectures on Ordinary Differential Equations.* Cambridge, Mass.: The Technology Press, and New York: John Wiley and Sons, Inc., 1958.
   This is a short book containing material on linear systems, and an introduction to the geometric theory of ordinary differential equations. A knowledge of advanced calculus and some linear algebra is required.

3. Coddington, Earl A., and Levinson, Norman, *Theory of Ordinary Differential Equations.* New York: McGraw-Hill, 1955.
   This is an advanced work which stresses the analytic aspects of the subject.

4. Lefschetz., S., *Differential Equations: Geometric Theory.* New York: Interscience, 1957.
   This is an advanced book which emphasizes the geometric approach.

5. Kamke, E., *Differentialgleichungen—Lösungsmethoden und Lösungen,* vol. I, 2nd ed. Leipzig: Akademische Verlagsgesellschaft, 1943; reprinted by J. W. Edwards, Ann Arbor, Mich., 1945.
   This book has a summary of many of the important results, and contains a large compilation of special equations and their solutions.

Answers to Exercises

CHAPTER 0

Chap. 0, Sec. 2

1. (a) $1 + i3$ (b) $-2 + i5$
   (c) $(12 + 4\sqrt{2}) + i(24 - 2\sqrt{2})$ (d) i
   (e) $\sqrt{41}$ (f) 4
   (g) 2

2. (a) $r = 2, \theta = \pi/3$ (b) $r = 2, \theta = \pi/2$
   (c) $r = 1, \theta = \pi/2$ (d) $r = 2, \theta = 0$

3. (a) circle of radius 1, center at $(2, 0)$
   (b) interior of the circle of radius 2, center at $(-2, 0)$
   (c) vertical strip consisting of all (x, y) such that $-3 \leqq x \leqq 3$, $-\infty < y < \infty$
   (d) two half-planes:
      (i) all (x, y) such that $-\infty < x < \infty, y > 1$
      (ii) all (x, y) such that $-\infty < x < \infty, y < -1$
   (e) ellipse with center $(-1/2, 0)$, foci $(1, 0)$ and $(-2, 0)$, major semi-axis 4, minor semi-axis $\sqrt{55}/2$

8. all z such that $|z| \leqq 1$

10. (a) $\sqrt{2}, -\sqrt{2}$ (b) $1, -\dfrac{1}{2} + i\dfrac{\sqrt{3}}{2}, -\dfrac{1}{2} - i\dfrac{\sqrt{3}}{2}$

Chap. 0, Sec. 3

1. (a) $2x$ (b) $3x - 2x^2$
   (c) $2 + i3 - i4x$ (d) $1 + i(5/6)$

2. (a) $F(x) = \dfrac{x^2}{2} + i\dfrac{x^4}{4}$ (b) $F'(x) = x + ix^3$

4. (a) $u(x, y) = x^2 - y^2$, $v(x, y) = 2xy$

6. (a) $F(x) = 5x^2 - 1 + 3ix$ (b) $F'(x) = 10x + 3i$
   (c) $\frac{2}{3} + \frac{3}{2}i$

Chap. 0, Sec. 4

1. (a) −3, multiplicity 1; 2, multiplicity 1

(b) $-\dfrac{1}{2} + i\dfrac{\sqrt{3}}{2}$, multiplicity 1; $-\dfrac{1}{2} - i\dfrac{\sqrt{3}}{2}$, multiplicity 1

(c) 2, multiplicity 2; 1, multiplicity 1

(d) 1, multiplicity 2; i, multiplicity 1

(e) $3^{1/4}$, $-3^{1/4}$, $i3^{1/4}$, $-i3^{1/4}$, all multiplicity 1

6. (a) i has multiplicity 3

(b) $-1 + i$, $-1 - i$

Chap. 0, Sec. 5

1. $1, -\dfrac{1}{2} + i\dfrac{\sqrt{3}}{2}, -\dfrac{1}{2} - i\dfrac{\sqrt{3}}{2}$

2. $\dfrac{\sqrt{2}}{2}(1 + i), -\dfrac{\sqrt{2}}{2}(1 + i)$

3. (a) $(24)^{1/3}\left(\dfrac{1}{2} + i\dfrac{\sqrt{3}}{2}\right), -(24)^{1/3}, (24)^{1/3}\left(\dfrac{1}{2} - i\dfrac{\sqrt{3}}{2}\right)$

(b) $2\sqrt{2}\left(\cos\dfrac{3\pi}{8} + i\sin\dfrac{3\pi}{8}\right),$

$2\sqrt{2}\left(\cos\dfrac{7\pi}{8} + i\sin\dfrac{7\pi}{8}\right),$

$2\sqrt{2}\left(\cos\dfrac{11\pi}{8} + i\sin\dfrac{11\pi}{8}\right),$

$2\sqrt{2}\left(\cos\dfrac{15\pi}{8} + i\sin\dfrac{15\pi}{8}\right)$

(c) $\sqrt{2}i, \sqrt{2}i, -\sqrt{2}i, -\sqrt{2}i$

(d) $\cos\left(\dfrac{2\pi k}{100}\right) + i\sin\left(\dfrac{2\pi k}{100}\right),$

$k = 0, 1, 2, \cdots, 99$

6. (b) (i) $\dfrac{e^r - 1}{r}$

(ii) $\dfrac{ae^a \cos b + be^a \sin b - a}{a^2 + b^2}$

(iii) $\dfrac{ae^a \sin b - be^a \cos b + b}{a^2 + b^2}$

8. $-\dfrac{3}{2} + \dfrac{\sqrt{17}}{2}, \quad -\dfrac{3}{2} - \dfrac{\sqrt{17}}{2}$

9. (a) $x = 0$ (b) all real x

(c) $x = 0$

10. (c) $s(z) = \dfrac{1}{1-z}$

Chap. 0, Sec. 6

1. (a) $-1 + 2i$ (b) $z_1 = -i, \quad z_2 = i$

2. $z_1 = \frac{5}{3}, \quad z_2 = -\frac{8}{3}, \quad z_3 = \frac{7}{3}$

3. Yes; $z_1 = 5, \quad z_2 = 1, \quad z_3 = -11$

CHAPTER 1

Chap. 1, Sec. 3

1. (a) $\phi(x) = \dfrac{e^{3x}}{3} - \cos x + c, \quad c$ any constant

(b) $\phi(x) = x^2 + \dfrac{x^3}{6} + c_1 x + c_2, \quad c_1, c_2$ any constants

(c) $\phi(x) = c_1 x^{k-1} + c_2 x^{k-2} + \cdots + c_k, \quad c_1, c_2, \cdots, c_k$ any constants

(d) $\phi(x) = \dfrac{x^5}{60} + c_1 x^2 + c_2 x + c_3, \quad c_1, c_2, c_3$ any constants

3. (b) $\phi(x) = \left(\dfrac{2}{5}\right)[1 + 4e^{5(1-x)}]$

(c) $\phi(x) = \left(\dfrac{2}{5}\right)\left[1 - \dfrac{2}{3 - e^{-5}} e^{-5x}\right]$

4. (a) $\phi(x) = \dfrac{x^3}{2} + \dfrac{x^2}{2} + c_1 x + c_2, \quad (0 \leqq x \leqq 1), \quad c_1, c_2$ any constants

(b) $\phi(x) = \dfrac{x^3}{2} + \dfrac{x^2}{2} + 2x + 1$

(c) $\phi(x) = \dfrac{x^3}{2} + \dfrac{x^2}{2} + 2x$

5. (d) $|\phi(x)| = |\phi(0)|$ for all x

Chap. 1, Sec. 6

1. (a) $\phi(x) = -\frac{1}{2} + ce^{2x}$, c any constant
   (b) $\phi(x) = \frac{1}{2}e^x + ce^{-x}$, c any constant
   (c) $\phi(x) = -\frac{1}{2}(x^2 + 2x + 1) + ce^{2x}$, c any constant
   (d) $\phi(x) = -e^{-x} + ce^{-x/3}$, c any constant
   (e) $\phi(x) = \left(\dfrac{3 - i}{10}\right)e^{ix} + ce^{-3x}$, c any constant

2. $\phi(\pi) = -i\pi$

3. (a) $\phi(x) = \dfrac{E}{R} + ce^{-Rx/L}$, c any constant

   (b) $\phi(x) = I_0 e^{-Rx/L} + \dfrac{E}{R}(1 - e^{-Rx/L})$

4. (a) $\phi(x) = \left(\dfrac{E}{R^2 + \omega^2 L^2}\right)(\omega L e^{-Rx/L} + R \sin \omega x - \omega L \cos \omega x)$

5. (a) $\phi(x) = \left(\dfrac{E}{R + i\omega L}\right)(e^{i\omega x} - e^{-Rx/L})$

   (b) $\phi_1(x) = \left(\dfrac{E}{R^2 + \omega^2 L^2}\right)(-R e^{-Rx/L} + R \cos \omega x + \omega L \sin \omega x)$

   (c) $\phi_2(x) = \left(\dfrac{E}{R^2 + \omega^2 L^2}\right)(\omega L e^{-Rx/L} + R \sin \omega x - \omega L \cos \omega x)$

6. (b) $\phi(x) = \frac{1}{2}\sin x - \frac{1}{2}\cos x + \frac{3}{5}\cos 2x + \frac{6}{5}\sin 2x - \frac{1}{10}e^{-x}$

7. (a) $\phi(x) = e^{-ax}\displaystyle\int_0^x e^{at}b(t)\,dt$

9. Given any $\epsilon > 0$ there is an $x_0 > 0$ such that $|b(x) - \beta| < \epsilon$ for $x \geqq x_0$. Write any solution ϕ as

$$\phi(x) = e^{-a(x-x_0)}\phi(x_0) + \int_{x_0}^x e^{-a(x-t)}b(t)\,dt$$

Chap. 1, Sec. 7

1. (a) $\phi(x) = \frac{1}{2} + ce^{-x^2}$, c any constant

(b) $\phi(x) = \frac{3}{4}x^3 - 1 + \frac{c}{x}$, $(x > 0)$, c any constant

(c) $\phi(x) = 3 + c \exp(-e^x)$

(d) $\phi(x) = (\sec x)e^{\sin x} + c \sec x$, c any constant

(e) $\phi(x) = \frac{x^2}{2} \exp(-x^2) + c \exp(-x^2)$, c any constant

2. (a) $\phi(x) = xe^{-\sin x}$

3. (b) $\phi(x) = \frac{1}{x} - \frac{6}{7x^2}$, $(x > 0)$

5. (b) $\phi(x) = (-\frac{1}{2} + ce^{-x^2})^{-1}$; also $\phi(x) = 0$

6. (c) $\int_0^{\xi} a(t)\, dt = 2\pi i k$, $(k = 0, \pm 1, \pm 2, \cdots)$;

$\int_0^{\xi} a(t)\, dt = \pi i k$, $(k = 0, \pm 1, \pm 2, \cdots)$;

$\int_0^{\xi} a(t)\, dt = 0$

(d) $a = \pi i k/\xi$, $(k = 0, \pm 1, \pm 2, \cdots)$

7. (d) (i) $\phi(x) = \frac{1}{10}(3 \cos x + \sin x)$

(ii) $\phi(x) = ce^{-\sin x} + 2(\sin x - 1)$, c any constant

8. $\phi(x) = (c - \frac{1}{4}e^{-2} - \frac{1}{4})e^{-2x}$, $(-\infty < x \leq -1)$,

$\phi(x) = (c - \frac{1}{4})e^{-2x} + \frac{1}{4}(1 + 2x)$, $(-1 < x \leq 0)$,

$\phi(x) = (c - \frac{3}{4})e^{-2x} + \frac{1}{4}(3 - 2x)$, $(0 < x \leq 1)$,

$\phi(x) = (c + \frac{1}{4}e^2 - \frac{3}{4})e^{-2x}$, $(1 < x < \infty)$, $c = \phi(0)$

9. If $a = 0$, $\psi(x) = x$, $(0 \leq x \leq \xi)$, and $\psi(x) = \xi$, $(x > \xi)$.

If $a \neq 0$, $\psi(x) = (1/a)(1 - e^{-ax})$, $(0 \leq x \leq \xi)$, and

$$\psi(x) = (e^{-ax}/a)(e^{a\xi} - 1), \quad (x > \xi)$$

12. (b) $\phi_k(x) = \exp[-i(2\pi k - \alpha)x]$

13. $g(x, y) = \dfrac{ie^{i(\alpha+l)}e^{-il(x-y)}}{1 - e^{i(\alpha+l)}}$, $(0 \leq y \leq x)$.

$g(x, y) = \dfrac{ie^{-il(x-y)}}{1 - e^{i(\alpha+l)}}$, $(x < y \leq 1)$

14. (a) $\phi(x) = e^x - 1$

CHAPTER 2

Chap. 2, Sec. 2

1. c_1, c_2 are any constants

  (a) $\phi(x) = c_1 e^{2x} + c_2 e^{-2x}$

  (b) $\phi(x) = c_1 + c_2 e^{-2x/3}$

  (c) $\phi(x) = c_1 e^{4ix} + c_2 e^{-4ix}$,   or $\phi(x) = c_1 \cos 4x + c_2 \sin 4x$

  (d) $\phi(x) = c_1 + c_2 x$

  (e) $\phi(x) = c_1 e^{i(\sqrt{2}-1)x} + c_2 e^{-i(\sqrt{2}+1)x}$

  (f) $\phi(x) = e^{2x}(c_1 \cos x + c_2 \sin x)$

  (g) $\phi(x) = c_1 e^{x} + c_2 e^{-3ix}$

2. (a) $\phi(x) = \frac{3}{5} e^{2x} + \frac{2}{5} e^{-3x}$

  (b) $\psi(x) = \frac{1}{5} e^{2x} - \frac{1}{5} e^{-3x}$

  (c) $\phi(1) = \frac{3}{5} e^2 + \frac{2}{5} e^{-3}$, $\psi(1) = \frac{1}{5} e^2 - \frac{1}{5} e^{-3}$

3. (a) $\phi(x) = \cos x + 2 \sin x$

  (b) $\phi(x) = c \sin x$, c any constant

  (c) $\phi(x) = c \sin x$, c any constant

  (d) $\phi(x) = 0$,   all x

5. (a)   (i) $\phi(x) = c_1 \exp\left[-\dfrac{Rx}{2L} + \dfrac{1}{2}\left(\dfrac{R^2}{L^2} - \dfrac{4}{LC}\right)^{1/2} x \right]$

$$+ c_2 \exp\left[-\dfrac{Rx}{2L} - \dfrac{1}{2}\left(\dfrac{R^2}{L^2} - \dfrac{4}{LC}\right)^{1/2} x \right], \quad c_1, c_2 \text{ any constants}$$

    (ii) $\phi(x) = (c_1 + c_2 x) \exp\left(-Rx/2L\right)$, c_1, c_2 any constants

    (iii) $\phi(x) = c_1 \exp\left[-\dfrac{Rx}{2L} + \dfrac{i}{2}\left(\dfrac{4}{LC} - \dfrac{R^2}{L^2}\right)^{1/2} x \right]$

$$+ c_2 \exp\left[-\dfrac{Rx}{2L} - \dfrac{i}{2}\left(\dfrac{4}{LC} - \dfrac{R^2}{L^2}\right)^{1/2} x \right], \quad c_1, c_2 \text{ any constants}$$

  (d) $\alpha = -\dfrac{R}{2L}$, $\beta = \dfrac{1}{2}\left(\dfrac{4}{LC} - \dfrac{R^2}{L^2}\right)^{1/2}$

8. (a)   (i) $k = 1, 2, 3, \cdots$

    (ii) $k = 0, 1, 2, \cdots$

    (iii) $k = 2n$, $n = 0, 1, 2, \cdots$

    (iv) $k = 2n - 1$, $n = 1, 2, 3, \cdots$

  (b)   (i) $\phi(x) = c \sin nx$, $c \neq 0$, $n = 1, 2, 3, \cdots$

    (ii) $\phi(x) = c \cos nx$, $c \neq 0$, $n = 0, 1, 2, \cdots$

    (iii) $\phi(x) = c_1 \cos 2nx + c_2 \sin 2nx$, $|c_1|^2 + |c_2|^2 \neq 0$, $n = 0, 1, 2, \cdots$

    (iv) $\phi(x) = c_1 \cos (2n - 1)x + c_2 \sin (2n - 1)x$, $|c_1|^2 + |c_2|^2 \neq 0$,

       $n = 1, 2, 3, \cdots$

9. $k = a_2 - (a_1^2/4)$

Chap. 2, Sec. 3

1. (a) $\phi(x) = \frac{1}{4}e^{3x} - \frac{1}{4}e^{-x}$

(b) $\phi(x) = 0$, all x

(c) $\phi(x) = \left(\dfrac{3i + 9}{5}\right)e^x + \left(\dfrac{1 - 3i}{5}\right)e^{-3ix}$

(d) $\phi(x) = \pi \cos \sqrt{10}x + \dfrac{\pi^2}{\sqrt{10}} \sin \sqrt{10}x$

3. $\phi(x) = \frac{1}{2}(e^x - e^{-x}), \quad (0 \le x \le 1)$,

$\phi(x) = \frac{1}{6}(2e^{3x-2} - e^{3x-4} + e^{-3x+4} - 2e^{-3x+2}), \quad (1 \le x \le 2)$

6. $A = \dfrac{k(k - 2i)}{k^2 + 4}, \quad B = \dfrac{2(2 + ik)}{k^2 + 4} e^{-ik}$,

$\phi(x) = \left(\dfrac{4ik - 2k^2}{k^2 + 4}\right)x + \left(\dfrac{4 + 2k^2 - 2ik}{k^2 + 4}\right) \quad \text{for } x \text{ in } I$

Chap. 2, Sec. 4

1. (a) independent   (b) independent
   (c) dependent   (d) independent
   (e) dependent   (f) independent

2. (a) false   (b) true
   (c) true   (d) true

3. (b) $W(\phi_1, \phi_2)(x) = 0$, all x

4. (b) $W(\phi_1, \phi_2)(x) = \beta e^{2\alpha x}$

7. $l = n^2\pi^2, \quad n = 1, 2, 3, \cdots; \quad \phi_n(x) = \sin n\pi x$

8. $W' = -a_1 W$

Chap. 2, Sec. 6

1. c_1, c_2 may be any constants

(a) $\phi(x) = c_1 \cos 2x + c_2 \sin 2x + \frac{1}{3} \cos x$

(b) $\phi(x) = c_1 \cos 3x + c_2 \sin 3x - \frac{1}{6}x \cos 3x$

(c) $\phi(x) = c_1 \cos x + c_2 \sin x - \cos x \log (\sec x + \tan x)$

(d) $\phi(x) = c_1 \exp [-i(\sqrt{2} + 1)x] + c_2 \exp [i(\sqrt{2} - 1)x] + x - 2i$

(e) $\phi(x) = e^{2x}(c_1 \cos x + c_2 \sin x) + \frac{3}{10}e^{-x} + \frac{44}{125} + \frac{16}{25}x + \frac{2}{5}x^2$

(f) $\phi(x) = c_1 e^{6x} + c_2 e^x + \frac{1}{74}(7 \cos x + 5 \sin x)$

(g) $\phi(x) = c_1 \cos x + c_2 \sin x + \dfrac{x}{2} \sin x - \dfrac{\sin x \sin 2x}{4}$

(h) $\phi(x) = (\cos x)(\log \cos x) + x \sin x + c_1 \cos x + c_2 \sin x$

(i) $\phi(x) = c_1 e^{x/2} + c_2 e^{-x/2} + \frac{1}{3}e^x$

(j) $\phi(x) = c_1 e^{2x/3} + c_2 e^{-3x/2} - \frac{1}{6}x - \frac{5}{36}$

2. (a) $B = \dfrac{A}{p(\alpha)}$

   (b) $\psi(x) = \dfrac{Axe^{\alpha x}}{p'(\alpha)}$,   if α has multiplicity one;

      $\psi(x) = \dfrac{A}{2}x^2 e^{\alpha x}$,   if α has multiplicity two

   (d) $\psi(x) = \dfrac{A_1}{p(\alpha_1)}e^{\alpha_1 x} + \dfrac{A_2}{p(\alpha_2)}e^{\alpha_2 x}$

3. (c) $B = \dfrac{E}{L}\left[\left(\dfrac{1}{LC} - \omega^2\right)^2 + \dfrac{\omega^2 R^2}{L^2}\right]^{-1/2}$, $\cos \alpha = \left(\dfrac{LB}{E}\right)\left(\dfrac{1}{LC} - \omega^2\right)$,

     $\sin \alpha = \left(\dfrac{LB}{E}\right)\left(\dfrac{\omega R}{L}\right)$

   (d) $\omega = \left(\dfrac{1}{LC} - \dfrac{R^2}{2L^2}\right)^{1/2}$

4. (a) $\phi(x) = c_1 \cos \omega x + c_2 \sin \omega x + \dfrac{A}{2\omega}x \sin \omega x$, c_1, c_2 any constants

Chap. 2, Sec. 7

1. (a) independent   (b) dependent
    (c) independent

3. (a) true   (b) false

4. (a) $\phi(x) = c_1 e^{2x} + c_2 \exp[(-1 + i\sqrt{3})x] + c_3 \exp[(-1 - i\sqrt{3})x]$,
     c_1, c_2, c_3 any constants
   (b) $\phi(x) = e^{\sqrt{2}x}(c_1 \cos \sqrt{2}x + c_2 \sin \sqrt{2}x) + e^{-\sqrt{2}x}(c_3 \cos \sqrt{2}x + c_4 \sin \sqrt{2}x)$,
     c_1, c_2, c_3, c_4 any constants
   (c) $\phi(x) = c_1 + c_2 e^{2x} + c_3 e^{3x}$, c_1, c_2, c_3 any constants
   (d) $\phi(x) = c_1 e^{ix} + c_2 e^{2ix} + c_3 e^{-2ix}$, c_1, c_2, c_3 any constants
   (e) $\phi(x) = \sum\limits_{k=1}^{100} c_k e^{r_k x}$, $r_k = (100)^{1/100} \exp i\left[\dfrac{(2k - 1)\pi}{100}\right]$, $k = 1, 2, \cdots, 100$;

     $c_1, \cdots, c_{100}$ any constants
   (f) $\phi(x) = c_1 e^{ix} + c_2 e^{-ix} + c_3 e^{2ix} + c_4 e^{-2ix}$, $c_1, \cdots, c_4$ any constants
   (g) $\phi(x) = c_1 e^{2x} + c_2 e^{-2x} + c_3 e^{2ix} + c_4 e^{-2ix}$, $c_1, \cdots, c_4$ any constants
   (h) $\phi(x) = (c_1 + c_2 x)e^{-x} + c_3 e^{2x}$, c_1, c_2, c_3 any constants
   (i) $\phi(x) = (c_1 + c_2 x + c_3 x^2)e^{ix}$, c_1, c_2, c_3 any constants

5. (b) $\phi(x) = \cos \sqrt{2}x \cosh \sqrt{2}x$

6. (a) $\phi_1(x) = 1$, $\phi_2(x) = x$, $\phi_3(x) = x^2$, $\phi_4(x) = x^3$

(b) $\phi_1(x) = \cosh kx \cos kx$,

$\phi_2(x) = \cosh kx \sin kx$,

$\phi_3(x) = \sinh kx \cos kx$,

$\phi_4(x) = \sinh kx \sin kx$, where $\lambda = 4k^4$

(c) $\phi_1(x) = \cosh kx$, $\phi_2(x) = \sinh kx$,

$\phi_3(x) = \cos kx$, $\phi_4(x) = \sin kx$, where $\lambda = -k^4$

Chap. 2, Sec. 8

1. (a) $\phi_1(x) = 1$, $\phi_2(x) = e^{2x}$, $\phi_3(x) = e^{-2x}$

(b) $W(\phi_1, \phi_2, \phi_3)(x) = 16$

(c) $\phi(x) = \dfrac{\sinh 2x}{2}$

2. (a) $\phi_1(x) = e^x$, $\phi_2(x) = xe^x$, $\phi_3(x) = e^{-x}$, $\phi_4(x) = \cos x$, $\phi_5(x) = \sin x$

(b) $W(\phi_1, \phi_2, \phi_3, \phi_4, \phi_5)(x) = 32e^x$

(c) $\phi(x) = \frac{5}{8}e^x - \frac{1}{4}xe^x + \frac{1}{8}e^{-x} + \frac{1}{4}\cos x - \frac{1}{4}\sin x$

Chap. 2, Sec. 9

1. (a) $\phi(x) = c_1 \cos x + c_2 \sin x$, c_1, c_2 any real constants

(b) $\phi(x) = c_1 e^x + c_2 e^{-x}$, c_1, c_2 any real constants

(c) $\phi(x) = c_1 e^x + c_2 e^{-x} + c_3 \cos x + c_4 \sin x$, c_1, c_2, c_3, c_4 any real constants

(d) $\phi(x) = \displaystyle\sum_{k=1}^{5} c_k \phi_k(x)$, where $c_1, \cdots, c_5$ are any real constants, and

$$\phi_1(x) = \exp\left[(2^{1/5}\cos(\pi/5))x\right]\cos\left[2^{1/5}\left(\sin\frac{\pi}{5}\right)x\right],$$

$$\phi_2(x) = \exp\left[(2^{1/5}\cos(\pi/5))x\right]\sin\left[2^{1/5}\left(\sin\frac{\pi}{5}\right)x\right],$$

$$\phi_3(x) = \exp\left[(2^{1/5}\cos(3\pi/5))x\right]\cos\left[2^{1/5}\left(\sin\frac{3\pi}{5}\right)x\right],$$

$$\phi_4(x) = \exp\left[(2^{1/5}\cos(3\pi/5))x\right]\sin\left[2^{1/5}\left(\sin\frac{3\pi}{5}\right)x\right],$$

$$\phi_5(x) = \exp(-2^{1/5}x)$$

(e) $\phi(x) = c_1 e^{2x} + c_2 e^{-2x} + c_3 e^x + c_4 e^{-x}$, c_1, c_2, c_3, c_4 any real constants

2. $\phi(x) = -\dfrac{1}{3}e^{-x} + \dfrac{1}{3}e^{x/2}\cos\dfrac{\sqrt{3}}{2}x + \dfrac{\sqrt{3}}{3}e^{x/2}\sin\dfrac{\sqrt{3}}{2}x$

3. (a) $\phi(x) = c_1 \cos x + c_2 \sin x$, c_1, c_2 any real constants

(b) $\phi(x) = 0$, all x

5. (c) $\phi(x) = c[(\cosh k - \cos k)(\sinh kx - \sin kx)$
$\qquad - (\sinh k - \sin k)(\cosh kx - \cos kx)]$, c any constant

(d) $k = 2n\pi$, $n = 0, 1, 2, \cdots$

(e) $\phi(x) = c_1 \cos 2n\pi x + c_2 \sin 2n\pi x$, c_1, c_2 any constants, $n = 0, 1, 2, \cdots$

Chap. 2, Sec. 10

1. (a) $\phi(x) = c_1 + c_2 e^x + c_3 e^{-x} - \dfrac{x^2}{2}$, c_1, c_2, c_3 any constants

(b) $\phi(x) = c_1 e^{2x} + e^{-x}(c_2 \cos \sqrt{3}x + c_3 \sin \sqrt{3}x) + \left(\dfrac{i - 8}{65}\right)e^{ix}$,

c_1, c_2, c_3 any constants

(c) $\phi(x) = c_1 e^{\sqrt{2}x} \cos x + c_2 e^{\sqrt{2}x} \sin x + c_3 e^{-\sqrt{2}x} \cos x + c_4 e^{-\sqrt{2}x} \sin x + \frac{1}{17} \cos x$,

$c_1, \cdots, c_4$ any constants

(d) $\phi(x) = (c_1 + c_2 x + c_3 x^2 + c_4 x^3)e^x + \dfrac{x^4}{24}e^x$, $c_1, \cdots, c_4$ any constants

(e) $\phi(x) = c_1 e^x + c_2 e^{-x} + c_3 \cos x + c_4 \sin x - \dfrac{x}{4} \sin x$,

$c_1, \cdots, c_4$ any constants

(f) $\phi(x) = (c_1 + c_2 x)e^{ix} + \frac{1}{2}x^2 e^{ix} + \frac{1}{2}e^{-ix}$, c_1, c_2 any constants

2. (a) $B = \dfrac{A}{p(\alpha)}$

(b) $\psi(x) = \dfrac{A}{p'(\alpha)} x e^{\alpha x}$

(c) $\psi(x) = \dfrac{A}{p^{(k)}(\alpha)} x^k e^{\alpha x}$

4. $h(x) = x e^{-kx}$,   if $k = \omega$,
$h(x) = (k^2 - \omega^2)^{-1/2} e^{-kx} \sinh [(k^2 - \omega^2)^{1/2}x]$,   if $k > \omega$,
$h(x) = (\omega^2 - k^2)^{-1/2} e^{-kx} \sin [(\omega^2 - k^2)^{1/2}x]$,   if $k < \omega$

5. $\psi(x) = 2 \cos x$, $(-\infty < x < -\pi)$,
$\psi(x) = \cos x - 1$, $(-\pi \leqq x < 0)$,
$\psi(x) = 1 - \cos x$, $(0 \leqq x \leqq \pi)$,
$\psi(x) = -2 \cos x$, $(\pi < x < \infty)$

Chap. 2, Sec. 11

1. (a) $\psi(x) = \frac{1}{3} \cos x$

(b) $\psi(x) = -\dfrac{x}{4} \cos 2x$

(c) $\psi(x) = \dfrac{3x}{4} e^{2x} - \dfrac{4}{3} e^{-x}$

(d) $\psi(x) = -\frac{1}{2}x^2 + \frac{1}{2}x - \frac{3}{4} - \frac{3}{10} \cos x - \frac{1}{10} \sin x$

(e) $\psi(x) = \frac{1}{162}(1 - 6x + 9x^2)e^{3x}$

(f) $\psi(x) = \frac{1}{50}[(11 - 5x)e^x \cos 2x + (-2 + 10x)e^x \sin 2x]$

(g) $\psi(x) = \left(\dfrac{3 - i}{20}\right)e^{2x} + \left(\dfrac{6 + 2i}{20}\right)e^{-2x}$

(h) $\psi(x) = \dfrac{x^5}{60} + \dfrac{1}{4} e^{-x} \sin x - \dfrac{1}{4} e^{-x} \cos x$

(i) $\psi(x) = \dfrac{x^5}{60} e^{-x}$

Chap. 2, Sec. 12

1. (a) use induction

CHAPTER 3

Chap. 3, Sec. 3

1. (a) $r = 1, -1$
   (b) $\phi_1(x) = x, \quad \phi_2(x) = 1/x$
   (c) $\phi_1(x) = \frac{1}{2}(x + x^{-1}),$
   $\phi_2(x) = \frac{1}{2}(x - x^{-1})$

2. $\phi_1(x) = 3x - 1, \quad \phi_2(x) = (3x - 1)^{-1}$

3. (a) $2u' + a_1 u = 0$

   (b) $u(x) = c \exp\left[-\dfrac{1}{2}\int_{x_0}^{x} a_1(t)\, dt\right], \quad c$ constant, x_0, x in I

6. (d) $a = \phi_1(\xi), \quad b = \phi_1'(\xi), \quad c = \phi_2(\xi), \quad d = \phi_2'(\xi)$

Chap. 3, Sec. 4

6. $\psi(x) = x^{1/2} \log x$

Chap. 3, Sec. 5

1. (a) $\phi_2(x) = x^5$
   (b) $\phi_2(x) = x \log x$
   (c) $\phi_2(x) = x \exp(x^2)$
   (d) $\phi_2(x) = -x - 1$

(e) $\phi_2(x) = \dfrac{x}{2} \log \left(\dfrac{1 + x}{1 - x}\right) - 1$

(f) $\phi_2(x) = x \displaystyle\int_1^x t^{-2} e^{t^2}\, dt$

2. $\phi_1(x) = x, \quad \phi_2(x) = x^2, \quad \phi_3(x) = x^3$

3. (a) $v = u'$ satisfies

$$\phi_1(x)v'' + [3\phi_1'(x) + a_1(x)\phi_1(x)]v' + [3\phi_1''(x) + 2a_1(x)\phi_1'(x) + a_2\phi_1(x)]v = 0$$

4. $\phi_3(x) = x^{-1}$

5. (a) let $\phi_2(x_0) = 0, \quad \phi_2'(x_0) = 1/\phi_1(x_0)$

(b) $\phi_2(x) = \phi_1(x) \displaystyle\int_{x_0}^x [\phi_1(t)]^{-2} A(t)\, dt, \quad$ where $A(t) = \exp\left[-\displaystyle\int_{x_0}^t a_1(s)\, ds\right]$

Chap. 3, Sec. 6

1. $\phi(x) = c_1 x^2 + c_2 x^{-1} + \frac{1}{2} - x, \quad c_1, c_2$ any constants

2. $\psi(x) = -2x \log x + x^2$

3. (a) $\phi_1(x) = x^{-2} \cos x, \quad \phi_2(x) = x^{-2} \sin x$

(b) $\phi(x) = c_1 x^{-2} \cos x + c_2 x^{-2} \sin x + 1 - 2x^{-2}, \quad c_1, c_2$ any constants

4. (b) $\phi(x) = c_1 x + c_2 x \log x + x^2, \cdot c_1, c_2$ any constants

Chap. 3, Sec. 7

1. (a) $\phi_1(x) = x,$

$$\phi_2(x) = \sum_{m=0}^{\infty} \frac{x^{2m}}{m!2^m(2m - 1)}$$

(b) $\phi_1(x) = 1 + \displaystyle\sum_{m=1}^{\infty} \frac{(-1)^m(-1)\cdot 8\cdot 17 \cdots (9m - 10)}{2\cdot 3\cdot 5\cdot 6 \cdots (3m - 1)(3m)} x^{3m},$

$$\phi_2(x) = x + \sum_{m=1}^{\infty} \frac{(-1)^m 2\cdot 11\cdot 20 \cdots (9m - 7)}{3\cdot 4\cdot 6\cdot 7 \cdots (3m)(3m + 1)} x^{3m+1}$$

(c) $\phi_1(x) = 1 + \displaystyle\sum_{m=1}^{\infty} \frac{x^{4m}}{3\cdot 4\cdot 7\cdot 8 \cdots (4m - 1)(4m)},$

$$\phi_2(x) = x + \sum_{m=1}^{\infty} \frac{x^{4m+1}}{4\cdot 5\cdot 8\cdot 9 \cdots (4m)(4m + 1)}$$

(d) $\phi_1(x) = 1 + \sum\limits_{m=1}^{\infty} \dfrac{(-1)^m 1 \cdot 5 \cdot 9 \cdots (4m-3)}{3 \cdot 4 \cdot 7 \cdot 8 \cdots (4m-1)(4m)} x^{4m},$

$\phi_2(x) = x + \sum\limits_{m=1}^{\infty} \dfrac{(-1)^m 2 \cdot 6 \cdot 10 \cdots (4m-2)}{4 \cdot 5 \cdot 8 \cdot 9 \cdots (4m)(4m+1)} x^{4m+1}$

(e) $\phi_1(x) = \sum\limits_{m=0}^{\infty} \dfrac{(-1)^m x^{2m}}{(2m)!} = \cos x,$

$\phi_2(x) = \sum\limits_{m=0}^{\infty} \dfrac{(-1)^m x^{2m+1}}{(2m+1)!} = \sin x$

All series converge for all real x.

2. $\phi(x) = \sum\limits_{m=0}^{\infty} \dfrac{(-1)^{m+1}(x-1)^{3m}}{3^m m!(3m-1)}$

3. $\phi(x) = x + \sum\limits_{m=1}^{\infty} \dfrac{(-1)^m [2 \cdot 3 + 1][4 \cdot 5 + 1] \cdots [(2m-2)(2m-1)+1]}{(2m+1)!} x^{2m+1},$

$r = 1$

4. $c_0 = 1$, $c_1 = 0$, $c_2 = -1/2$, $c_3 = -1/6$, $c_4 = 0$, $c_5 = 1/40$

5. $\phi(x) = 1 + \sum\limits_{m=1}^{\infty} \dfrac{5 \cdot 9 \cdots (4m-3)}{(4m)!} x^{4m}$

6. (b) see Sec. 8

(c) see Sec. 8

7. (a) $\phi_1(x) = 1 + \sum\limits_{m=1}^{\infty} \dfrac{(-\alpha^2)(2^2 - \alpha^2) \cdots [(2m-2)^2 - \alpha^2]}{(2m)!} x^{2m},$

$\phi_2(x) = x + \sum\limits_{m=1}^{\infty} \dfrac{(1^2 - \alpha^2)(3^2 - \alpha^2) \cdots [(2m-1)^2 - \alpha^2]}{(2m+1)!} x^{2m+1}$

(b) ϕ_1 is a polynomial if α is an even integer, ϕ_2 is a polynomial if α is an odd integer.

8. (a) $\phi_1(x) = 1 + \sum\limits_{m=1}^{\infty} \dfrac{2^m(-\alpha)(2-\alpha) \cdots (2m-2-\alpha)}{(2m)!} x^{2m},$

$\phi_2(x) = x + \sum\limits_{m=1}^{\infty} \dfrac{2^m(1-\alpha)(3-\alpha) \cdots (2m-1-\alpha)}{(2m+1)!} x^{2m+1}$

(b) ϕ_1 is a polynomial if α is an even integer, ϕ_2 is a polynomial if α is an odd integer

(d) $H_0(x) = 1$, $H_1(x) = 2x$, $H_2(x) = 4x^2 - 2$, $H_3(x) = 8x^3 - 12x$

Chap. 3, Sec. 8

9. $\phi(x) = \frac{1}{2} \log \left(\dfrac{1 + x}{1 - x} \right)$

10. $Q_1(x) = -\phi_1(x)$

CHAPTER 4

Chap. 4, Sec. 2

1. (a) $\phi(x) = c_1 x^{-3} + c_2 x^2$, c_1, c_2 any constants
  (b) $\phi(x) = c_1 x^{-1/2} + c_2 x$, c_1, c_2 any constants

  (c) $\phi(x) = c_1 x^2 + c_2 x^{-2} - \dfrac{x}{3}$, c_1, c_2 any constants

  (d) $\phi(x) = c_1 x^3 + c_2 x^3 \log x + \frac{1}{2} x^3 \log^2 x$, c_1, c_2 any constants
  (e) $\phi(x) = c_1 x + c_2 x \log x + c_3 x^{-1}$, c_1, c_2, c_3 any constants

2. c_1, c_2 are any constants but may be different for $x > 0$ and for $x < 0$
  (a) $\phi(x) = c_1 |x|^{2i} + c_2 |x|^{-2i} + \frac{1}{4}$
  (b) $\phi(x) = x^2(c_1 |x|^i + c_2 |x|^{-i})$
  (c) $\phi(x) = c_1 |x|^i + c_2 |x|^3$

  (d) $\phi(x) = c_1 |x|^{2\sqrt{\pi}} + c_2 |x|^{-2\sqrt{\pi}} + \dfrac{x}{1 - 4\pi}$

6. (a) $c = 1/q(k)$
  (b) $c = 1/q'(k)$

  (c) $\psi(x) = \dfrac{x^k}{2} \log^2 x$.

Chap. 4, Sec. 3

1. (a) $x = 0$, regular
  (b) $x = 0$, regular
  (c) $x = 0$, not regular
  (d) $x = 0$, regular
  (e) $x = 1$, regular; $x = -1$, regular
  (f) $x = -2$ regular; $x = 1$, not regular
  (g) $x = 0$, regular

2. (a) $q(r) = r^2 - 1$; $r_1 = 1$, $r_2 = -1$
  (b) $q(r) = r^2 - \frac{1}{4}$; $r_1 = \frac{1}{2}$, $r_2 = -\frac{1}{2}$
  (c) $q(r) = r^2 - \frac{9}{4} r + \frac{1}{2}$; $r_1 = 2$, $r_2 = \frac{1}{4}$
  (d) $q(r) = r^2 + 1$; $r_1 = i$, $r_2 = -i$
  (e) $q(r) = r^2 + 1$; $r_1 = i$, $r_2 = -i$

3. (b) $q(r) = r^2$; $r_1 = 0$, $r_2 = 0$

4. (a) $\phi_1(x) = x \sum\limits_{k=0}^{\infty} \dfrac{(-1)^k x^k}{1\cdot3\cdot5\cdot7\,\cdots\,(2k+1)}$,

$\phi_2(x) = x^{1/2} \sum\limits_{k=0}^{\infty} \dfrac{(-1)^k x^k}{2^k k!} = x^{1/2} e^{-x/2}$

(b) $\phi_1(x) = x^i \sum\limits_{k=0}^{\infty} \dfrac{i(1+i)\,\cdots\,(k-1+i)}{k!(1+2i)(2+2i)\,\cdots\,(k+2i)}\, x^k$,

$\phi_2(x) = x^{-i} \sum\limits_{k=0}^{\infty} \dfrac{(-i)(1-i)\,\cdots\,(k-1-i)}{k!(1-2i)(2-2i)\,\cdots\,(k-2i)}\, x^k$

Chap. 4, Sec. 4

1. (a) any constant times ϕ_1 or ϕ_2, where

$\phi_1(x) = \sum\limits_{k=0}^{\infty} \dfrac{(-1)^k 3^k x^k}{k!5\cdot8\cdot11\,\cdots\,(3k+2)}$,

$\phi_2(x) = |x|^{-2/3} \sum\limits_{k=0}^{\infty} \dfrac{(-1)^k 3^k x^k}{k!1\cdot4\cdot7\,\cdots\,(3k-2)}$

(b) any constant times ϕ_1, where $\phi_1(x) = \sum\limits_{m=0}^{\infty} \dfrac{(-1)^m}{(m!)^2} \left(\dfrac{x}{2}\right)^{2m}$

(c) $\phi(x) = |x|^{-1/2}(c_1 \cos x + c_2 \sin x)$, c_1, c_2 any constants

2. (a) $q(r) = r^2 + 1$

(b) $c_1 = \dfrac{-i}{(1+2i)}$,

$c_2 = \dfrac{i}{2\cdot(1+2i)\cdot2\cdot(2+2i)}$,

$c_3 = \left(\dfrac{-9-2i}{6}\right) \dfrac{1}{(1+2i)\cdot2\cdot(2+2i)\cdot3\cdot(3+2i)}$

3. (a) $\phi(x) = 1 + \sum\limits_{k=1}^{\infty} c_k(x-1)^k$, where

$c_k = \dfrac{(\alpha+k)(\alpha+k-1)\,\cdots\,(\alpha+1)\alpha(\alpha-1)\,\cdots\,(\alpha-k+2)(\alpha-k+1)}{2^k(k!)^2}$;

converges for $|x-1| < 2$

4. (b) $q(r) = r^2$; $r_1 = 0$, $r_2 = 0$

(c) $\phi(x) = 1 + \sum\limits_{k=1}^{\infty} \dfrac{(0 - \alpha)(1 - \alpha) \cdots (k - 1 - \alpha)}{(k!)^2} x^k$

5. (b) $L_0(x) = 1$, $L_1(x) = 1 - x$, $L_2(x) = 2 - 4x + x^2$

Chap. 4, Sec. 6

1. (a)   (i) $r_1 = 1/2$, $r_2 = -2$

(ii) $r_1 = 3/2$, $r_2 = 1/2$

(iii) $r_1 = -1$, $r_2 = -1$

(b)   (i) $\phi_1(x) = |x|^{1/2}\sigma_1(x)$, $\phi_2(x) = |x|^{-2}\sigma_2(x)$, σ_1, σ_2 power series convergent for $|x| < \infty$

(ii) $\phi_1(x) = |x|^{3/2}\sigma_1(x)$, $\sigma_1(0) = 1$, $\phi_2(x) = |x|^{1/2}\sigma_2(x) + c(\log|x|)\phi_1(x)$, σ_1, σ_2 power series convergent for $|x| < \infty$, $c \neq 0$

(iii) $\phi_1(x) = |x|^{-1}\sigma_1(x)$, $\sigma_1(0) = 1$, $\phi_2(x) = \sigma_2(x) + (\log|x|)\phi_1(x)$, σ_1, σ_2 power series convergent for $|x| < 1$, $\sigma_2(0) = 3$

2. (a) $q(r) = r^2 - \alpha^2$; $r_1 = \alpha$, $r_2 = -\alpha$

(b) if $\alpha = 0$ there are two solutions of the form $\phi_1(x) = \sigma_1(x)$, $\phi_2(x) = x\sigma_2(x) + (\log|x|)\phi_1(x)$, where σ_1, σ_2 are power series convergent for $|x| < \infty$; if $\alpha > 0$, 2α not a positive integer, two solutions are of the form $\phi_1(x) = |x|^{\alpha}\sigma_1(x)$, $\phi_2(x) = |x|^{-\alpha}\sigma_2(x)$, where σ_1, σ_2 are two power series convergent for $|x| < \infty$; if 2α is a positive integer two solutions have the form $\phi_1(x) = |x|^{\alpha}\sigma_1(x)$, $\phi_2(x) = |x|^{-\alpha}\sigma_2(x) + c(\log|x|)\phi_1(x)$, where σ_1, σ_2 are power series convergent for $|x| < \infty$

3. (a) $\phi_1(x) = |x|^{-1}\sum\limits_{k=0}^{\infty} \dfrac{(-1)^k x^k}{(k!)^2}$

$\phi_2(x) = -2|x|^{-1}\sum\limits_{k=1}^{\infty} \dfrac{(-1)^k}{(k!)^2}\left(1 + \tfrac{1}{2} + \cdots + \dfrac{1}{k}\right)x^k + (\log|x|)\phi_1(x)$

(b) $\phi_1(x) = |x|^{-1}(1 - x)$,

$\phi_2(x) = x^2\sum\limits_{k=0}^{\infty} \dfrac{(-1)^k 2^k(k + 1)}{(k + 3)!} x^k$

(c) $\phi_1(x) = x^{-3}\left[1 + \sum\limits_{m=1}^{\infty} \dfrac{1}{1 \cdot 3 \cdot 4 \cdot 6 \cdots (3m - 2)(3m)} x^{3m}\right]$;

$\phi_2(x) = x^{-1}\left[1 + \sum\limits_{m=1}^{\infty} \dfrac{1}{3 \cdot 5 \cdot 6 \cdot 8 \cdots (3m)(3m + 2)} x^{3m}\right]$

(d) $\phi_1(x) = x$, $\phi_2(x) = x\,e^{2x}$

(e) $\phi_1(x) = \left(\dfrac{x}{2}\right) \displaystyle\sum_{m=0}^{\infty} \dfrac{(-1)^m}{m!(m+1)!} \left(\dfrac{x}{2}\right)^{2m}$,

$$\phi_2(x) = -\dfrac{1}{2}\left(\dfrac{x}{2}\right)^{-1}\left\{1 + \left(\dfrac{x}{2}\right)^2\right.$$

$$+ \left(\dfrac{x}{2}\right)^2 \sum_{m=1}^{\infty} \dfrac{(-1)^m}{m!(m+1)!}\left[\left(1 + \tfrac{1}{2} + \cdots + \dfrac{1}{m}\right) + \left(1 + \tfrac{1}{2} + \cdots + \dfrac{1}{m+1}\right)\right]\left(\dfrac{x}{2}\right)^{2m}\right\}$$

$$+ (\log |x|)\phi_1(x); \quad \text{see Sec. 8}$$

(f) $\phi_1(x) = x^2$,

$$\phi_2(x) = x^{-1}\left[1 + 3x + 6x^2 - 3\sum_{k=4}^{\infty} \dfrac{2^k x^k}{(k-3)k!}\right] - 4x^2 \log |x|$$

5. $\Psi(x, r_2) = C_m(r_2)\Phi(x, r_1)$

6. (a) $c_k = -d_k/k$, where $d_k = \displaystyle\sum_{j=0}^{k-1} c_j \alpha_{k-j}, \quad k = 1, 2, \cdots$

Chap. 4, Sec. 9

3. (b) $t^2 y'' + \left(\dfrac{2t^2}{t^2-1}\right)ty' + \dfrac{\alpha(\alpha+1)}{t^2-1}\, y = 0$

(c) $q(r) = r^2 - r - \alpha^2 - \alpha; \quad r_1 = \alpha + 1, \quad r_2 = -\alpha$

4. $\phi_1(x) = x, \quad \phi_2(x) = x^{-2} \displaystyle\sum_{m=0}^{\infty} \dfrac{x^{-2m}}{2m+3}$

6. (a) $q(r) = r^2 + (\gamma - 1)r$

(b) $\phi(x) = 1 + \displaystyle\sum_{k=1}^{\infty} c_k x^k, \quad (|x| < 1)$,

$$\text{where} \quad c_k = \dfrac{\alpha(\alpha+1)\cdots(\alpha+k-1)\beta(\beta+1)\cdots(\beta+k-1)}{k!\gamma(\gamma+1)\cdots(\gamma+k-1)}$$

7. (b) $\phi_1(x) = x^n, \quad \phi_2(x) = x^{-(n+1)}$

CHAPTER 5

Chap. 5, Sec. 2

1. any real-valued differentiable function defined implicitly by the relation:

(a) $y = c \exp (x^3/3)$

(b) $y^2 = x^2 + c$

(c) $3y^2 - 2y^3 = 3x^2 + 2x^3 + c$

(d) $e^y = \log|1 + e^x| + c$

(e) $\log\left|\dfrac{y-2}{y+2}\right| = \dfrac{4x^2}{3} + c$ (*c* is a real constant); also $y = 2,\ y = -2$

2. (b) if $y_0 = 0$, all real x;
    if $y_0 \neq 0$, all real $x \neq x_0 + (1/y_0)$
  (c) if $y_0 = 0$, all x;
    if $y_0 > 0$, $-\infty < x < x_0 + (1/y_0)$;
    if $y_0 < 0$, $x_0 + (1/y_0) < x < \infty$

3. (a) $\phi(x) = (x - x_0 + \sqrt{y_0})^2,\ \ (x \geq x_0 - \sqrt{y_0})$,
    $\phi(x) = 0, \quad (x < x_0 - \sqrt{y_0})$

4. any real-valued differentiable function defined by the relation:
  (a) $2 \tan^{-1}(y/x) = \log(x^2 + y^2) + c$
  (b) $x \log y + y = cx$
  (c) $\tan^{-1}(y/x) = \log x + c$
  (d) $\exp(2y/x) = 2 \log x + c$ (*c* is a real constant)

5. any real-valued differentiable function defined by the relation:
  (a) $(y - \frac{3}{2})^2 + 2(y - \frac{3}{2})(x + \frac{1}{2}) - (x + \frac{1}{2})^2 = c$

  (b) $4 \tan^{-1}\left[\dfrac{14y + 7x - 5}{\sqrt{3}(7x - 1)}\right]$

      $= \sqrt{3} \log\left[(x - \tfrac{1}{7})^2 + (x - \tfrac{1}{7})(y + \tfrac{3}{7}) + (y + \tfrac{3}{7})^2\right] + c$
  (c) $\log|x + y| + x - 2y = c$ (*c* is a real constant)

6. (b) any real-valued differentiable function defined by the relation

$$2 \tan^{-1}\!\left(\frac{y-3}{x+2}\right) = \log|x + 2| + c, \quad c \text{ any real constant}$$

7. (a) $y = cx$,   *c* any real constant
  (b) $x^2 + y^2 = c$, $c > 0$
  (c) $2y^2 + x^2 = c$, $c > 0$
  (d) $y^3 = cx^2$,   *c* any real constant
  (e) $xy = c$,   *c* any real constant
  (f) $x = c \exp(-y^2)$,   *c* any real constant

Chap. 5, Sec. 3

1. solutions are any real-valued differentiable functions satisfying the relations given
  (a) exact, $x^2 y + y^3 = c$
  (b) not exact
  (c) exact, $e^z + ye^y = c$
  (d) exact, $\sin x \cos^2 y = c$
  (e) not exact
  (f) exact, $x^2 + 2xy - y^2 = c$
  (g) exact, $ye^{2x} + x^2 \cos y = c$
  (h) exact, $xy + x^3 \log|x| = c$ (*c* is a real constant)

2. any real-valued differentiable function defined by the relation:

    (a) $x^2(y^3 + 1) = c$ (b) $\sin x \cos^2 y = c$

    (c) $x^5 y^3 + x^2 y^2 = c$ (d) $x \exp (x + y) = c$

6. (a) $u(x) = \exp \left[\displaystyle\int_{x_0}^{x} a(t)\, dt \right], \quad x_0 \text{ in } I$

    (b) $\phi(x) = ce^{-A(x)} + e^{-A(x)} \displaystyle\int_{x_0}^{x} e^{A(t)} b(t)\, dt, \quad \text{where } A' = a$

Chap. 5, Sec. 4

1. (a) use induction

    (b) $\phi_0(x) = 2, \quad \phi_1(x) = 2 + 7x, \quad \phi_2(x) = 2 + 7x + \frac{21}{2}x^2,$

        $\phi_3(x) = 2 + 7x + \frac{21}{2}x^2 + \frac{21}{2}x^3$

    (c) $\phi(x) = \frac{1}{3}(7e^{3x} - 1)$

    (d) use the series for e^{3x}

2. (a) $\phi_0(x) = 0, \quad \phi_1(x) = \dfrac{x^3}{3}, \quad \phi_2(x) = \dfrac{x^3}{3} + \dfrac{x^7}{63},$

    $\phi_3(x) = \dfrac{x^3}{3} + \dfrac{x^7}{63} + \dfrac{2x^{11}}{11 \cdot 189} + \dfrac{x^{15}}{15 \cdot 63^2}$

    (b) $\phi_0(x) = 1, \quad \phi_1(x) = 1 + x + \dfrac{x^2}{2},$

    $\phi_2(x) = 1 + x + \dfrac{x^2}{2} + \dfrac{x^3}{3} + \dfrac{x^4}{8},$

    $\phi_3(x) = 1 + x + \dfrac{x^2}{2} + \dfrac{x^3}{3} + \dfrac{x^4}{8} + \dfrac{x^5}{15} + \dfrac{x^6}{48}$

    (c) $\phi_0(x) = 0, \quad \phi_1(x) = 0, \quad \phi_2(x) = 0, \quad \phi_3(x) = 0$

    (d) $\phi_0(x) = 1, \quad \phi_1(x) = 1 + x,$

    $\phi_2(x) = 1 + x + x^2 + \dfrac{x^3}{3},$

    $\phi_3(x) = 1 + x + x^2 + x^3 + \dfrac{2x^4}{3} + \dfrac{x^5}{3} + \dfrac{x^6}{9} + \dfrac{x^7}{63}$

3. (a) all ϕ_k are polynomials

    (b) $\phi(x) = 1/(1 - x), \quad \text{exists for } -\infty < x < 1$

    (c) solution does not exist for $x \geq 1$

4. (a) $M = 2$ (b) $|x| \leq 1/2$

6. (a) $\phi_0(x) = y_0$,

$$\phi_{k+1}(x) = y_0 + (x - x_0)y_1 + \int_{x_0}^x (x - t)f(t, \phi_k(t)) \, dt, \quad (k = 0, 1, 2, \cdots)$$

7. (a) $\phi_k(x) = 1 - \dfrac{x^2}{2!} + \dfrac{x^3}{3!} + \dfrac{x^4}{4!} - \dfrac{x^5}{5!} - \dfrac{x^6}{6!} + \cdots + (-1)^{k+1} \dfrac{x^{2k+1}}{(2k + 1)!}$,

$(k = 1, 2, 3, \cdots)$

(b) $\phi(x) = \cos x - \sin x + x$

8. (c) $\phi_0(x) = 0$,

$$\phi_{k+1}(x) = \frac{\sin \lambda x}{\lambda} + \int_0^x \frac{\sin \lambda(x - t)}{\lambda} f(t, \phi_k(t)) \, dt, \quad (k = 0, 1, 2, \cdots)$$

Chap. 5, Sec. 5

1. (a) $K = 2$

(b) $K = 3$

(c) $K = $ maximum $\{2a^3, 2a^4\}$

(d) $K = 4M_a + M_b$, where

$M_a = \max_{|x| \leq 1} |a(x)|, \quad M_b = \max_{|x| \leq 1} |b(x)|$

(e) $K = \max_{|x| \leq 1} |a(x)|$

3. (a) $K = 1$

Chap. 5, Sec. 6

1. (a) $\phi(x) = e^{-x^2} \displaystyle\int_0^x e^{t^2} \, dt$

(d) use Theorem 8

(e) $\phi_3(x) = x - \dfrac{2x^3}{3} + \dfrac{4x^5}{15}$

2. (a) $\phi(x) = \tan x$, exists on $-\pi/2 < x < \pi/2$

(b) all ϕ_k are polynomials

(c) $K = 2$

Chap. 5, Sec. 7

1. $K = 2a^3 + 3a^2 + 4a + 1$

2. (a) $K = 1/(1 - a^2)$

8. (a) if $\lambda = 0$, $\phi(x) = x + \displaystyle\int_0^x (x - t)q(t)\phi(t) \, dt$;

if $\lambda > 0$, $\phi(x) = \dfrac{\sin \lambda x}{\lambda} + \displaystyle\int_0^x \frac{\sin \lambda(x - t)}{\lambda} q(t)\phi(t) \, dt$

Chap. 5, Sec. 8

1. (b) $\phi(x) = \frac{1}{10} \exp(x^2/2)$

4. (b) $\phi_\lambda(x) = y_0 \exp\left[-\lambda^2(x - x_0) + \int_{x_0}^{x} q(t) \, dt \right]$

CHAPTER 6

Chap. 6, Sec. 2

1. (a) path of particle is the ellipse

$$\frac{x^2}{a^2} + \frac{k^2 y^2}{b^2} = 1$$

(b) period $2\pi/k$

2. (a) $x(t) = (v_0 \cos \alpha)t, \quad y(t) = -\frac{gt^2}{2} + (v_0 \sin \alpha)t$

(b) particle path is the parabola

$$y = -\frac{g}{2} \frac{x^2}{(v_0 \cos \alpha)^2} + (\tan \alpha)x$$

(c) vertex: $x = \frac{v_0^2 \sin 2\alpha}{2g}, \quad y = \frac{v_0^2 \sin^2 \alpha}{2g}$;

time required to reach vertex: $\frac{v_0 \sin \alpha}{g}$

(d) $x(T) = \frac{v_0^2 \sin 2\alpha}{g}$; maximum $x(T)$ when $\alpha = \pi/4$

4. (a) $x'' = -\epsilon x', \quad y'' = -g - \epsilon y'$

(b) $x(t) = \frac{v_0 \cos \alpha}{\epsilon} (1 - e^{-\epsilon t})$,

$$y(t) = \left(v_0 \sin \alpha + \frac{g}{\epsilon}\right)\left(\frac{1 - e^{-\epsilon t}}{\epsilon}\right) - \frac{xg}{\epsilon}$$

5. $r(t_0), r'(t_0), \theta(t_0), \theta'(t_0),$ for some t_0

Chap. 6, Sec. 3

1. (a) $\phi(x) = c_1 + c_2 e^{-x} + x, \quad c_1, c_2$ any constants

(b) $\phi(x) = x + c_1 + c_2 \int_0^t \exp(-e^s) \, ds, \quad c_1, c_2$ any constants

(c) $\phi(x) = (c_1 x + c_2)^{1/5}$, c_1, c_2 any constants

(d) $\phi(x) = c_1 \cos kx + c_2 \sin kx$, c_1, c_2 any constants

(e) $\phi(x) = c$,   any constant;

$\phi(x) = 2a \tan [a(x + b)]$ $a > 0, b$ constant;

$$\phi(x) = \frac{-2}{x + c}, \quad c \text{ any constant;}$$

$\phi(x) = -2a \coth [a(x + b)]$ $a > 0, b$ constant

(f) $\phi(x) = c_1 + c_2 x^3 + \dfrac{x^4}{4}$, c_1, c_2 any constants

(g) $\phi(x) = c$,   any constant; any twice differentiable function ϕ defined implicitly by

$$y + a \log |x - a| = \frac{x}{a} + b, \quad a, b \text{ constants, } a \neq 0;$$

$\phi(x) = \sqrt{2}(a - x)^{1/2}$, a constant

2. $\phi(x) = -\log (\cos x)$, $(-\pi/2 < x < \pi/2)$

3. $\phi(x) = \left(1 - \dfrac{3x}{2}\right)^{2/3}$

5. (c) $\phi(x) = 4 \tan^{-1} (e^x) - \pi$

(d) look up the subject of elliptic integrals

Chap. 6, Sec. 4

1. (a) $\mathbf{y} + \mathbf{z} = (8 + 2i, 2i, 0)$

(b) $\mathbf{y} - \mathbf{z} = (8, 4i, -4)$

(c) $s = 1/4$

4. (a) $\boldsymbol{\phi}(1) = (1, 1, i)$

(b) $\boldsymbol{\phi}'(x) = (1, 2x, 4ix^3)$, $\boldsymbol{\phi}'(2) = (1, 4, 32i)$

(c) $\displaystyle\int_{-1}^{1} \boldsymbol{\phi}(x)\, dx = \left(0, \frac{2}{3}, \frac{2i}{5}\right)$

(d) $\left| \displaystyle\int_{-1}^{1} \boldsymbol{\phi}(x)\, dx \right| = \dfrac{16}{15}$, $\displaystyle\int_{-1}^{1} |\boldsymbol{\phi}(x)|\, dx = \dfrac{31}{15}$

Chap. 6, Sec. 5

1. (a) $M = 3$ (b) $K = 2$

2. (a) $\mathbf{f}(x, \mathbf{y}) = (ay_1 + by_2, cy_1 + dy_2)$

(b) $K = |a| + |b| + |c| + |d|$

3. $\boldsymbol{\phi}(x) = (e^x, (x + 2)e^x)$

4. $\phi(x) = \frac{1}{5}(e^{3x} + 4e^{-2x}, 3e^{3x} - 8e^{-2x})$

5. $\phi(x) = \left(\dfrac{e^{3x}}{3} - \dfrac{e^{2x}}{2} + \dfrac{1}{6}, \dfrac{2e^{3x}}{3} - \dfrac{e^{2x}}{2} - \dfrac{1}{6}\right)$

Chap. 6, Sec. 6

1. (a) $\mathbf{f}(x, \mathbf{y}) = (y_2^2 + 1, y_1^2)$, $\mathbf{y}_0 = (0, 0)$

   (b) $M = 2$, $K = 2$, $\alpha = 1/2$

   (c) $\phi_0(x) = (0, 0)$, $\phi_1(x) = (x, 0)$, $\phi_2(x) = \left(x, \dfrac{x^3}{3}\right)$

2. $\mathbf{f}$ is continuous and satisfies a Lipschitz condition on each strip $\mid x \mid \; \leqq a, \mid \mathbf{y} \mid \, < \, \infty$ with Lipschitz constant $K = 3 + 2a + a^3 + e^a$

4. (a) $\mathbf{f}$ is continuous and satisfies a Lipschitz condition on $\mid x \mid \; < \, \infty, \mid \mathbf{y} \mid \, < \, \infty$ with Lipschitz constant $K = 1 + \epsilon$

   (c) $\phi(x) = c_1 e^{(1+\epsilon)x}(1, 1) + c_2 e^{(1-\epsilon)x}(1, -1)$, $\quad c_1, c_2$ any constants

   (d) $\phi(x) = e^{(1-\epsilon)x}(1, -1)$, $\quad \psi(x) = e^x(1, -1)$

Chap. 6, Sec. 7

3. (c) $\phi(x) = e^{r_1 x}(\sqrt{89} - 3, 10)$, where $r_1 = \dfrac{9 + \sqrt{89}}{2}$; another solution is

$$\psi(x) = e^{r_2 x}(\sqrt{89} + 3, -10), \quad \text{where } r_2 = \dfrac{9 - \sqrt{89}}{2}$$

5. (b) $p(r) = r^2 - (a + d)r + (ad - bc)$

   (c) if r_1, r_2 are the roots of p, then if $r_1 \neq r_2$, $\phi(x) = c_1 e^{r_1 x} + c_2 e^{r_2 x}$, $\quad c_1, c_2$ any constants; $\phi(x) = (c_1 + c_2 x)e^{r_1 x}$ if $r_1 = r_2$, $\quad c_1, c_2$ any constants

   (d) (i) if $r_1 \neq r_2$ all solutions have the form $\phi(x)' = e^{r_1 x}\boldsymbol{\alpha} + e^{r_2 x}\boldsymbol{\beta}$, where $\boldsymbol{\alpha} = (\alpha_1, \alpha_2)$, $\boldsymbol{\beta} = (\beta_1, \beta_2)$ are such that

$$(a - r_1)\alpha_1 + b\alpha_2 = 0, \qquad c\alpha_1 + (d - r_1)\alpha_2 = 0,$$
$$(a - r_2)\beta_1 + b\beta_2 = 0, \qquad c\beta_1 + (d - r_2)\beta_2 = 0$$

     (ii) if $r_1 = r_2$ all solutions have the form $\phi(x) = e^{r_1 x}(\boldsymbol{\alpha} + x\boldsymbol{\beta})$, where $\boldsymbol{\alpha} = (\alpha_1, \alpha_2)$, $\boldsymbol{\beta} = (\beta_1, \beta_2)$ satisfy

$$(a - r_1)\beta_1 + b\beta_2 = 0, \qquad c\beta_1 + (d - r_1)\beta_2 = 0,$$
$$(a - r_1)\alpha_1 + b\alpha_2 = \beta_1, \qquad c\alpha_1 + (d - r_1)\alpha_2 = \beta_2$$

   (e) $\phi(x) = c_1 e^x(1, 1) + c_2 e^{2x}(3, 2)$, $\quad c_1, c_2$ any constants

Chap. 6, Sec. 8

1. (a) $y_1' = y_2$, $\quad y_2' = -a_2 y_1 - a_1 y_2$

   (d) apply Theorem 5 to ϕ and $\psi(x) = 0, \mathbf{f} = \mathbf{g}$

2. Apply Theorem 5 to the first order system associated with the equation

3. See Note to Ex. 3, Sec. 6, and corollary to Theorem 4

4. (a) $y_1' = y_2$, $y_2' = -\epsilon y_2$, $y_3' = y_4$, $y_4' = -g - \epsilon y_4$; $y_1' = y_2$, $y_2' = 0$, $y_3' = y_4$, $y_4' = -g$; both with initial conditions

$$y_1(0) = 0, \quad y_2(0) = v_0 \cos \alpha, \quad y_3(0) = 0, \quad y_4(0) = v_0 \sin \alpha$$

(b) $\boldsymbol{\phi} = (\phi_1, \phi_2, \phi_3, \phi_4)$, where

$$\phi_1(x) = \frac{v_0 \cos \alpha}{\epsilon} (1 - e^{-\epsilon x}),$$

$$\phi_2(x) = (v_0 \cos \alpha)e^{-\epsilon x},$$

$$\phi_3(x) = \left(v_0 \sin \alpha + \frac{g}{\epsilon}\right)\left(\frac{1 - e^{-\epsilon x}}{\epsilon}\right) - \frac{xg}{\epsilon},$$

$$\phi_4(x) = (v_0 \sin \alpha)e^{-\epsilon x} + \frac{g}{\epsilon}(e^{-\epsilon x} - 1); \quad \boldsymbol{\psi} = (\psi_1, \psi_2, \psi_3, \psi_4), \quad \text{where}$$

$$\psi_1(x) = (v_0 \cos \alpha)x, \qquad\qquad \psi_2(x) = v_0 \cos \alpha,$$

$$\psi_3(x) = (v_0 \sin \alpha)x - \frac{gx^2}{2}, \qquad \psi_4(x) = v_0 \sin \alpha - gx$$

Index

A

Analytic coefficients, 126
Analytic function, 126
Annihilator method, 92
Approximation to solution:
  of first order equation, 222
  of system, 253

B

Basis for linear space, 108
Bernoulli's equation, 46 (Ex. 5)
Bessel equation, 168, 172, 182 (Ex. 2)
Bessel function:
  order α, first kind, 174
  order n, second kind, 178
  order zero:
    first kind, 169
    second kind, 170
  zeros of, 171 (Ex. 3), 172 (Ex. 6),
    179 (Exs. 5, 6, 9, 10), 180 (Ex. 14)
Birkhoff, G., 17
Boundary conditions, 37
Brahe, T., 234

C

Cauchy-Riemann equations, 16 (Ex. 5)
Central force, 231
Characteristic polynomial, 50, 71
Chebyshev equation, 131 (Ex. 7)
Chebyshev polynomial, 131 (Ex. 7)
Complex number:
  conjugate, 4
  definition, 1

difference, 2
equality, 2
imaginary part, 4
magnitude, 4
negative, 2
product, 2
quotient, 3
real part, 4
reciprocal, 3
sum, 2
unit, 2
zero, 2
Complex n-dimensional space, 240
Complex plane, 4
Constant coefficient operator, 94

D

Derivative, 9, 12
Determinant, 27
Differential equation:
  first order, 33, 35
  nth order, 36
  solution of, 33, 36
  with periodic coefficients, 46 (Exs.
    6, 7)
Differential operator, 49
Dimension of linear space, 108
Direction field, 34

E

Euclidean length of vector, 245 (Ex. 6)
Euler equation:
  nth order, 148
  second order, 145

A CATALOG OF SELECTED
DOVER BOOKS
IN SCIENCE AND MATHEMATICS

Astronomy

BURNHAM'S CELESTIAL HANDBOOK, Robert Burnham, Jr. Thorough guide to the stars beyond our solar system. Exhaustive treatment. Alphabetical by constellation: Andromeda to Cetus in Vol. 1; Chamaeleon to Orion in Vol. 2; and Pavo to Vulpecula in Vol. 3. Hundreds of illustrations. Index in Vol. 3. 2,000pp. 6⅛ x 9¼.
Vol. I: 0-486-23567-X
Vol. II: 0-486-23568-8
Vol. III: 0-486-23673-0

EXPLORING THE MOON THROUGH BINOCULARS AND SMALL TELESCOPES, Ernest H. Cherrington, Jr. Informative, profusely illustrated guide to locating and identifying craters, rills, seas, mountains, other lunar features. Newly revised and updated with special section of new photos. Over 100 photos and diagrams. 240pp. 8¼ x 11. 0-486-24491-1

THE EXTRATERRESTRIAL LIFE DEBATE, 1750–1900, Michael J. Crowe. First detailed, scholarly study in English of the many ideas that developed from 1750 to 1900 regarding the existence of intelligent extraterrestrial life. Examines ideas of Kant, Herschel, Voltaire, Percival Lowell, many other scientists and thinkers. 16 illustrations. 704pp. 5⅜ x 8½. 0-486-40675-X

THEORIES OF THE WORLD FROM ANTIQUITY TO THE COPERNICAN REVOLUTION, Michael J. Crowe. Newly revised edition of an accessible, enlightening book re-creates the change from an earth-centered to a sun-centered conception of the solar system. 242pp. 5⅜ x 8½. 0-486-41444-2

ARISTARCHUS OF SAMOS: The Ancient Copernicus, Sir Thomas Heath. Heath's history of astronomy ranges from Homer and Hesiod to Aristarchus and includes quotes from numerous thinkers, compilers, and scholasticists from Thales and Anaximander through Pythagoras, Plato, Aristotle, and Heraclides. 34 figures. 448pp. 5⅜ x 8½. 0-486-43886-4

A COMPLETE MANUAL OF AMATEUR ASTRONOMY: TOOLS AND TECHNIQUES FOR ASTRONOMICAL OBSERVATIONS, P. Clay Sherrod with Thomas L. Koed. Concise, highly readable book discusses: selecting, setting up and maintaining a telescope; amateur studies of the sun; lunar topography and occultations; observations of Mars, Jupiter, Saturn, the minor planets and the stars; an introduction to photoelectric photometry; more. 1981 ed. 124 figures. 25 halftones. 37 tables. 335pp. 6½ x 9¼. 0-486-42820-8

AMATEUR ASTRONOMER'S HANDBOOK, J. B. Sidgwick. Timeless, comprehensive coverage of telescopes, mirrors, lenses, mountings, telescope drives, micrometers, spectroscopes, more. 189 illustrations. 576pp. 5⅝ x 8¼. (Available in U.S. only.) 0-486-24034-7

STAR LORE: Myths, Legends, and Facts, William Tyler Olcott. Captivating retellings of the origins and histories of ancient star groups include Pegasus, Ursa Major, Pleiades, signs of the zodiac, and other constellations. "Classic."–Sky & Telescope. 58 illustrations. 544pp. 5⅜ x 8½. 0-486-43581-4

Chemistry

THE SCEPTICAL CHYMIST: THE CLASSIC 1661 TEXT, Robert Boyle. Boyle defines the term "element," asserting that all natural phenomena can be explained by the motion and organization of primary particles. 1911 ed. viii+232pp. 5⅜ x 8½.
0-486-42825-7

RADIOACTIVE SUBSTANCES, Marie Curie. Here is the celebrated scientist's doctoral thesis, the prelude to her receipt of the 1903 Nobel Prize. Curie discusses establishing atomic character of radioactivity found in compounds of uranium and thorium; extraction from pitchblende of polonium and radium; isolation of pure radium chloride; determination of atomic weight of radium; plus electric, photographic, luminous, heat, color effects of radioactivity. ii+94pp. 5⅜ x 8½.
0-486-42550-9

CHEMICAL MAGIC, Leonard A. Ford. Second Edition, Revised by E. Winston Grundmeier. Over 100 unusual stunts demonstrating cold fire, dust explosions, much more. Text explains scientific principles and stresses safety precautions. 128pp. 5⅜ x 8½.
0-486-67628-5

MOLECULAR THEORY OF CAPILLARITY, J. S. Rowlinson and B. Widom. History of surface phenomena offers critical and detailed examination and assessment of modern theories, focusing on statistical mechanics and application of results in mean-field approximation to model systems. 1989 edition. 352pp. 5⅜ x 8½.
0-486-42544-4

CHEMICAL AND CATALYTIC REACTION ENGINEERING, James J. Carberry. Designed to offer background for managing chemical reactions, this text examines behavior of chemical reactions and reactors; fluid-fluid and fluid-solid reaction systems; heterogeneous catalysis and catalytic kinetics; more. 1976 edition. 672pp. 6⅛ x 9¼.
0-486-41736-0 $31.95

ELEMENTS OF CHEMISTRY, Antoine Lavoisier. Monumental classic by founder of modern chemistry in remarkable reprint of rare 1790 Kerr translation. A must for every student of chemistry or the history of science. 539pp. 5⅜ x 8½. 0-486-64624-6

MOLECULES AND RADIATION: An Introduction to Modern Molecular Spectroscopy. Second Edition, Jeffrey I. Steinfeld. This unified treatment introduces upper-level undergraduates and graduate students to the concepts and the methods of molecular spectroscopy and applications to quantum electronics, lasers, and related optical phenomena. 1985 edition. 512pp. 5⅜ x 8½.
0-486-44152-0

A SHORT HISTORY OF CHEMISTRY, J. R. Partington. Classic exposition explores origins of chemistry, alchemy, early medical chemistry, nature of atmosphere, theory of valency, laws and structure of atomic theory, much more. 428pp. 5⅜ x 8½. (Available in U.S. only.)
0-486-65977-1

GENERAL CHEMISTRY, Linus Pauling. Revised 3rd edition of classic first-year text by Nobel laureate. Atomic and molecular structure, quantum mechanics, statistical mechanics, thermodynamics correlated with descriptive chemistry. Problems. 992pp. 5⅜ x 8½.
0-486-65622-5

ELECTRON CORRELATION IN MOLECULES, S. Wilson. This text addresses one of theoretical chemistry's central problems. Topics include molecular electronic structure, independent electron models, electron correlation, the linked diagram theorem, and related topics. 1984 edition. 304pp. 5⅜ x 8½.
0-486-45879-2

Engineering

DE RE METALLICA, Georgius Agricola. The famous Hoover translation of greatest treatise on technological chemistry, engineering, geology, mining of early modern times (1556). All 289 original woodcuts. 638pp. 6¾ x 11.　　0-486-60006-8

FUNDAMENTALS OF ASTRODYNAMICS, Roger Bate et al. Modern approach developed by U.S. Air Force Academy. Designed as a first course. Problems, exercises. Numerous illustrations. 455pp. 5⅜ x 8½.　　0-486-60061-0

DYNAMICS OF FLUIDS IN POROUS MEDIA, Jacob Bear. For advanced students of ground water hydrology, soil mechanics and physics, drainage and irrigation engineering and more. 335 illustrations. Exercises, with answers. 784pp. 6⅛ x 9¼.
　　0-486-65675-6

THEORY OF VISCOELASTICITY (SECOND EDITION), Richard M. Christensen. Complete consistent description of the linear theory of the viscoelastic behavior of materials. Problem-solving techniques discussed. 1982 edition. 29 figures. xiv+364pp. 6⅛ x 9¼.　　0-486-42880-X

MECHANICS, J. P. Den Hartog. A classic introductory text or refresher. Hundreds of applications and design problems illuminate fundamentals of trusses, loaded beams and cables, etc. 334 answered problems. 462pp. 5⅜ x 8½.　　0-486-60754-2

MECHANICAL VIBRATIONS, J. P. Den Hartog. Classic textbook offers lucid explanations and illustrative models, applying theories of vibrations to a variety of practical industrial engineering problems. Numerous figures. 233 problems, solutions. Appendix. Index. Preface. 436pp. 5⅜ x 8½.　　0-486-64785-4

STRENGTH OF MATERIALS, J. P. Den Hartog. Full, clear treatment of basic material (tension, torsion, bending, etc.) plus advanced material on engineering methods, applications. 350 answered problems. 323pp. 5⅜ x 8½.　　0-486-60755-0

A HISTORY OF MECHANICS, René Dugas. Monumental study of mechanical principles from antiquity to quantum mechanics. Contributions of ancient Greeks, Galileo, Leonardo, Kepler, Lagrange, many others. 671pp. 5⅜ x 8½. 0-486-65632-2

STABILITY THEORY AND ITS APPLICATIONS TO STRUCTURAL MECHANICS, Clive L. Dym. Self-contained text focuses on Koiter postbuckling analyses, with mathematical notions of stability of motion. Basing minimum energy principles for static stability upon dynamic concepts of stability of motion, it develops asymptotic buckling and postbuckling analyses from potential energy considerations, with applications to columns, plates, and arches. 1974 ed. 208pp. 5⅜ x 8½.
　　0-486-42541-X

BASIC ELECTRICITY, U.S. Bureau of Naval Personnel. Originally a training course; best nontechnical coverage. Topics include batteries, circuits, conductors, AC and DC, inductance and capacitance, generators, motors, transformers, amplifiers, etc. Many questions with answers. 349 illustrations. 1969 edition. 448pp. 6½ x 9¼.
　　0-486-20973-3

ROCKETS, Robert Goddard. Two of the most significant publications in the history of rocketry and jet propulsion: "A Method of Reaching Extreme Altitudes" (1919) and "Liquid Propellant Rocket Development" (1936). 128pp. 5⅜ x 8½. 0-486-42537-1

STATISTICAL MECHANICS: PRINCIPLES AND APPLICATIONS, Terrell L. Hill. Standard text covers fundamentals of statistical mechanics, applications to fluctuation theory, imperfect gases, distribution functions, more. 448pp. 5⅜ x 8½.
0-486-65390-0

ENGINEERING AND TECHNOLOGY 1650–1750: ILLUSTRATIONS AND TEXTS FROM ORIGINAL SOURCES, Martin Jensen. Highly readable text with more than 200 contemporary drawings and detailed engravings of engineering projects dealing with surveying, leveling, materials, hand tools, lifting equipment, transport and erection, piling, bailing, water supply, hydraulic engineering, and more. Among the specific projects outlined-transporting a 50-ton stone to the Louvre, erecting an obelisk, building timber locks, and dredging canals. 207pp. 8⅜ x 11¼.
0-486-42232-1

THE VARIATIONAL PRINCIPLES OF MECHANICS, Cornelius Lanczos. Graduate level coverage of calculus of variations, equations of motion, relativistic mechanics, more. First inexpensive paperbound edition of classic treatise. Index. Bibliography. 418pp. 5⅜ x 8½. 0-486-65067-7

PROTECTION OF ELECTRONIC CIRCUITS FROM OVERVOLTAGES, Ronald B. Standler. Five-part treatment presents practical rules and strategies for circuits designed to protect electronic systems from damage by transient overvoltages. 1989 ed. xxiv+434pp. 6⅛ x 9¼. 0-486-42552-5

ROTARY WING AERODYNAMICS, W. Z. Stepniewski. Clear, concise text covers aerodynamic phenomena of the rotor and offers guidelines for helicopter performance evaluation. Originally prepared for NASA. 537 figures. 640pp. 6⅛ x 9¼.
0-486-64647-5

INTRODUCTION TO SPACE DYNAMICS, William Tyrrell Thomson. Comprehensive, classic introduction to space-flight engineering for advanced undergraduate and graduate students. Includes vector algebra, kinematics, transformation of coordinates. Bibliography. Index. 352pp. 5⅜ x 8½. 0-486-65113-4

HISTORY OF STRENGTH OF MATERIALS, Stephen P. Timoshenko. Excellent historical survey of the strength of materials with many references to the theories of elasticity and structure. 245 figures. 452pp. 5⅜ x 8½. 0-486-61187-6

ANALYTICAL FRACTURE MECHANICS, David J. Unger. Self-contained text supplements standard fracture mechanics texts by focusing on analytical methods for determining crack-tip stress and strain fields. 336pp. 6⅛ x 9¼. 0-486-41737-9

STATISTICAL MECHANICS OF ELASTICITY, J. H. Weiner. Advanced, self-contained treatment illustrates general principles and elastic behavior of solids. Part 1, based on classical mechanics, studies thermoelastic behavior of crystalline and polymeric solids. Part 2, based on quantum mechanics, focuses on interatomic force laws, behavior of solids, and thermally activated processes. For students of physics and chemistry and for polymer physicists. 1983 ed. 96 figures. 496pp. 5⅜ x 8½.
0-486-42260-7

Mathematics

FUNCTIONAL ANALYSIS (Second Corrected Edition), George Bachman and Lawrence Narici. Excellent treatment of subject geared toward students with background in linear algebra, advanced calculus, physics and engineering. Text covers introduction to inner-product spaces, normed, metric spaces, and topological spaces; complete orthonormal sets, the Hahn-Banach Theorem and its consequences, and many other related subjects. 1966 ed. 544pp. 6⅛ x 9¼. 0-486-40251-7

DIFFERENTIAL MANIFOLDS, Antoni A. Kosinski. Introductory text for advanced undergraduates and graduate students presents systematic study of the topological structure of smooth manifolds, starting with elements of theory and concluding with method of surgery. 1993 edition. 288pp. 5⅜ x 8½. 0-486-46244-7

VECTOR AND TENSOR ANALYSIS WITH APPLICATIONS, A. I. Borisenko and I. E. Tarapov. Concise introduction. Worked-out problems, solutions, exercises. 257pp. 5⅝ x 8¼. 0-486-63833-2

AN INTRODUCTION TO ORDINARY DIFFERENTIAL EQUATIONS, Earl A. Coddington. A thorough and systematic first course in elementary differential equations for undergraduates in mathematics and science, with many exercises and problems (with answers). Index. 304pp. 5⅜ x 8½. 0-486-65942-9

FOURIER SERIES AND ORTHOGONAL FUNCTIONS, Harry F. Davis. An incisive text combining theory and practical example to introduce Fourier series, orthogonal functions and applications of the Fourier method to boundary-value problems. 570 exercises. Answers and notes. 416pp. 5⅜ x 8½. 0-486-65973-9

COMPUTABILITY AND UNSOLVABILITY, Martin Davis. Classic graduate-level introduction to theory of computability, usually referred to as theory of recurrent functions. New preface and appendix. 288pp. 5⅜ x 8½. 0-486-61471-9

AN INTRODUCTION TO MATHEMATICAL ANALYSIS, Robert A. Rankin. Dealing chiefly with functions of a single real variable, this text by a distinguished educator introduces limits, continuity, differentiability, integration, convergence of infinite series, double series, and infinite products. 1963 edition. 624pp. 5⅜ x 8½. 0-486-46251-X

METHODS OF NUMERICAL INTEGRATION (SECOND EDITION), Philip J. Davis and Philip Rabinowitz. Requiring only a background in calculus, this text covers approximate integration over finite and infinite intervals, error analysis, approximate integration in two or more dimensions, and automatic integration. 1984 edition. 624pp. 5⅜ x 8½. 0-486-45339-1

INTRODUCTION TO LINEAR ALGEBRA AND DIFFERENTIAL EQUATIONS, John W. Dettman. Excellent text covers complex numbers, determinants, orthonormal bases, Laplace transforms, much more. Exercises with solutions. Undergraduate level. 416pp. 5⅜ x 8½. 0-486-65191-6

RIEMANN'S ZETA FUNCTION, H. M. Edwards. Superb, high-level study of landmark 1859 publication entitled "On the Number of Primes Less Than a Given Magnitude" traces developments in mathematical theory that it inspired. xiv+315pp. 5⅜ x 8½. 0-486-41740-9

CALCULUS OF VARIATIONS WITH APPLICATIONS, George M. Ewing. Applications-oriented introduction to variational theory develops insight and promotes understanding of specialized books, research papers. Suitable for advanced undergraduate/graduate students as primary, supplementary text. 352pp. 5⅜ x 8½.
0-486-64856-7

MATHEMATICIAN'S DELIGHT, W. W. Sawyer. "Recommended with confidence" by *The Times Literary Supplement,* this lively survey was written by a renowned teacher. It starts with arithmetic and algebra, gradually proceeding to trigonometry and calculus. 1943 edition. 240pp. 5⅜ x 8½.
0-486-46240-4

ADVANCED EUCLIDEAN GEOMETRY, Roger A. Johnson. This classic text explores the geometry of the triangle and the circle, concentrating on extensions of Euclidean theory, and examining in detail many relatively recent theorems. 1929 edition. 336pp. 5⅜ x 8½.
0-486-46237-4

COUNTEREXAMPLES IN ANALYSIS, Bernard R. Gelbaum and John M. H. Olmsted. These counterexamples deal mostly with the part of analysis known as "real variables." The first half covers the real number system, and the second half encompasses higher dimensions. 1962 edition. xxiv+198pp. 5⅜ x 8½. 0-486-42875-3

CATASTROPHE THEORY FOR SCIENTISTS AND ENGINEERS, Robert Gilmore. Advanced-level treatment describes mathematics of theory grounded in the work of Poincaré, R. Thom, other mathematicians. Also important applications to problems in mathematics, physics, chemistry and engineering. 1981 edition. References. 28 tables. 397 black-and-white illustrations. xvii + 666pp. 6⅛ x 9¼.
0-486-67539-4

COMPLEX VARIABLES: Second Edition, Robert B. Ash and W. P. Novinger. Suitable for advanced undergraduates and graduate students, this newly revised treatment covers Cauchy theorem and its applications, analytic functions, and the prime number theorem. Numerous problems and solutions. 2004 edition. 224pp. 6½ x 9¼.
0-486-46250-1

NUMERICAL METHODS FOR SCIENTISTS AND ENGINEERS, Richard Hamming. Classic text stresses frequency approach in coverage of algorithms, polynomial approximation, Fourier approximation, exponential approximation, other topics. Revised and enlarged 2nd edition. 721pp. 5⅜ x 8½.
0-486-65241-6

INTRODUCTION TO NUMERICAL ANALYSIS (2nd Edition), F. B. Hildebrand. Classic, fundamental treatment covers computation, approximation, interpolation, numerical differentiation and integration, other topics. 150 new problems. 669pp. 5⅜ x 8½.
0-486-65363-3

MARKOV PROCESSES AND POTENTIAL THEORY, Robert M. Blumental and Ronald K. Getoor. This graduate-level text explores the relationship between Markov processes and potential theory in terms of excessive functions, multiplicative functionals and subprocesses, additive functionals and their potentials, and dual processes. 1968 edition. 320pp. 5⅜ x 8½.
0-486-46263-3

ABSTRACT SETS AND FINITE ORDINALS: An Introduction to the Study of Set Theory, G. B. Keene. This text unites logical and philosophical aspects of set theory in a manner intelligible to mathematicians without training in formal logic and to logicians without a mathematical background. 1961 edition. 112pp. 5⅜ x 8½.
0-486-46249-8

INTRODUCTORY REAL ANALYSIS, A.N. Kolmogorov, S. V. Fomin. Translated by Richard A. Silverman. Self-contained, evenly paced introduction to real and functional analysis. Some 350 problems. 403pp. 5³⁄₈ x 8½. 0-486-61226-0

APPLIED ANALYSIS, Cornelius Lanczos. Classic work on analysis and design of finite processes for approximating solution of analytical problems. Algebraic equations, matrices, harmonic analysis, quadrature methods, much more. 559pp. 5³⁄₈ x 8½. 0-486-65656-X

AN INTRODUCTION TO ALGEBRAIC STRUCTURES, Joseph Landin. Superb self-contained text covers "abstract algebra": sets and numbers, theory of groups, theory of rings, much more. Numerous well-chosen examples, exercises. 247pp. 5³⁄₈ x 8½. 0-486-65940-2

QUALITATIVE THEORY OF DIFFERENTIAL EQUATIONS, V. V. Nemytskii and V.V. Stepanov. Classic graduate-level text by two prominent Soviet mathematicians covers classical differential equations as well as topological dynamics and ergodic theory. Bibliographies. 523pp. 5³⁄₈ x 8½. 0-486-65954-2

THEORY OF MATRICES, Sam Perlis. Outstanding text covering rank, nonsingularity and inverses in connection with the development of canonical matrices under the relation of equivalence, and without the intervention of determinants. Includes exercises. 237pp. 5³⁄₈ x 8½. 0-486-66810-X

INTRODUCTION TO ANALYSIS, Maxwell Rosenlicht. Unusually clear, accessible coverage of set theory, real number system, metric spaces, continuous functions, Riemann integration, multiple integrals, more. Wide range of problems. Undergraduate level. Bibliography. 254pp. 5³⁄₈ x 8½. 0-486-65038-3

MODERN NONLINEAR EQUATIONS, Thomas L. Saaty. Emphasizes practical solution of problems; covers seven types of equations. ". . . a welcome contribution to the existing literature. . . ."–*Math Reviews.* 490pp. 5³⁄₈ x 8½. 0-486-64232-1

MATRICES AND LINEAR ALGEBRA, Hans Schneider and George Phillip Barker. Basic textbook covers theory of matrices and its applications to systems of linear equations and related topics such as determinants, eigenvalues and differential equations. Numerous exercises. 432pp. 5³⁄₈ x 8½. 0-486-66014-1

LINEAR ALGEBRA, Georgi E. Shilov. Determinants, linear spaces, matrix algebras, similar topics. For advanced undergraduates, graduates. Silverman translation. 387pp. 5³⁄₈ x 8½. 0-486-63518-X

MATHEMATICAL METHODS OF GAME AND ECONOMIC THEORY: Revised Edition, Jean-Pierre Aubin. This text begins with optimization theory and convex analysis, followed by topics in game theory and mathematical economics, and concluding with an introduction to nonlinear analysis and control theory. 1982 edition. 656pp. 6⅛ x 9¼. 0-486-46265-X

SET THEORY AND LOGIC, Robert R. Stoll. Lucid introduction to unified theory of mathematical concepts. Set theory and logic seen as tools for conceptual understanding of real number system. 496pp. 5⅝ x 8¼. 0-486-63829-4

TENSOR CALCULUS, J.L. Synge and A. Schild. Widely used introductory text covers spaces and tensors, basic operations in Riemannian space, non-Riemannian spaces, etc. 324pp. 5⅝ x 8¼. 0-486-63612-7

ORDINARY DIFFERENTIAL EQUATIONS, Morris Tenenbaum and Harry Pollard. Exhaustive survey of ordinary differential equations for undergraduates in mathematics, engineering, science. Thorough analysis of theorems. Diagrams. Bibliography. Index. 818pp. 5⅜ x 8½. 0-486-64940-7

INTEGRAL EQUATIONS, F. G. Tricomi. Authoritative, well-written treatment of extremely useful mathematical tool with wide applications. Volterra Equations, Fredholm Equations, much more. Advanced undergraduate to graduate level. Exercises. Bibliography. 238pp. 5⅜ x 8½. 0-486-64828-1

FOURIER SERIES, Georgi P. Tolstov. Translated by Richard A. Silverman. A valuable addition to the literature on the subject, moving clearly from subject to subject and theorem to theorem. 107 problems, answers. 336pp. 5⅜ x 8½. 0-486-63317-9

INTRODUCTION TO MATHEMATICAL THINKING, Friedrich Waismann. Examinations of arithmetic, geometry, and theory of integers; rational and natural numbers; complete induction; limit and point of accumulation; remarkable curves; complex and hypercomplex numbers, more. 1959 ed. 27 figures. xii+260pp. 5⅜ x 8½. 0-486-42804-8

THE RADON TRANSFORM AND SOME OF ITS APPLICATIONS, Stanley R. Deans. Of value to mathematicians, physicists, and engineers, this excellent introduction covers both theory and applications, including a rich array of examples and literature. Revised and updated by the author. 1993 edition. 304pp. 6⅛ x 9¼. 0-486-46241-2

CALCULUS OF VARIATIONS, Robert Weinstock. Basic introduction covering isoperimetric problems, theory of elasticity, quantum mechanics, electrostatics, etc. Exercises throughout. 326pp. 5⅜ x 8½. 0-486-63069-2

THE CONTINUUM: A CRITICAL EXAMINATION OF THE FOUNDATION OF ANALYSIS, Hermann Weyl. Classic of 20th-century foundational research deals with the conceptual problem posed by the continuum. 156pp. 5⅜ x 8½. 0-486-67982-9

CHALLENGING MATHEMATICAL PROBLEMS WITH ELEMENTARY SOLUTIONS, A. M. Yaglom and I. M. Yaglom. Over 170 challenging problems on probability theory, combinatorial analysis, points and lines, topology, convex polygons, many other topics. Solutions. Total of 445pp. 5⅜ x 8½. Two-vol. set. Vol. I: 0-486-65536-9 Vol. II: 0-486-65537-7

INTRODUCTION TO PARTIAL DIFFERENTIAL EQUATIONS WITH APPLICATIONS, E. C. Zachmanoglou and Dale W. Thoe. Essentials of partial differential equations applied to common problems in engineering and the physical sciences. Problems and answers. 416pp. 5⅜ x 8½. 0-486-65251-3

STOCHASTIC PROCESSES AND FILTERING THEORY, Andrew H. Jazwinski. This unified treatment presents material previously available only in journals, and in terms accessible to engineering students. Although theory is emphasized, it discusses numerous practical applications as well. 1970 edition. 400pp. 5⅜ x 8½. 0-486-46274-9

Math–Decision Theory, Statistics, Probability

INTRODUCTION TO PROBABILITY, John E. Freund. Featured topics include permutations and factorials, probabilities and odds, frequency interpretation, mathematical expectation, decision-making, postulates of probability, rule of elimination, much more. Exercises with some solutions. Summary. 1973 edition. 247pp. 5⅜ x 8½. 0-486-67549-1

STATISTICAL AND INDUCTIVE PROBABILITIES, Hugues Leblanc. This treatment addresses a decades-old dispute among probability theorists, asserting that both statistical and inductive probabilities may be treated as sentence-theoretic measurements, and that the latter qualify as estimates of the former. 1962 edition. 160pp. 5⅜ x 8½. 0-486-44980-7

APPLIED MULTIVARIATE ANALYSIS: Using Bayesian and Frequentist Methods of Inference, Second Edition, S. James Press. This two-part treatment deals with foundations as well as models and applications. Topics include continuous multivariate distributions; regression and analysis of variance; factor analysis and latent structure analysis; and structuring multivariate populations. 1982 edition. 692pp. 5⅜ x 8½. 0-486-44236-5

LINEAR PROGRAMMING AND ECONOMIC ANALYSIS, Robert Dorfman, Paul A. Samuelson and Robert M. Solow. First comprehensive treatment of linear programming in standard economic analysis. Game theory, modern welfare economics, Leontief input-output, more. 525pp. 5⅜ x 8½. 0-486-65491-5

PROBABILITY: AN INTRODUCTION, Samuel Goldberg. Excellent basic text covers set theory, probability theory for finite sample spaces, binomial theorem, much more. 360 problems. Bibliographies. 322pp. 5⅜ x 8½. 0-486-65252-1

GAMES AND DECISIONS: INTRODUCTION AND CRITICAL SURVEY, R. Duncan Luce and Howard Raiffa. Superb nontechnical introduction to game theory, primarily applied to social sciences. Utility theory, zero-sum games, n-person games, decision-making, much more. Bibliography. 509pp. 5⅜ x 8½. 0-486-65943-7

INTRODUCTION TO THE THEORY OF GAMES, J. C. C. McKinsey. This comprehensive overview of the mathematical theory of games illustrates applications to situations involving conflicts of interest, including economic, social, political, and military contexts. Appropriate for advanced undergraduate and graduate courses; advanced calculus a prerequisite. 1952 ed. x+372pp. 5⅜ x 8½. 0-486-42811-7

FIFTY CHALLENGING PROBLEMS IN PROBABILITY WITH SOLUTIONS, Frederick Mosteller. Remarkable puzzlers, graded in difficulty, illustrate elementary and advanced aspects of probability. Detailed solutions. 88pp. 5⅜ x 8½. 0-486-65355-2

PROBABILITY THEORY: A CONCISE COURSE, Y. A. Rozanov. Highly readable, self-contained introduction covers combination of events, dependent events, Bernoulli trials, etc. 148pp. 5⅝ x 8¼. 0-486-63544-9

THE STATISTICAL ANALYSIS OF EXPERIMENTAL DATA, John Mandel. First half of book presents fundamental mathematical definitions, concepts and facts while remaining half deals with statistics primarily as an interpretive tool. Well-written text, numerous worked examples with step-by-step presentation. Includes 116 tables. 448pp. 5⅜ x 8½. 0-486-64666-1

Math–Geometry and Topology

ELEMENTARY CONCEPTS OF TOPOLOGY, Paul Alexandroff. Elegant, intuitive approach to topology from set-theoretic topology to Betti groups; how concepts of topology are useful in math and physics. 25 figures. 57pp. 5⅜ x 8½. 0-486-60747-X

A LONG WAY FROM EUCLID, Constance Reid. Lively guide by a prominent historian focuses on the role of Euclid's Elements in subsequent mathematical developments. Elementary algebra and plane geometry are sole prerequisites. 80 drawings. 1963 edition. 304pp. 5⅜ x 8½. 0-486-43613-6

EXPERIMENTS IN TOPOLOGY, Stephen Barr. Classic, lively explanation of one of the byways of mathematics. Klein bottles, Moebius strips, projective planes, map coloring, problem of the Koenigsberg bridges, much more, described with clarity and wit. 43 figures. 210pp. 5⅜ x 8½. 0-486-25933-1

THE GEOMETRY OF RENÉ DESCARTES, René Descartes. The great work founded analytical geometry. Original French text, Descartes's own diagrams, together with definitive Smith-Latham translation. 244pp. 5⅜ x 8½. 0-486-60068-8

EUCLIDEAN GEOMETRY AND TRANSFORMATIONS, Clayton W. Dodge. This introduction to Euclidean geometry emphasizes transformations, particularly isometries and similarities. Suitable for undergraduate courses, it includes numerous examples, many with detailed answers. 1972 ed. viii+296pp. 6⅛ x 9¼. 0-486-43476-1

EXCURSIONS IN GEOMETRY, C. Stanley Ogilvy. A straightedge, compass, and a little thought are all that's needed to discover the intellectual excitement of geometry. Harmonic division and Apollonian circles, inversive geometry, hexlet, Golden Section, more. 132 illustrations. 192pp. 5⅜ x 8½. 0-486-26530-7

THE THIRTEEN BOOKS OF EUCLID'S ELEMENTS, translated with introduction and commentary by Sir Thomas L. Heath. Definitive edition. Textual and linguistic notes, mathematical analysis. 2,500 years of critical commentary. Unabridged. 1,414pp. 5⅜ x 8½.. Three-vol. set.
Vol. I: 0-486-60088-2 Vol. II: 0-486-60089-0 Vol. III: 0-486-60090-4

SPACE AND GEOMETRY: IN THE LIGHT OF PHYSIOLOGICAL, PSYCHOLOGICAL AND PHYSICAL INQUIRY, Ernst Mach. Three essays by an eminent philosopher and scientist explore the nature, origin, and development of our concepts of space, with a distinctness and precision suitable for undergraduate students and other readers. 1906 ed. vi+148pp. 5⅜ x 8½. 0-486-43909-7

GEOMETRY OF COMPLEX NUMBERS, Hans Schwerdtfeger. Illuminating, widely praised book on analytic geometry of circles, the Moebius transformation, and two-dimensional non-Euclidean geometries. 200pp. 5⅝ x 8¼. 0-486-63830-8

DIFFERENTIAL GEOMETRY, Heinrich W. Guggenheimer. Local differential geometry as an application of advanced calculus and linear algebra. Curvature, transformation groups, surfaces, more. Exercises. 62 figures. 378pp. 5⅜ x 8½.
0-486-63433-7

History of Math

THE WORKS OF ARCHIMEDES, Archimedes (T. L. Heath, ed.). Topics include the famous problems of the ratio of the areas of a cylinder and an inscribed sphere; the measurement of a circle; the properties of conoids, spheroids, and spirals; and the quadrature of the parabola. Informative introduction. clxxxvi+326pp. 5⅜ x 8½.
0-486-42084-1

A SHORT ACCOUNT OF THE HISTORY OF MATHEMATICS, W. W. Rouse Ball. One of clearest, most authoritative surveys from the Egyptians and Phoenicians through 19th-century figures such as Grassman, Galois, Riemann. Fourth edition. 522pp. 5⅜ x 8½. 0-486-20630-0

THE HISTORY OF THE CALCULUS AND ITS CONCEPTUAL DEVELOP-MENT, Carl B. Boyer. Origins in antiquity, medieval contributions, work of Newton, Leibniz, rigorous formulation. Treatment is verbal. 346pp. 5⅜ x 8½. 0-486-60509-4

THE HISTORICAL ROOTS OF ELEMENTARY MATHEMATICS, Lucas N. H. Bunt, Phillip S. Jones, and Jack D. Bedient. Fundamental underpinnings of modern arithmetic, algebra, geometry and number systems derived from ancient civilizations. 320pp. 5⅜ x 8½. 0-486-25563-8

THE HISTORY OF THE CALCULUS AND ITS CONCEPTUAL DEVELOP-MENT, Carl B. Boyer. Fluent description of the development of both the integral and differential calculus—its early beginnings in antiquity, medieval contributions, and a consideration of Newton and Leibniz. 368pp. 5⅜ x 8½. 0-486-60509-4

GAMES, GODS & GAMBLING: A HISTORY OF PROBABILITY AND STATISTICAL IDEAS, F. N. David. Episodes from the lives of Galileo, Fermat, Pascal, and others illustrate this fascinating account of the roots of mathematics. Features thought-provoking references to classics, archaeology, biography, poetry. 1962 edition. 304pp. 5⅜ x 8½. (Available in U.S. only.) 0-486-40023-9

OF MEN AND NUMBERS: THE STORY OF THE GREAT MATHEMATICIANS, Jane Muir. Fascinating accounts of the lives and accomplishments of history's greatest mathematical minds—Pythagoras, Descartes, Euler, Pascal, Cantor, many more. Anecdotal, illuminating. 30 diagrams. Bibliography. 256pp. 5⅜ x 8½.
0-486-28973-7

HISTORY OF MATHEMATICS, David E. Smith. Nontechnical survey from ancient Greece and Orient to late 19th century; evolution of arithmetic, geometry, trigonometry, calculating devices, algebra, the calculus. 362 illustrations. 1,355pp. 5⅜ x 8½. Two-vol. set.   Vol. I: 0-486-20429-4   Vol. II: 0-486-20430-8

A CONCISE HISTORY OF MATHEMATICS, Dirk J. Struik. The best brief history of mathematics. Stresses origins and covers every major figure from ancient Near East to 19th century. 41 illustrations. 195pp. 5⅜ x 8½. 0-486-60255-9

Physics

OPTICAL RESONANCE AND TWO-LEVEL ATOMS, L. Allen and J. H. Eberly. Clear, comprehensive introduction to basic principles behind all quantum optical resonance phenomena. 53 illustrations. Preface. Index. 256pp. 5⅜ x 8½.
0-486-65533-4

QUANTUM THEORY, David Bohm. This advanced undergraduate-level text presents the quantum theory in terms of qualitative and imaginative concepts, followed by specific applications worked out in mathematical detail. Preface. Index. 655pp. 5⅜ x 8½.
0-486-65969-0

ATOMIC PHYSICS (8th EDITION), Max Born. Nobel laureate's lucid treatment of kinetic theory of gases, elementary particles, nuclear atom, wave-corpuscles, atomic structure and spectral lines, much more. Over 40 appendices, bibliography. 495pp. 5⅜ x 8½.
0-486-65984-4

A SOPHISTICATE'S PRIMER OF RELATIVITY, P. W. Bridgman. Geared toward readers already acquainted with special relativity, this book transcends the view of theory as a working tool to answer natural questions: What is a frame of reference? What is a "law of nature"? What is the role of the "observer"? Extensive treatment, written in terms accessible to those without a scientific background. 1983 ed. xlviii+172pp. 5⅜ x 8½.
0-486-42549-5

AN INTRODUCTION TO HAMILTONIAN OPTICS, H. A. Buchdahl. Detailed account of the Hamiltonian treatment of aberration theory in geometrical optics. Many classes of optical systems defined in terms of the symmetries they possess. Problems with detailed solutions. 1970 edition. xv + 360pp. 5⅜ x 8½. 0-486-67597-1

PRIMER OF QUANTUM MECHANICS, Marvin Chester. Introductory text examines the classical quantum bead on a track: its state and representations; operator eigenvalues; harmonic oscillator and bound bead in a symmetric force field; and bead in a spherical shell. Other topics include spin, matrices, and the structure of quantum mechanics; the simplest atom; indistinguishable particles; and stationary-state perturbation theory. 1992 ed. xiv+314pp. 6⅛ x 9¼.
0-486-42878-8

LECTURES ON QUANTUM MECHANICS, Paul A. M. Dirac. Four concise, brilliant lectures on mathematical methods in quantum mechanics from Nobel Prize-winning quantum pioneer build on idea of visualizing quantum theory through the use of classical mechanics. 96pp. 5⅜ x 8½.
0-486-41713-1

THIRTY YEARS THAT SHOOK PHYSICS: THE STORY OF QUANTUM THEORY, George Gamow. Lucid, accessible introduction to influential theory of energy and matter. Careful explanations of Dirac's anti-particles, Bohr's model of the atom, much more. 12 plates. Numerous drawings. 240pp. 5⅜ x 8½. 0-486-24895-X

ELECTRONIC STRUCTURE AND THE PROPERTIES OF SOLIDS: THE PHYSICS OF THE CHEMICAL BOND, Walter A. Harrison. Innovative text offers basic understanding of the electronic structure of covalent and ionic solids, simple metals, transition metals and their compounds. Problems. 1980 edition. 582pp. 6⅛ x 9¼.
0-486-66021-4

HYDRODYNAMIC AND HYDROMAGNETIC STABILITY, S. Chandrasekhar. Lucid examination of the Rayleigh-Benard problem; clear coverage of the theory of instabilities causing convection. 704pp. 5⅝ x 8¼. 0-486-64071-X

INVESTIGATIONS ON THE THEORY OF THE BROWNIAN MOVEMENT, Albert Einstein. Five papers (1905–8) investigating dynamics of Brownian motion and evolving elementary theory. Notes by R. Fürth. 122pp. 5⅜ x 8½. 0-486-60304-0

THE PHYSICS OF WAVES, William C. Elmore and Mark A. Heald. Unique overview of classical wave theory. Acoustics, optics, electromagnetic radiation, more. Ideal as classroom text or for self-study. Problems. 477pp. 5⅜ x 8½. 0-486-64926-1

GRAVITY, George Gamow. Distinguished physicist and teacher takes reader-friendly look at three scientists whose work unlocked many of the mysteries behind the laws of physics: Galileo, Newton, and Einstein. Most of the book focuses on Newton's ideas, with a concluding chapter on post-Einsteinian speculations concerning the relationship between gravity and other physical phenomena. 160pp. 5⅜ x 8½. 0-486-42563-0

PHYSICAL PRINCIPLES OF THE QUANTUM THEORY, Werner Heisenberg. Nobel Laureate discusses quantum theory, uncertainty, wave mechanics, work of Dirac, Schroedinger, Compton, Wilson, Einstein, etc. 184pp. 5⅜ x 8½. 0-486-60113-7

ATOMIC SPECTRA AND ATOMIC STRUCTURE, Gerhard Herzberg. One of best introductions; especially for specialist in other fields. Treatment is physical rather than mathematical. 80 illustrations. 257pp. 5⅜ x 8½. 0-486-60115-3

AN INTRODUCTION TO STATISTICAL THERMODYNAMICS, Terrell L. Hill. Excellent basic text offers wide-ranging coverage of quantum statistical mechanics, systems of interacting molecules, quantum statistics, more. 523pp. 5⅜ x 8½. 0-486-65242-4

THEORETICAL PHYSICS, Georg Joos, with Ira M. Freeman. Classic overview covers essential math, mechanics, electromagnetic theory, thermodynamics, quantum mechanics, nuclear physics, other topics. First paperback edition. xxiii + 885pp. 5⅜ x 8½. 0-486-65227-0

PROBLEMS AND SOLUTIONS IN QUANTUM CHEMISTRY AND PHYSICS, Charles S. Johnson, Jr. and Lee G. Pedersen. Unusually varied problems, detailed solutions in coverage of quantum mechanics, wave mechanics, angular momentum, molecular spectroscopy, more. 280 problems plus 139 supplementary exercises. 430pp. 6½ x 9¼. 0-486-65236-X

THEORETICAL SOLID STATE PHYSICS, Vol. 1: Perfect Lattices in Equilibrium; Vol. II: Non-Equilibrium and Disorder, William Jones and Norman H. March. Monumental reference work covers fundamental theory of equilibrium properties of perfect crystalline solids, non-equilibrium properties, defects and disordered systems. Appendices. Problems. Preface. Diagrams. Index. Bibliography. Total of 1,301pp. 5⅜ x 8½. Two volumes. Vol. I: 0-486-65015-4 Vol. II: 0-486-65016-2

WHAT IS RELATIVITY? L. D. Landau and G. B. Rumer. Written by a Nobel Prize physicist and his distinguished colleague, this compelling book explains the special theory of relativity to readers with no scientific background, using such familiar objects as trains, rulers, and clocks. 1960 ed. vi+72pp. 5⅜ x 8½. 0-486-42806-0

A TREATISE ON ELECTRICITY AND MAGNETISM, James Clerk Maxwell. Important foundation work of modern physics. Brings to final form Maxwell's theory of electromagnetism and rigorously derives his general equations of field theory. 1,084pp. 5⅜ x 8½. Two-vol. set. Vol. I: 0-486-60636-8 Vol. II: 0-486-60637-6

MATHEMATICS FOR PHYSICISTS, Philippe Dennery and Andre Krzywicki. Superb text provides math needed to understand today's more advanced topics in physics and engineering. Theory of functions of a complex variable, linear vector spaces, much more. Problems. 1967 edition. 400pp. 6½ x 9¼. 0-486-69193-4

INTRODUCTION TO QUANTUM MECHANICS WITH APPLICATIONS TO CHEMISTRY, Linus Pauling & E. Bright Wilson, Jr. Classic undergraduate text by Nobel Prize winner applies quantum mechanics to chemical and physical problems. Numerous tables and figures enhance the text. Chapter bibliographies. Appendices. Index. 468pp. 5⅜ x 8½. 0-486-64871-0

METHODS OF THERMODYNAMICS, Howard Reiss. Outstanding text focuses on physical technique of thermodynamics, typical problem areas of understanding, and significance and use of thermodynamic potential. 1965 edition. 238pp. 5⅜ x 8½. 0-486-69445-3

THE ELECTROMAGNETIC FIELD, Albert Shadowitz. Comprehensive undergraduate text covers basics of electric and magnetic fields, builds up to electromagnetic theory. Also related topics, including relativity. Over 900 problems. 768pp. 5⅝ x 8¼. 0-486-65660-8

GREAT EXPERIMENTS IN PHYSICS: FIRSTHAND ACCOUNTS FROM GALILEO TO EINSTEIN, Morris H. Shamos (ed.). 25 crucial discoveries: Newton's laws of motion, Chadwick's study of the neutron, Hertz on electromagnetic waves, more. Original accounts clearly annotated. 370pp. 5⅜ x 8½. 0-486-25346-5

EINSTEIN'S LEGACY, Julian Schwinger. A Nobel Laureate relates fascinating story of Einstein and development of relativity theory in well-illustrated, nontechnical volume. Subjects include meaning of time, paradoxes of space travel, gravity and its effect on light, non-Euclidean geometry and curving of space-time, impact of radio astronomy and space-age discoveries, and more. 189 b/w illustrations. xiv+250pp. 8⅜ x 9¼. 0-486-41974-6

THE VARIATIONAL PRINCIPLES OF MECHANICS, Cornelius Lanczos. Philosophic, less formalistic approach to analytical mechanics offers model of clear, scholarly exposition at graduate level with coverage of basics, calculus of variations, principle of virtual work, equations of motion, more. 418pp. 5⅜ x 8½. 0-486-65067-7

Paperbound unless otherwise indicated. Available at your book dealer, online at **www.doverpublications.com**, or by writing to Dept. GI, Dover Publications, Inc., 31 East 2nd Street, Mineola, NY 11501. For current price information or for free catalogues (please indicate field of interest), write to Dover Publications or log on to **www.doverpublications.com** and see every Dover book in print. Dover publishes more than 400 books each year on science, elementary and advanced mathematics, biology, music, art, literary history, social sciences, and other areas.